W9-AAE-528

Physical and Thermal Technologies

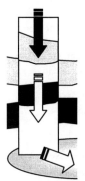

Remediation of Chlorinated and Recalcitrant Compounds

EDITORS
Godage B. Wickramanayake
and Arun R. Gavaskar
Battelle

The Second International Conference on Remediation of Chlorinated and Recalcitrant Compounds

Monterey, California, May 22–25, 2000

 BATTELLE PRESS
Columbus • Richland

Library of Congress Cataloging-in-Publication Data

International Conference on Remediation of Chlorinated and Recalcitrant Compounds
(2nd : 2000 : Monterey, Calif.)
    Physical and thermal technologies : remediation of chlorinated and recalcitrant
compounds (C2-5) / edited by Godage B. Wickramanayake, Arun R. Gavaskar.
        p.   cm.
    "The Second International Conference on Remediation of Chlorinated and
Recalcitrant Compounds, Monterey, California, May 22–25, 2000."
    Includes bibliographical references and index.
    ISBN 1-57477-099-3 (alk. paper)
        1. Organochlorine compounds--Environmental aspects--Congresses. 2. Hazardous
waste site remediation--Congresses. I. Wickramanayake, Godage B., 1953–
II. Gavaskar, Arun R., 1962–   III. Title.

TD1066.O73 I58 2000e
628.5'2--dc21

                                                                    00-034243

Printed in the United States of America

Battelle Press
505 King Avenue
Columbus, Ohio 43201-2693, USA
614-424-6393 or 1-800-451-3543
Fax: 1-614-424-3819
E-mail: press@battelle.org
Website: www.battelle.org/bookstore

For information on future environmental remediation meetings, contact:
Remediation Conferences
Battelle
505 King Avenue
Columbus, Ohio 43201-2693, USA
Fax: 614-424-3667
Website: www.battelle.org/conferences

CONTENTS

Multiphase Extraction/Bioslurping

Hot Fluids Injection for In Situ Soil Treatment

Soil Heating Technologies

Emerging and Innovative Technologies

Applying Multiple Remediation Technologies

FOREWORD

Physical and thermal technologies for the remediation of sites contaminated with chlorinated compounds and other difficult-to-treat contaminants are in the mainstream of contemporary remediation practice. Because of the continual need to find cost-effective ways to remediate highly contaminated sites, innovative applications of these technologies are being developed by remediation professionals, and there are cases where combinations of technologies may prove to be synergistic solutions. *Physical and Thermal Technologies: Remediation of Chlorinated and Recalcitrant Compounds* provides recent case studies that cover air sparging, groundwater recirculating wells, multiphase extraction/bioslurping, hot fluids injection for in situ soil treatment, soil heating technologies, emerging and innovative technologies for site remediation, and the application of multiple remediation technologies.

This is one of seven volumes resulting from the Second International Conference on Remediation of Chlorinated and Recalcitrant Compounds (May 22–25, 2000, Monterey, California). Like the first meeting in the series, which was held in May 1998, the 2000 conference focused on the more problematic contaminants—chlorinated solvents, pesticides/herbicides, PCBs/dioxins, MTBE, DNAPLs, and explosives residues—in all environmental media and on physical, chemical, biological, thermal, and combined technologies for dealing with these compounds. The conference was attended by approximately 1,450 environmental professionals involved in the application of environmental assessment and remediation technologies at private- and public-sector sites around the world.

A short paper was invited for each presentation accepted for the program. Each paper submitted was reviewed by a volume editor for general technical content. Because of the need to complete publication shortly after the Conference, no in-depth peer review, copy-editing, or detailed typesetting was performed for the majority of the papers in these volumes. Papers for 60% of the presentations given at the conference appear in the proceedings. Each section in this and the other six volumes corresponds to a technical session at the Conference. Most papers are printed as submitted by the authors, with resulting minor variations in word usage, spelling, abbreviations, the manner in which numbers and measurements are presented, and formatting.

We would like to thank the people responsible for the planning and conduct of the Conference and the production of the proceedings. Valuable input to our task of defining the scope of the technical program and delineating sessions was provided by a steering committee made up of several Battelle scientists and engineers – Bruce Alleman, Abraham Chen, James Gibbs, Neeraj Gupta, Mark Kelley, and Victor Magar. The committee members, along with technical reviewers from Battelle and many other organizations, reviewed more than 600 abstracts submitted for the Conference and determined the content of the individual sessions. Karl Nehring provided valuable advice on the development of the program schedule and the organization of the proceedings volumes. Carol Young, with assistance from Gina Melaragno, maintained program data, corresponded with speakers and authors, and compiled the final program and abstract books. Carol and Lori Helsel were responsible for the proceedings production effort, receiving assistance on specific aspects from Loretta Bahn, Tom Wilk, and Mark Hendershot. Lori, in particular, spent many hours examining papers for format and contacting authors as necessary to obtain revisions. Joe Sheldrick, the manager of Battelle Press, provided valuable production-planning advice; he and Gar Dingess designed the volume covers.

Battelle organizes and sponsors the Conference on Remediation of Chlorinated and Recalcitrant Compounds. Several organizations made financial contributions toward the 2000 Conference. The co-sponsors were EnviroMetal Technologies Inc. (ETI); Geomatrix Consultants, Inc.; the Naval Facilities Engineering Command (NAVFAC); Parsons Engineering Science, Inc.; and Regenesis.

As stated above, each article submitted for the proceedings was reviewed by a volume editor for basic technical content. As necessary, authors were asked to provide clarification and additional information. However, it would have been impossible to subject more than 300 papers to a rigorous peer review to verify the accuracy of all data and conclusions. Therefore, neither Battelle nor the Conference co-sponsors or supporting organizations can endorse the content of the materials published in these volumes, and their support for the Conference should not be construed as such an endorsement.

Godage B. Wickramanayake and Arun R. Gavaskar
Conference Chairs

NUMERICAL MODELING OF KINETIC INTERPHASE MASS TRANSFER DURING AIR SPARGING

Ronald W. Falta (Clemson University, Clemson, South Carolina)

ABSTRACT: A dual media multiphase flow approach is proposed for modeling the local interphase mass transfer that occurs during in-situ air sparging. The method is applied to two- or three-phase flow in porous media to simulate the small gas channels that form during air sparging, allowing resolution of the local diffusive mass transfer of contaminants between the flowing gas phase and nearby stagnant liquid filled zones. This approach provides a good match with laboratory column experiments in which dissolved trichloroethylene (TCE) is removed by air sparging, and it is shown that the simulation results are very sensitive to the nature of the local mass transfer regime.

INTRODUCTION

Recent laboratory studies have demonstrated that dissolved volatile organic compound (VOC) removal during air sparging is limited by the mass transfer of the VOC into the flowing gas phase [Braida and Ong, 1998; J.S. Gierke, personal communication, 1999]. Mass transfer limitations may occur at several scales due to the heterogeneous nature of gas distributions during air sparging. At a large scale, the air sparging zone is very strongly influenced by heterogeneities, which form capillary and permeability barriers to the gas flow. If the geometry and locations of these heterogeneities are well known or if the media is homogeneous, it is possible to accurately model the sparge gas flow field using a conventional multiphase flow numerical approach [Hein et al., 1997; McCray and Falta, 1997]. This type of simulator, however, cannot resolve local (subgridblock) mass transfer effects that arise due to the millimeter to centimeter scale gas channels which form during sparging [Clayton, 1998; Elder and Benson, 1999]. Because these channels typically occur at a scale smaller than that of a numerical model gridblock, compositional multiphase flow simulators that assume local chemical equilibrium between the phases may overestimate the rate of interphase mass transfer during air sparging.

While models have been developed which can account for the kinetic interphase mass transfer during sparging [Braida and Ong, 1998; Elder et al., 1999], these codes do not model the transient two- or three-dimensional multiphase flow that occurs during sparging. On the other hand almost all of the current two- or three-dimensional compositional multiphase flow simulators assume local equilibrium between phases, and there have been only limited efforts to model kinetic interphase mass transfer in multi-dimensional, field scale multiphase contaminant transport [Sleep and Sykes, 1989]. It is significant to note that the kinetic interphase mass transfer approach has never been applied to a field scale multiphase flow air sparging simulation.

MODELING INTERPHASE MASS TRANSFER WITH A DUAL MEDIA APPROACH

An alternative method for modeling the local mass transfer during air sparging involves a dual media formulation. This technique is fairly straightforward, and can be easily implemented in existing compositional multiphase flow simulators that assume local phase equilibrium. The method is based on a dual media formulation which is widely used for modeling both single-phase and multiphase flow and transport processes in fractured media (see, for example, Barenblatt et al. [1960]; Warren and Root [1963]; Grisak and Pickens [1980]; Pruess and Narasimhan [1985]). The method is also commonly used to describe single-phase solute transport in which the solute is affected by physical or chemical nonequilibrium effects due to structured soils, adsorption, or dead end pore effects [Coats and Smith, 1964; van Genuchten et al., 1974; Rao et al., 1979]. Recently, the method has also been applied to large field scale single phase solute transport in highly heterogeneous aquifers by Harvey [1996] and Feehley et al. [2000].

With this approach, each overall gridblock is considered to consist of two volume fractions, a volume associated with the fractures, and a volume associated with the porous matrix. The two volumes are assigned different properties, corresponding to the respective media. Each of these domains are globally connected in a normal way, and the overall grid structure can be one-, two-, or three-dimensional with any coordinate system. The global connections retain their normal (single media) nodal distances, but the global gridblock connection areas are divided into two area fractions corresponding to the average fracture area between gridblocks, and to the average matrix area between gridblocks. These global area fractions are not necessarily equal to the media volume fractions, and they depend on the assumed fracture geometry inside the gridblock.

Locally, at the scale of a single gridblock, the two domains are attached with a simple one-dimensional connection. This connection is characterized by an interfacial area (per unit total volume) between the two media, A_{12}, and by the average nodal distance between the two media, $(d_1 + d_2 = d_{12})$. Here, the subscripts 1 and 2 refer to the fracture and matrix parts of the overall gridblock and the local dual media connection. It is important to understand that with this approach, individual fractures and matrix blocks within the gridblock are not explicitly modeled. Rather the average fracture and matrix block responses are simulated. The ratio of A_{12} / d_{12} determines the magnitude of the conductance between the media, and it is often called the dual media interaction parameter. Large values of the interfacial area, or small nodal separation distances correspond to rapid fluid, chemical, and heat fluxes between the media. In the limit of very large A_{12} or very small d_{12}, the two media remain at equilibrium with each other, and the dual media formulation responds as if it were a normal single media formulation with local equilibrium in each gridblock.

In the present work, the method is applied to porous media to simulate the effect of local gas channels that form during air sparging. The local (subgridblock) geometry is assumed to consist of two regions: a volume fraction characterized by a high capillary pressure and a volume fraction characterized by a lower capillary pressure. These volume fractions, the global area fractions, the interfacial area (A_{12}), and the interfacial distances (d_1 and d_2) are estimated using an appropriate conceptual model of the local two-phase flow and are inputs to the numerical model. The Multiple Interacting Continua (MINC) subroutines in the T2VOC code (see Pruess [1991] or Falta et al. [1995]) were modified to allow direct input of these quantities for this purpose.

Under two-phase flow conditions, the nonwetting phase (gas) will flow preferentially through the low capillary pressure media, while the wetting phase (aqueous) will remain in the high capillary pressure media. This local contrast in gas phase saturation between the two media results in a aqueous phase diffusional limitation to the mass transfer of a contaminant between the two media. Thus the numerical model allows resolution of the local mass transfer kinetics between the flowing gas phase, and nearby stagnant water filled zones. Compared to the usual local equilibrium approach, the dual media approach doubles the number of equations to be solved at each time-step. The two media are assumed to have the same intrinsic permeability and porosity, so that single phase flows of gas and water are not affected.

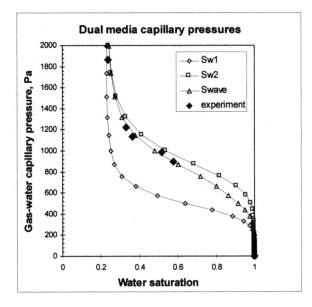

FIGURE 1. Comparison of dual media capillary pressures with experimental data. The center curve is a volume weighted average of the two media curves. The dual media capillary parameters were adjusted so that the weighted average would give a best fit with the experimental data.

The dual media capillary pressure and relative permeability properties must be estimated for this method. One reasonable approach is to choose the two media properties so that the weighted averages (with local capillary pressure equilibrium) duplicate the single media properties. Figure 1 illustrates this approach. Here, the measured capillary pressure from sand used in Michigan Technological University column sparging experiments (J.S. Gierke, personal communication, 1999) provides the single media property to be matched. Assuming dual media volume fractions of 0.2 for the low capillary pressure media, and 0.8 for the higher capillary pressure media, the capillary curve parameters were adjusted until the volume weighted average matched the experimental data.

The two capillary pressure functions shown in Figure 1 give rise to a large contrast in water saturation in the two media. For example, at an overall water saturation of 0.9 (a gas saturation of 0.1), the high capillary pressure media has a water saturation of about 0.97, while the low capillary pressure media has a water saturation of only about 0.62. A matching approach can also be used to determine the appropriate relative permeability functions. By adjusting the parameters in the function, the global area fraction weighed average of the dual media gas relative permeabilities is similar to the single media one. On the basis of the above discussions, it is expected that the dual media formulation can be made to duplicate "normal" bulk porous media characteristics, while providing a mechanism for local kinetic interphase mass transfer.

The fact that substantial water saturation remains in the low capillary pressure media is significant. Within each media, in each gridblock, local equilibrium is assumed. Thus as the gas flows through the low capillary pressure region, it equilibrates with the contaminant dissolved in the pore water (or present as a NAPL) in that volume. At the same time, there is a diffusive exchange with the high capillary pressure region, so the model acts like a dual region model, with local equilibrium in one region, and kinetic mass transfer with the other region. The relative contributions of these regions are determined by the media volume fractions and by the water saturations in the two adjacent media.

SIMULATION OF LABORATORY COLUMN EXPERIMENTS

A series of laboratory column air sparging experiments were performed at Michigan Technological University during the summer of 1998 (J.S. Gierke, personal communication, 1999). These were performed in a vertical 5 cm diameter column, with a packed length of 28.3 cm. The column was packed with a 20x30 mesh Ottawa sand, having a porosity of 0.34 and an intrinsic permeability of 2.6×10^{-10} m^2. Capillary pressure data for this sand is shown in Figure 1 TCE was dissolved in water to solubility (1100 mg/l) and 128.5 ml of this solution was injected into the bottom of the initially dry column. Filling the column in this manner resulted in about 10 cm of unsaturated sand above about 18 cm of water saturated sand. Air sparging was begun immediately after filling, and continued until the effluent gas concentration was below detection limits. Experiments were performed at several gas flow rates.

The 2.21 cm/min and 0.55 cm/min (not shown here) injection rate experiments were modeled numerically using T2VOC with the new dual media approach [Falta, 2000]. A dual media grid consisting of 114 gridblocks was generated using the built-in subroutines in T2VOC. As mentioned earlier, these subroutines were modified to allow direct entry of the dual media volume fractions, global area fractions, the dual media nodal separation distances, and the dual media interfacial area per unit volume. Unfortunately, we do not have a method for estimating these dual media parameters based on standard porous media measurements. Therefore, the dual media geometrical parameters were estimated using a conceptual model described by Falta [2000] and were refined slightly by attempting to match the experiments. This situation is analogous to the problem of determining a kinetic interphase mass transfer coefficient – interfacial area product for standard kinetic mass transfer modeling.

Figure 2 shows a dual media T2VOC simulation of the 2.21 cm/min experiment. Here, a value of 300 m^2/m^3 was used for A_{12}, d_1 was set equal to 1 mm, and d_2 was set equal to 4 mm. The other multiphase flow parameters (capillary pressure curves and relative permeability curves) were determined by matching the single media values when possible. Although the numerical model overpredicts the gas concentration in the first few minutes, the overall match with the experimental data is excellent, especially compared to the earlier local equilibrium simulation. The initial overprediction of gas concentration could probably be eliminated by reducing the volume fraction of the low permeability

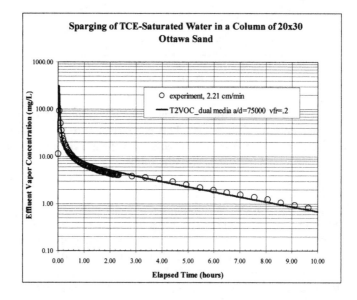

FIGURE 2. Comparison of a numerical simulation using the dual media mass transfer approach with experimental data from a column air sparging test operated at a darcy velocity of 2.21 cm/min.

media (in which local equilibrium with the water is assumed), but this type of fine tuning was not attempted.

A series of simulations were conducted using different A_{12}/d_2 ratios in order to evaluate the sensitivity of the results to the rate of mass transfer. Figure 3 shows the effect of increasing the A_{12}/d_2 ratio. As the ratio is increased by factors of 10 and 100, to 750,000 and 7,500,000 the rate of mass transfer greatly increases, and begins to approach the pure local equilibrium simulation (equivalent to $A_{12}/d_2 = \infty$). It is apparent that these would be unrealistically high rates of interphase mass transfer for this experiment.

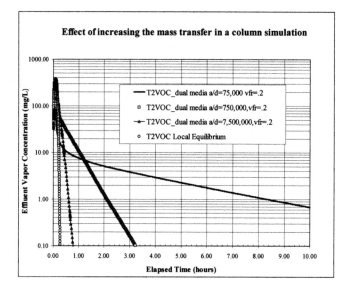

FIGURE 3. Effect of increasing the interaction parameter (A/d) between the dual media in a column simulation. This is equivalent to increasing the mass transfer coefficient between the two media, and the curves approach the local equilibrium curve.

CONCLUSIONS

It is possible to adapt an existing 3-D local equilibrium multiphase flow and transport simulator to model the problem of locally rate limited mass transfer by using a dual media approach. This method is completely analogous to the technique used in fractured porous media. Because the dual media approach is simply a mesh generation step, it could be readily adapted to other multiphase flow and transport codes that use the integral finite difference method.

The dual media approach allows for a fraction of the system to be at local phase equilibrium, with first order rate limited transport with the other fraction. This model achieved a good match with laboratory column air sparging data, and the column scale simulations were very sensitive to the rates of local mass transfer. The results of field scale modeling using this approach are presented by Falta [2000]. These show a substantially lower sensitivity to the local mass transfer regime.

ACKNOWLEDGEMENTS

This work was funded by the Michigan Technological University through a grant from the US Department of Energy. All experimental work referred to here was conducted at Michigan Tech by J.S. Gierke and co-workers. Their collaboration is gratefully acknowledged.

REFERENCES

Barenblatt, G.E., I.P. Zheltov, and I.N. Kochina. 1960. "Basic concept of the theory of seepage of homogeneous liquids in fissured rocks." *Journal of Applied Math Mech., English Translation*, 24, 1286-1303.

Braida, W.J., and S.K. Ong. 1998. "Air sparging: air-water mass transfer coefficients." *Water Resources Research*, Vol. 34, No. 12, p. 3245-3253.

Brusseau, M.L., Z. Gerstl, D. Augustijn, and P.S.C. Rao. 1994. "Simulating solute transport in an aggregated soil with the dual-porosity model: measured and optimized parameter values." *Journal of Hydrology*, 163, 187-193.

Clayton, W.S. 1998. "A field and labaoratory investigation of air fingering during air sparging." *Ground Water Monitoring and Remediation*, Vol. 18, No. 3, p. 134-145.

Coats, K.H., and B.D. Smith. 1964. "Dead-end pore volume and dispersion in porous media." *Society of Petroleum Engineers Journal*, 4, 73-84.

Elder, C.R., and C.H. Benson. 1999. "Air channel formation, size, spacing, and tortuosity during air sparging." *Ground Water Monitoring and Remediation*, Vol. 19, No. 3, p.171-181.

Elder, C.R., C.H. Benson, and G. Eykholt. 1999. "Factors affecting mass removal during in situ air sparging." *J. of Geotech. And Geoenviron. Engin.*, Vol. 125, No. 11.

Falta, R.W. 2000. "Numerical Modeling of Kinetic Interphase Mass Transfer During Air Sparging using a Dual Media Approach." Submitted to *Water Resources Research*.

Falta, R.W., K. Pruess, S. Finsterle, and A. Battistelli. 1995. "T2VOC User's Guide." *Lawrence Berkeley Laboratory Report LBL-36400*.

Feehley, C.E., C. Zheng, and F.J. Molz. 2000. "A dual-domain mass transfer approach for modeling solute transport in heterogeneous aquifers: Application to the MADE site." *Water Resources Research*, in press.

Grisak, G.E., and J.F. Pickens. 1980. "Solute transport through fractured media, 1. The effect of matrix diffusion." *Water Resources Research*, Vol. 16, No. 4, p 719-730.

Harvey, C.F. 1996. *Solute transport in spatially heterogeneous aquifers: Mapping large-scale structures and modeling small-scale effects* Ph.D. dissertation, Stanford University, Stanford, CA.

Hein, G.L., N.J. Hutzler, J.S. Gierke, and R.W. Falta. 1997. "Three-dimensional experimental testing of a two-phase flow-modeling approach to air sparging." *Ground Water Monitoring and Remediation*, Vol. 17, No. 3.

McCray, J.E. and R.W. Falta. 1997. "Numerical simulations of air sparging for subsurface NAPL remediation." *Ground Water* Vol. 35, No. 1.

Pruess, K. 1991. "TOUGH2-A general purpose simulator for multiphase fluid and heat flow." *Report No. LBL-29400, Lawrence Berkeley Laboratory*.

Pruess, K. and T. N. Narasimhan. 1985. "A practical method for modeling fluid flow and heat flow in fractured porous media." *Society of Petroleum Engineers Journal*, 25(1), 14-26.

Rao, P.S.C., J.M. Davidson, R.E. Jessup, and H.M. Selim. 1979. "Evaluation of conceptual models for describing non-equilibrium adsorption-desorption of pesticides during steady-flow in soils." *Soil Science Soc. of America J.*, 43, 22-28.

Sleep, B.E., and J.F. Sykes. 1989. "Modeling the transport of volatile organics in variably saturated media." *Water Resources Research, Vol. 25, No. 1*, 81-92.

van Genuchten, M.Th., J.M. Davidson, and P.J. Wierenga. 1974. "An evaluation of kinetic and equilibrium equations for the prediction of pesticide movement through porous media." *Soil Sci. of America Proceedings*, 38, 29-35.

Warren, J.E., and P.J. Root. 1963. "The behavior of naturally fractured reservoirs." *Soc. Pet. Eng. J., Transactions, AIME*, 228, 245-255.

PILOT TESTING AND IMPLEMENTATION OF BIOSPARGING FOR VINYL CHLORIDE TREATMENT

Marc A. Cannata, Scott W. Hoxworth, and Todd P. Swingle, P.E.,
Parsons Engineering Science, Winter Park, Florida
Ed Carver, 45th Space Wing Cape Canaveral Air Station, Florida

ABSTRACT: Field scale pilot testing through full-scale implementation of a biosparge system has taken place at Hangar K, Cape Canaveral Air Station (CCAS). Initially, a pilot test was performed to determine the effectiveness of a biosparge system and evaluate key design parameters. Discrete depth monitoring wells, neutron density tubes, and slug tests were used to evaluate the performance of biosparging. The pilot study also evaluated several modes of air injection including multi-level sparging and pulsed injection sequences. The study yielded sufficient results to support the design of a full-scale system and to generate observations related to various potential monitoring techniques. A 1200-foot (365.8 m) long biosparge curtain was constructed, which consisted of two rows of 60 sparge wells (120 total). The wells in each row are spaced on 20-ft (6.1 m) centers and the two rows are offset yielding an effective 10-ft (3.1 m) spacing. The sparge wells have an air injection depth of 40-45 ft (12.2-13.7 m) below ground surface (bgs) (immediately above intermediate clays), and have well flow rates of 5 scfm (0.14 std. m³/min.) at 14 psig (96.5 kPa). The biosparge wall was designed and installed using creative methods that minimized interference with utilities and substantially reduced the construction costs. This paper focuses on the use of a variety of test methods for evaluating sparge system performance as well as the use of creative design/construction techniques to enhance the value of the approach.

INTRODUCTION

The purpose of this project was to evaluate the performance of biosparging to support full-scale implementation at Hangar K, a facility at Cape Canaveral Air Station (CCAS). Various chemical compounds, which included chlorinated solvents such as trichloroethylene (TCE), were utilized at locations inside and outside Hangar K during historic launch support operations. Historic releases of these solvents led to a groundwater contamination plume, consisting of TCE and its daughter products dichloroethylene (DCE) and vinyl chloride (VC).

Analytical data and modeling activities suggested that the VC and/or DCE portions of the plume had the potential to reach a protected surface water body. Through development of an alternatives analysis, it was determined that biosparging would be the best approach to achieve control of the dissolved plume and prevent migration of contaminants to the surface water.

Biosparging is an effective mechanism for removal of VC and to a lesser extent DCE. Biosparging functions by injecting air at a low rate into the aquifer below the zone of contamination. At a relatively close well spacing, the injected air promotes oxygenation of the aquifer as necessary to promote aerobic biodegradation. The system is located to focus upon VC and DCE for the purpose of stimulating aerobic biodegradation of contaminants.

BIOSPARGING PILOT TESTING

The objectives of the field scale pilot test were to (1) evaluate the effectiveness of in-situ biosparging as a method for controlling migration of VC and DCE, and (2) develop design data to support the implementation of a full scale biosparging system to achieve this control of plume migration. The pilot test was performed in November and December 1998.

Testing Setup. Biosparge testing was accomplished using a vertical nested sparge well (SW) pair, given the designation SW1-25 and SW1-40. These depths focus on the primary area of contamination, which extended from approximately 20 to 40 ft (6.1 to 12.2 m) bgs.

The biosparge pilot test configuration consisted of a well head connection, pressure gauge, air hose, galvanized steel piping, inline regulator, flow indicator, gate valve, oil filter, shut-off valve, and an air compressor. A process flow diagram of this configuration and the biosparge well configuration is shown in Figure 1.

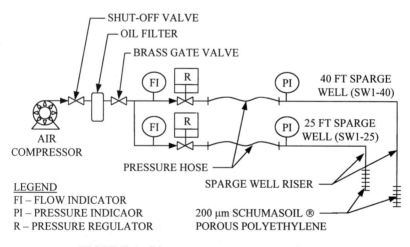

FIGURE 1. Biosparge system process flow diagram.

The following methods were used to evaluate the biosparge system's performance (See Figure 2 for pilot study layout):
- *Discrete Depth Monitoring Wells* – Nine sets of discrete groundwater monitoring point clusters were installed to monitor dissolved oxygen and volatile organic compounds. The monitoring point clusters were located on

three radials from the biosparge wells. Each radial included a well cluster at distances of 5, 10, and 15 feet (1.5, 3.1 and 4.6 m) from the injection wells. Each cluster consisted of six monitoring points completed at 5 ft (1.5 m) intervals between 10 and 35 ft (3.1 and 10.7 m) bgs. Each well point was constructed using a 6-inch (15.2 cm) long 0.010-inch (0.25 mm) slotted screen.

- *Neutron Density Tubes* – Nine neutron probe access tubes were installed in association with each of the discrete sparge monitoring well clusters. The access tubes were constructed using PVC capped on the bottom and were completed to at depths of 37 ft (11.3 m) bgs. These tubes provided access for the neutron density probe as means to assess the level of air saturation in the aquifer.
- *Slug Test Wells* – Two piezometer pairs were installed for the purpose of conducting slug tests to monitor changes in permeability due to biosparging. These piezometer pairs were located at 5 and 10 ft (1.5 and 3.1 m) radial distances from the sparge points and completed at depths of approximately 20 and 38 ft (6.1 and 11.6 m) bgs with 10 ft (3.1 m) screened intervals.
- *Additional Monitoring Wells* – Eight additional "typical" monitoring wells were also installed to verify the pilot test location and provide future long-term monitoring of the expanded biosparge system. These wells were completed with a 5-ft. (1.5 m) screened interval above the clay layer at depths ranging from 41 to 44 ft (12.5 and 13.4 m) bgs. The wells were located at 250-foot (76.2 m) intervals along the planned biosparge wall. Each location included two wells, each approximately 50 ft (15.2 m) up-gradient and down-gradient of the planned full-scale biosparge system.

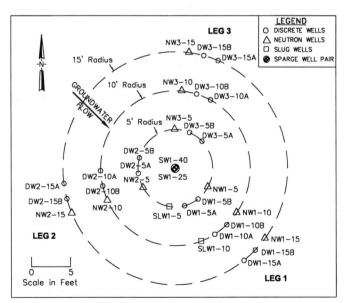

FIGURE 2. Biosparge pilot study layout.

Testing Methods. The pilot test data collection activities and procedures were developed and implemented in an effort to define the following parameters: initial site conditions, zone/radius of influence, air injection pressure, injection flow rate, and contaminant removal efficiency and effectiveness.

The pilot system was evaluated at several different operating modes, which included SW1-25 and SW1-40 operating independently, and SW1-25 and SW1-40 operating concurrently at flow rates of 2 and 5 scfm (0.06 and 0.14 std. m^3/min.). For each mode of operation, air was injected continuously and using pulsed air injection at 45 to 60 minute intervals. The variation of injection depths, flow rates and cycle (pulsed or continuous) helped evaluate the most effective method for maximizing the radius of influence. The tests that were performed and the measurements that were taken during the operation of each mode are as follows:

- *Discrete Depth Monitoring Wells* – Groundwater samples for VOC analyses (Method SW8260) were collected from each of the discrete sampling points prior to and upon completion of the pilot study. The only VOC detections in the groundwater occurred at the 20, 25, 30, and 35 ft (6.1, 7.6, 9.1 and 10.7 m) discrete monitoring well depths. Dissolved oxygen was also measured prior to each sparging activity to acquire baseline data and then on continuous rotations between each monitoring point. Baseline measurements ranged from 0 to 0.45 mg/L dissolved oxygen. Due to the quantity of wells, the D.O. sampling was limited to a full round of measurements twice a day.
- *Neutron Density Tubes* – A neutron probe was used to monitor changes in air saturation due to air sparging. The neutron probe measures water content by neutron moderation, or energy loss. Neutron probe testing was completed before and during air sparge activities at the same frequency as the D.O. sampling. The neutron probe measured changes in density before and during sparging activities. These changes in density are directly related to levels of air saturation resulting from the air sparging activities. Neutron probe measurements were recorded at 5 ft (1.5 m) intervals between 8 and 30 ft (2.4 and 9.1 m) bgs. A change in weight of water measurements during sparging activities was the primary indicator and was used to aid in evaluating the sparging effectiveness.
- *Slug Test Wells* – Slug tests were performed before and during each sparge mode. These tests were based on the principal that hydraulic conductivity is a function of available pore space, and to the extent pore space in the vicinity of a well is occupied by air, hydraulic conductivity will be reduced. Thus, reduced conductivity might be an indication of increased formation air content, which correlates to sparging effectiveness. Pressure transducers were installed with no greater than 30 ft (9.1 m) of water above the instrument. Data was collected with a multi-channel data logger calibrated to take water level measurements in increments of 1/10 seconds for a minimum of the first 10 seconds or a maximum of the first 30 seconds according to subsurface

sediment characteristics observed during the piezometer installation. Both slug-in and slug-out tests were performed in order to counter check and verify results.

Results and Conclusions. Upon completing the testing and evaluation for each of the operational modes previously discussed, it was decided the optimum operating condition was to pulse the intermediate well (SW1-40) at 5 scfm (0.14 std. m^3/min). This mode of operation was then tested for 7 days. Graphical representation of data collected in the final mode of operation is presented in Figure 3.

The results from each of the biosparge test methods are as follows:

Groundwater VOC and Dissolved Oxygen Results – Upon completing the pilot test, the discrete wells were immediately sampled and analyzed. There were virtually no VOC detections in the 15 ft (4.6 m) wells and minimal detections in the 20 ft (6.1 m) wells. Groundwater monitoring results produced an overall reduction in groundwater contamination with a significant reduction (approximately 50% or greater) in vinyl chloride at depths greater than 25 ft (7.6 m). The reduction in groundwater contamination had a cone effect, with an approximate 5 to 10 ft (1.5 to 3.1 m) radius of influence at the 35 ft (10.7 m) depth and a 10 to 15 ft (3.1 to 4.6 m) radius of influence at the 25 and 30 ft (7.6 and 9.1 m) depths. This reduction in VOC concentrations is a sign that biosparging continued to promote biodegradation of the chlorinated hydrocarbons throughout the pilot test.

Baseline D.O. measurements indicated anaerobic conditions existed prior to pilot study activities. This supported the conclusion that biological activity was taking place and depleted all available oxygen in the aquifer. According to Nyer *et al* (1996), aqueous environments can switch from aerobic to anaerobic with oxygen concentrations of approximately 1.0 mg/L. This 1.0 mg/L value was used as the screening level for D.O. samples (effective oxygenation) taken during the pilot test. D.O. influence became apparent at deeper depths with extended operation of the injection system. Operating the system for approximately 5 to 7 days was required to begin observing influence near the injection depth. The same cone effect as previously seen in the groundwater VOC results was apparent in the D.O. samples, with a 5 ft (1.5 m) radius of influence at the 35 ft (10.7 m) depth, a 5 to 10 ft (1.5 to 3.1 m) radius of influence at the 25 to 30 ft (7.6 to 9.1 m) depths, and a 10 to 15 ft (3.1 to 4.6 m) radius of influence at the 15 to 20 ft (4.6 to 6.1 m) depths. This D.O. data established a minimum 5 ft (1.5 m) radius of influence to a potential 10 ft (3.1 m) radius of influence. Radius of Influence for each discrete sampling depth for the final mode (5 cfm [0.14 std. m^3/min] pulsed operation for SW1-40) are shown on Figure 3.

The observations associated with the D.O. and vinyl chloride reduction data indicates a strong relationship between D.O. and VOC reduction. In addition, a VOC reduction can be seen as a shadow of D.O. distribution on the down gradient side of the test cell. This event supports the conclusion that treatment continues to occur as water flows beyond the zone of influence and is biologically facilitated.

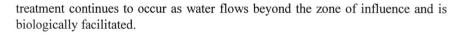
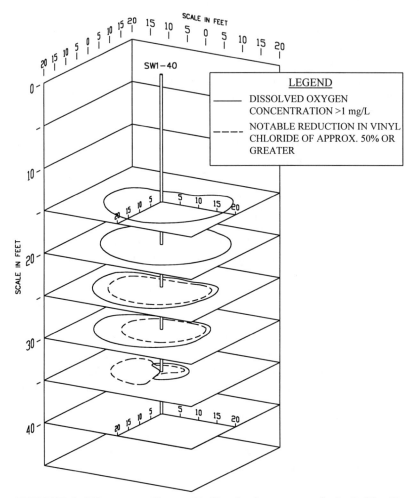

FIGURE 3. Biosparge pilot study dissolved oxygen and vinyl chloride observations.

Neutron Probe Results – The neutron probe provides an additional means for evaluating the effectiveness of the injection system. The weight of water is the primary parameter that was evaluated because as air is injected into the aquifer it displaces the water, thus reducing the weight of water in the zone of detection. A weight of water difference greater than 1.0 lb/ft^3 (16 kg/m^3) between baseline and observed measurements was used to evaluate the effectiveness of the sparging activities. The neutron probes ability to take accurate measurements seems to be affected by its sensitivity and limited radius of influence. Since biosparging has low flow rates, the weight of water difference is not as greatly affected as it would be if higher flow rates were being produced. As a result, the neutron density measurements did not produce a definable radius of influence; however, there was evidence of predominant flow direction in the pilot study. This predominant flow direction provides additional evidence that the biosparging has achieved substantial formation influence in at least the down gradient direction.

Slug Test Results – Slug test results illustrate the difficulty of performing this analysis in sandy soils present at the pilot test site and typical of surficial aquifer soils at Cape Canaveral. Well recovery times were very rapid (transient times are less than 18 seconds) with usable data only generated during the first six seconds of slug insertion or withdrawal. Slug tests in the two intermediate wells (35 feet [10.7 m]) indicated an increase in hydraulic conductivity during pilot testing from initial conditions. This was inconsistent with expected results (no change or a decrease in hydraulic conductivity). Results in the two shallow wells conflicted with each other and failed to show a correlation with neutron density measurements and dissolved oxygen concentrations at similar points in the aquifer. The slug tests produced results that were unsuccessful, due to the difficulty in determining hydraulic conductivity in sandy soils. Therefore, the slug test results were deemed ineffective for this type of aquifer and application.

Based on all the results evaluated from this pilot study, it is apparent that the most optimum operating conditions was to pulse the intermediate well (SW1-40) at 5 cfm (0.14 std. m³/min). These results effectively support biosparging as an appropriate in-situ treatment for dissolved chlorinated hydrocarbons at the Hangar K Area, and a means to prevent discharge to various surface waters. This pilot study showed biosparging enhanced aerobic biodegradation by providing oxygen to the aquifer. The following are design parameters that were determined from the pilot study results:

- Well :   Single screened well above clay layer at approximately 42 ft (12.8 m) bls
- Air Flow :   Approximately 5 cfm (0.14 std. m³/min) per well, pulsed
- Radius of Influence : 5 to 10 ft (1.5 to 3.1 m)

In regards to monitoring techniques, it was determined that discrete depth monitoring wells continue to provide invaluable information. Neutron density measurements are believed to offer promise and are probably applicable for higher flow rate applications. The slug tests were unsuccessful in this application. It is possible that this monitoring technique could still be applicable for higher air flows in lower permeability soils.

BIOSPARGE SYSTEM IMPLEMENTATION
Based on the sampling results of area wells and the pilot study, a 1200 foot (365.8 m) biosparge wall was constructed to prevent migration of VOC

contaminants. The following summarizes key parameters used for design of the full-scale biosparge system:

- Length of treatment wall: 1200 ft (365.8 m)
- Required air flow rate per well: 5 scfm @14 psig, nominal
                                        (0.14 std. m³/min. at 96.5 kPa).
- Radius of influence: 5 ft (1.5 m) minimum effective
- Sparge air injection depth: 40-45 ft (12.2-13.7 m) bgs
                                        (immediately above intermediate
                                        clays)

Results from the biosparge pilot study demonstrated the effectiveness of biosparging to remove vinyl chloride and dichloroethylene contaminants at a minimum radius of influence of 5 ft (1.5 m) using a pulsed injection rate of 5 scfm (0.14 std. m³/min) at 40 ft (12.2 m) bgs. Some evidence exists that there was potential for a 10 ft (3.1 m) radius of influence. Based on the length required for the wall and well spacing requirements, 120 sparge wells were installed to complete the treatment system. The treatment wall consisted of two rows of sparge wells each spaced 20 ft (6.1 m) on center. The second row was staggered 10 ft (3.1 m) in comparison to the first. This configuration provided an adequate treatment zone and accommodates the 5 ft (1.5 m) radius of influence without causing wells to conflict in the event a 10 ft (3.1 m) radius of influence exists. A nominal injection pressure of 14 psig (96.5 kPa) (designed maximum of 17 psig [117.2 kPa]) was required at the well head of the sparge well for 5 scfm (0.14 std. m³/min) flow rate into the aquifer.

Based on a flow rate of 5 scfm (0.14 std. m³/min) per well, the complete system required a total air flow of 600 scfm (17 std. m³/min). To meet these airflow requirements, the system required a compressor rated for 600 scfm (17 std. m³/min) at 40 psig ([275.8 kPa] accommodate piping losses). A Quincy Model QSLP 100 low-pressure rotary screw air compressor was selected to deliver compressed air to the system. This compressor was installed on an equipment pad with a fence enclosure. In addition, an air receiver tank, oil filters, pressure relief valve, regulator valve, solenoid valves, and controls for the sparge system were installed at the pad as additional components for the air compressor system.

Creative Design Approach. A 3-inch (7.6 cm) diameter below grade header connects the sparge compressor to the biosparge wall above grade header piping. The biosparge wall above grade header pipe is mounted to the back of a traffic guardrail to allow ready access to control valves for each individual well and provide protection for the system piping. This above ground piping approach eliminated any conflict with the existing below grade ground utilities at the site and substantially reduced construction costs due to the elimination of below grade well vaults and controls. The biosparge wall header consists of 1 ½, 2, 2 ½, and 3 inch (3.8, 5.1, 6.4 and 7.6 cm) diameter galvanized steel pipe with a ¾ inch (1.9 cm) diameter drop leg pipe from the biosparge wall header, to each individual sparge well completed below grade. Each drop leg is equipped with connections to enable measurement and adjustment of individual well flow and pressure using portable pressure gauges and air flow meters. With this configuration, header piping was sized to deliver the required flow to each drop leg while providing sufficient pressure and flow to the most remote well at the end of the treatment wall.

Design Effectiveness. Sampling performed from the up gradient and down gradient monitoring wells, since the implementation of the full-scale biosparge system, has shown a significant reduction in vinyl chloride (See Tables 1 and 2).

TABLE 1. Reduction in Vinyl Chloride down-gradient of biosparge wall.

Well ID	Vinyl Chloride (ug/L)		
	Pre-Sparging	Post-Sparging	% Reduction
INDA-MWI25	690	0	100%
INDA-MWI27	52	11	79%
INDA-MWI29	1.5	0	100%
INDA-MWI31	ND	ND	Not applicable

TABLE 2. Reduction in Vinyl Chloride up-gradient of biosparge wall.

Well ID	Vinyl Chloride		
	Pre-Sparging	Post-Sparging	% Reduction
INDA-MWI26	270	4.6	98%
INDA-MWI28	650	720	-11%
INDA-MWI30	35	1.7	95%
INDA-MWI32	6.6	0	100%

Reduction of VOCs in upgradient wells may be a result of air migration past the anticipated radius of influence. Monitoring wells at a greater distance upgradient continue to verify migration of the contaminant plume toward the biosparge wall.

SUMMARY AND CONCLUSIONS

The following conclusions can be drawn from the pilot test and full-scale system installation:

- Biosparging can be an effective method for controlling migration of VC and DCE;
- Discrete depth monitoring wells, VOC sampling, and DO sampling continue to be a reasonable method of data collection for testing activities;
- Neutron probes offer a potentially viable technique for higher air flow applications;
- Slug tests were ineffective in this application but may have applicability for higher flow and/or lower permeability applications;
- Creative design and construction techniques can ease the financial and logistical burden of implementing such systems;
- Full-scale systems such as this can be and are effective.

REFERENCES

Nyer, E.K., D.F. Kidd, P.L. Palmer, T.L. Crossman, S. Fam, F.J. Johns II, G. Boettcher, and S.S. Suthersan. 1996. *In Situ Treatment Technology.* Geraghty and Miller Environmental Science and Engineering Series. Lewis Publishers, Boca Raton, FL.

Parsons ES. 1998. *Corrective Measures Study/Interim measures Work Plan for Hangar K Area, SWMU 22 (DP-35).* Prepared for U.S. Air Force, Installation Restoration Program, 45th Space Wing Facilities at Cape Canaveral Air Station, Florida.

Parsons ES. 1999. *Technical Memorandum for Testing/Interim Measures for 1798 Canal Aeration System and Hangar K Biosparge System for Hangar K Area, SWMU 22 (DP-35).* Prepared for U.S. Air Force, Installation Restoration Program, 45th Space Wing Facilities at Cape Canaveral Air Station, Florida.

EXPERIENCE-BASED MODIFICATIONS TO A GCW TO IMPROVE OPERATIONAL RELIABILITY

Chris Hood, P.E. (CH2M HILL, Navarre, FL USA)
Susanne Borchert, P.G. (CH2M HILL, Chicago, IL USA)

ABSTRACT: A Groundwater Circulation Well (GCW) system has been operating as an Interim Measure (IM) since April 1998 to reduce source area chlorinated solvent contamination in both unconfined and semi-confined aquifer conditions at the subject site. Four Density-Driven-Convection (DDC)-type treatment wells were installed; all are operated by a single mechanical system. In order to effectively reduce the dissolved phase solvent concentrations, system performance data was frequently monitored and evaluated resulting in a variety of system optimizations. These included 1) reconfiguring the internal parts of the treatment wells to balance hydraulic flow, 2) reversing the flow direction of the circulation cell to accommodate varying yields of inhomogeneous aquifers, 3) shocking the treatment wells with chlorine to mitigate iron bio-fouling that reduced yield efficiencies, 4) adjusting the system start-up method to continue utilizing the existing pressure blower, and 5) evaluation of flow measurement in the treatment wells. Due to the inherent flexibility of this unique DDC-system, most changes could be carried out without major system redesign; however, some limitations were found. Overall concentrations were reduced two orders of magnitude in the first 18 months of operation in the semi-confined intermediate zone and a single order of magnitude in the unconfined shallow zone.

INTRODUCTION

In an attempt to find alternatives for the costly implementation of pump and treat systems for chlorinated solvent impacted groundwater, several innovative *in situ* technologies have been implemented across the U.S. and Europe over the last decade (USEPA 1998). Closely monitored remedial activity at these sites provides insight and knowledge on the advantages and limitations of each technology. This knowledge is valuable when converted to improved system design. System improvement requires continuous review of performance data and a closely coordinated relationship between monitoring results and system optimizations. It also requires the remediation system be flexible to allow system changes in response to changing remediation conditions.

An *in situ* groundwater circulation well (GCW) technology was chosen for a site in the Southeast U.S. due to its proven flexibility in responding to changing conditions and its effectiveness in addressing impacted groundwater (USEPA 1998; Miller and Roote 1997). Following a summary of conditions at the subject site, this paper briefly describes the remediation design and implementation of four GCWs. The focus shifts to the relationship between the monitoring program and subsequent system optimizations, showing how an intimate relationship between these two steps of a remediation project leads to cost savings. The discussion addresses the advantages and limitations of the unique GCW system revealed as adaptations were made based on monitoring data.

SITE DESCRIPTION

Maintenance activities have taken place for approximately 30 years at the subject site. In 1989 a spill of PD-680 solvent was reported and resulted in the excavation of a tank and some contaminated soil. After the spill response, several investigations were conducted in accordance with the Resource Conservation and Recovery Act (RCRA). Multiple groundwater investigative techniques were used to define the nature and extent of the impact (BCM, 1994; BCM, 1995; CH2M HILL, 1997a). The information from these studies provided a basis for selecting the remediation technology to be used as an interim measure (IM) to address chlorinated solvents in groundwater at the site (CH2M HILL, 1997b).

Geologic and Hydrogeologic Setting. The focus of the site geology was on two zones of the Sand-and-Gravel Aquifer, which is defined as a single unit consisting of undifferentiated fine to coarse sands and gravels, interspersed with relatively discontinuous layers of silt and clay (Barr et al., 1981). The interspersed lower permeability lenses vary in size and extent and, at the subject site, create an unconfined surficial zone and a semi-confined, intermediate zone. The surficial zone is approximately 1.5 to 10.5 m below ground surface (bgs), while an intermediate zone extends from 12 to 26 m bgs. A relatively thin ($\pm$ 15-cm-thick) semi-confining clay layer was encountered between 10 and 12 m bgs. This clay layer was part of a 0.6- to 2.0-m-thick restrictive unit of silt and silty sand that dips slightly to the north. A denser sand layer was found at approximately 17 to 24 m bgs at all cone penetrometer technology (CPT) locations. Figure 1 presents a stratigraphic profile in a generalized cross-section showing unconfined and semi-confined aquifer zones.

Clusters of three permanent monitoring wells (MWs) were installed across the site with the screened intervals between 3 and 6 m bgs for shallow wells, 12 and 17.5 m bgs for intermediate wells, and 23 and 26 m bgs for deep wells. Both intermediate and deep MWs were used to monitor conditions in the intermediate zone. The potentiometric surface in the surficial zone typically is about 3 m higher than that of the intermediate zone, and a slight downward vertical gradient was measured from the upper intermediate to the lower intermediate zone of the aquifer. The horizontal hydraulic gradient in the shallow aquifer varies from 0.005 to 0.00015 m/m to the northeast, and the gradient in the intermediate zone is 0.0005 m/m to the west to southwest. The horizontal hydraulic conductivity at the site is 16 m/day based on slug testing data. The hydraulic conductivity anisotropy ranged from an estimated 1:10 to 1:50 in the aquifer, depending on the amount of silt layers encountered.

Chlorinated Solvent Concentrations. Although a low permeability silty-clayey layer separates the Sand-and-Gravel Aquifer's shallow and intermediate zones, both portions of the aquifer have been impacted by the solvents. The extent of the initial plumes are shown in Figure 2. The lateral extent of chlorinated solvents in

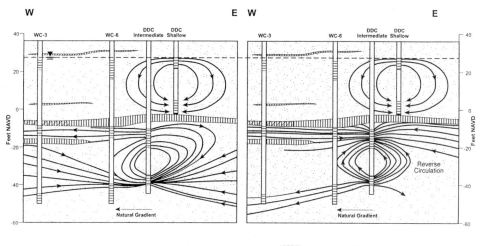

FIGURE 1. Cross-Section of the Site.

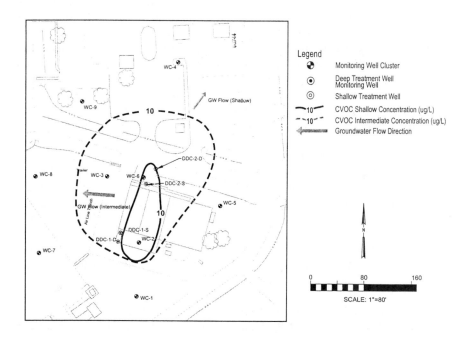

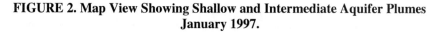

**FIGURE 2. Map View Showing Shallow and Intermediate Aquifer Plumes
January 1997.**

the intermediate zone was larger than in the shallow zone. The shallow zone (1.5 to 10.5 m bgs) was impacted primarily by trichloroethene (TCE) and dichloroethene (DCE) (48 and 34 μg/L, respectively); while the intermediate zone (12 to 26 m bgs) contained almost exclusively PCE (4,500 μg/L maximum concentration).

REMEDIATION SYSTEM DESIGN AND INSTALLATION

Objectives. The remediation design objectives were to achieve remediation goals of 3 μg/L TCE, 3 μg/L PCE, and 70 μg/L cis-1,2 DCE within a 5-year period or sooner, reduce the source of the contamination to meet the remediation goals, mitigate off-site plume migration, and limit the annual operation and maintenance and reporting budget to less than $100,000 per year. Based on the propriety nature of GCWs, the treatment wells and mechanical system were designed by subcontractors. CH2M HILL oversaw system installation and start-up and conducts system operation, maintenance, monitoring, and reporting.

Design. In order to meet remediation objectives and address the contamination in the shallow and intermediate zones, four GCWs were designed for the IM: two in the unconfined shallow zone and two in the semi-confined intermediate zone. All four dually-screened Density-Driven-Convection (DDC)-type GCWs were designed to use induced air-lift as the *in situ* stripping and standard flow groundwater pumping mechanism. In the unconfined aquifer the wells were designed to pull water from just above the restrictive layer and discharge treated water through the upper screen and infiltration gallery. In the semi-confined zone, an eductor pipe DDC-well design was implemented. This design utilizes packers and multiple screens to regulate hydrostatic pressures associated with the semi-confined zone during treatment of groundwater from this zone. Air flow rates were chosen to produce air to water ratios ranging from 1:20 to 1:40 with predicted single pass *in situ* stripping efficiencies of between 80 and 95%.

In addition to the four treatment wells (TWs), the remediation system design consisted of piping to and from an equipment trailer with the following major components: a pressure blower and vacuum blower each with particle filters and with multiple influent and effluent air regulation valves, pressure gauges, a water-knockout drum with a small pump, three 55-gallon drum granular-activated carbon (GAC) vessels, a carbon dioxide (CO_2) injection system (for pH balance), a trailer cooling system, and telemetry to regulate equipment remotely. The system was designed as a closed-loop system with respect to air flow and contaminant removal.

Installation. System construction began in February 1998 and was completed in April 1998 (CH2M HILL, 1998). First, two lithologic borings were advanced to characterize subsurface units at the locations of the DDC TWs. Data gathered from these borings were used to place the TW influent and effluent screens. Using mud rotary drilling, a 12-inch (0.3-m) diameter borehole was advanced for the construction of each shallow TW, and a 17-inch (0.4-m) borehole was advanced for the construction of each intermediate TW. Two monitoring piezometers were installed with screens adjacent to the influent and effluent screens of the TWs. Well development is critical for effective operation of any GCW; therefore separate development of each screen was conducted using inflatable packers, surging, and over pumping with centrifugal, submersible, and airlift pumping.

Monitoring. The physical parameters of the operational system were monitored as were chemical data reflecting the effectiveness of the system on reducing chlorinated solvent concentrations in groundwater. Air flow rates to and from the TWs, temperature inside the trailer, pH levels in groundwater, and groundwater level measurements are the primary operational parameters.

RESULTS AND DISCUSSION

Data collected during system performance monitoring was continuously reviewed to determine possible changes for system optimization. Table 1 summarizes various observations from the monitoring data, the conclusions drawn, and the remedy chosen to overcome the problem and optimize system performance. Each system modification is addressed in greater detail in the text following the table.

TABLE 1. Observations, Conclusions, and System Modifications

Observations	Conclusions	System Optimizations
- GW hydraulics not balanced for continuous flow - Schedule 40 and 80 PVC bends - Need to use screen for discharge - Holes can be used for intake	- Redesign eductor for better effluent water distribution and increased mounding pressure - Improve water deflector on the air line	- Reconfigured eductor pipe for optimum flow and stripping
- Minimal hydraulic head differential between upper and lower part of TW - Water levels and pressure transducer tests showed smaller circulation size than expected in the intermediate zone	- Leaking packer causing partial short circuiting of circulation cell	- Replaced shale trap packer with inflatable type and initiated packer tightness testing (pressure differential)
- Downgradient VOC migration - Tighter, finer soils in upper intermediate zone limiting efficient release of stripped GW into the aquifer - Water levels and pressure transducer tests showed limited pressure differential radially outward from an intermediate TW	- Conventional GCW circulation cell was not developing in intermediate aquifer	- Reversed flow in the two intermediate DDC TWs to pull groundwater from the former effluent screen and release it from the former influent screen.
- Iron biofilm observed on submerged air line and eductor pipe - GW flow through TWs reduced - Iron particulate on blower filter - Increased oxygen levels in GW due to in situ stripping	- Iron fouling of the screen increased oxygen in GW enabling increased iron bacteria growth	- Chlorine shock in TWs to eliminate iron fouling bacteria

TABLE 1. Observations, Conclusions, and System Modifications (Continued)

Observations	Conclusions	System Optimizations
- Adjusting air line submergence to change GW flow rates through the TW physically difficult - Empirical GW flow rates tables based on drawdown curves from well development, do not take circulation hydraulic dynamics into consideration - Unable to calculate exact mass removal rates with estimated GW flow rates	- Cannot determine exact GW flow rates which influence how to optimize system in many ways	- Install exact flow measuring devices in eductors/TWs (not carried out to date)
- Required start-up pressure is greater than the operational one - Limited total air volume was sufficient to start-up 2 to 3 of the 4 TWs at once, depending on seasonal water level	-Insufficient pressure blower capacity to overcome initial water displacement in 4 TWs with air distributed to all TWs at once	-Developed new start-up instructions, starting operations on one TW and successively adding other wells by pressure adjustments through regulating valves

GCW = Groundwater Circulation Well; GW = Groundwater; TW = Treatment Well; VOC = Volatile Organic Compounds

Since the packer and eductor pipe changes are closely related (both manipulate groundwater flow through the TW) these two are discussed jointly. Water level data from the TW piezometers were collected frequently and were used to help evaluate the operation of the DDC system. Over time, a diminishing difference between the influent and effluent water levels of the intermediate TW piezometers was recorded. The re-evaluation concluded additional "free board" was required within the TW to create an increased hydraulic head needed to force the treated water into the less permeable upper intermediate zone. At about the same time of this evaluation, an eight-hour pressure transducer test (in one of the intermediate TWs) showed that intense circulation pumping only influenced the MWs around this intermediate TW to a small degree. Based on these results, it was concluded the shale trap packer was leaking causing partial short-circuiting inside the TW. Subsequent modifications to the system included adding a section of casing to the eductor pipe for increased hydraulic head, modifying the 125 slot screen for the diffusion of water, and installing an inflatable packer around the eductor pipe. These modifications allowed groundwater to accumulate to a higher hydrostatic level inside the TW exerting more pressure onto the aquifer and thus producing an efficient circulation cell.

Although measurable intermediate zone circulation cells were developing, this change caused operational problems with the treatment trailer. Water was pulled through the vacuum lines at a rate higher than the knockout drum system could handle. This resulted in automatic system shutdowns. As a consequence, the eductor pipe design was reconfigured again. The reconfiguration added air to the annular space between the well casing and the eductor pipe. This optimization of the eductor pipe and air lines reversed the flow in the intermediate TWs. The intermediate TWs have been successfully operating with reverse flow over the past eight months.

As part of the normal operational system monitoring, pressure and air flow rates are monitored at each of the TWs. Over time the difference between the pressure in the air supply lines for the two shallow TWs became pronounced, prompting a review of the air line submergence depth. While taking the air lines out of the TW casings, a film of biological substance was observed on the air line as well as on the eductor pipe on the sections that were in contact with the air-water mixture. Monitoring data from the CO_2 system, which maintains the pH in the wells equal to that of the aquifer, showed continuous treatment. pH levels measured in the groundwater remained stable and below levels at which metals (particularly iron) would precipitate. A sample of the biological material was collected and sent to a laboratory for analysis. Results indicated high organic content and concentrations of iron, vanadium, manganese, nickel, and other metals. Based on this information, it was determined that the material was primarily biological in nature, thus shock chlorination was performed on the TWs. This treatment was successful and cleared the screens thereby increasing the recharge rate of the TWs.

One factor that directly influences the size and effectiveness of the circulation cell, and therefore the remediation time requirements, is the rate at which the groundwater is pumped through the TW. The DDC-type systems simultaneously use the air-lift effect of injected air as the *in situ* stripping mechanism and the pumping mechanism. Primarily the submergence depth of the air line determines the rate of pumping. Unfortunately, it is physically cumbersome to alter the air line depth and thereby the groundwater flow rate. In addition, the empirically determined curve that correlates the submergence with the pump rate is site-specific and is determined during TW development. Hydraulics during this one-way extraction, however, are not the same as those governing conditions during groundwater circulation, which uses pressure differentials versus one extraction pressure. The inability to determine the exact groundwater flow rate furthermore prohibits contaminant mass removal rates to be calculated, although monitoring of the contaminant concentrations in the groundwater passing through the TWs is conducted regularly. Flow-measuring devices have been evaluated for their applicability to the DDC wells; however, to date an adequate solution for accurately measuring groundwater flow rates has not been implemented.

The capacity of the pressure blower (compressor) was primarily based on three factors: 1) 3-m of hydrostatic pressure to overcome for air line submergence, 2) air flow rates for efficient stripping with an air to water ratio of 1:20 to 1:40, and 3) head loss in air distribution piping and fittings. Airflow rates and the pressure in the air lines were monitored using both digital and analog output. Data showed blower capacity did not limit on-going operations of the DDC remediation system; however, system start-up was difficult due to differences in start-up and operation air distribution requirements. Overcoming the hydrostatic pressure with the compressed air distributed across the four TWs often failed, resulting in system shutdown. In lieu of installing a higher capacity blower, new system start-up instructions were developed. TWs were successively started up one at a time, enabling the larger volume of air to overcome the hydrostatic pressure and initiate air-lift pumping at each well. While requiring additional time to re-balance the overall system after start-up, using the sequential method optimized use of the existing equipment.

CONCLUSIONS

By continuously interpreting data gathered during the operation of a unique DDC-type GCW remediation system, required system optimizations could be identified and carried out promptly. We were able to take advantage of the inherent flexibility of the DDC-type remediation system to optimize its effectiveness under unconfined and semi-confined conditions. In addition, we were able to successfully adapt the circulation cell to the stratigraphically heterogeneous conditions in the intermediate zone by reversing the groundwater flow direction in two TWs and their surrounding matrix. The optimizations ultimately led to successful operation of the two intermediate TWs. Chlorinated solvent concentrations decreased by two orders of magnitude in the intermediate aquifer. As a result of the system optimizations, the IM appears to be the final remedy for the site and will result in overall time and cost savings in meeting regulatory requirements for site closure.

ACKNOWLEDGEMENTS

We would like to thank Mathilda Cox for patiently editing the text. Mr. Ben Coulter provided an insightful review.

REFERENCES

Barr, D., A. Maristany, and T. Kwader. 1981. *Water Resources of Southern Okaloosa and Walton Counties, Northwest Florida.* Northwest Florida Water Management District Water Resources Assessment 81-1.

BCM Engineers Inc. May 9, 1994. *Hurlburt Field Preliminary Assessment (PA) Report for the Installation Restoration Program.* Prepared for the U.S. Air Force, Center for Environmental Excellence (AFCEE).

BCM Engineers Inc. August 1995. *Site Inspection Report: AOC-125 UST Spill (Building 90731).* Prepared for AFCEE.

CH2M HILL. 1997a. *RCRA Facility Investigation Report for SS-125, UST Spill at Weapons Maintenance Facility (90731).* Prepared for AFCEE.

CH2M HILL. 1997b. *Interim Measure Alternatives Evaluation for SS-125.* Prepared for AFCEE.

CH2M HILL. May 1998 – November 1999. *Monitoring Reports for SS-125, UST Spill at Weapons Maintenance Facility (90731).* Prepared for AFCEE.

Miller, Ralinda R. and Roote, Diane S. February 1997. *Technology Overview Report, TO-97-01, In-well Vapor Stripping* GWRTAC, Ground-Water Remediation Technologies Analysis Center.

USEPA, Solid Waste and Emergency Response. October 1998. *Field Applications of in situ Remediation Technologies: Ground-Water Circulation Wells.* EPA 542-R-98-009.

OZONE SUPERSPARGING FOR CHLORINATED
AND FLUORINATED HVOC REMOVAL

William B. Kerfoot (K-V Associates, Inc., Mashpee, Massachusetts)

ABSTRACT: Ozone supersparging with Criegee oxidation is demonstrated for chlorinated and fluorinated ethenes and ethanes. Independent testing has verified that Criegee oxidation dominates in the combined gas/aqueous fractions. No partial breakdown products have been detected in laboratory or pilot tests. The reaction proceeds through a low molecular ratio of ozone to attack the central bond of the ethene molecule, when ethene derivatives are targeted, resulting in the decomposition of the halogenated fragments clearing into a carbonyl compound and a hydroxy hydroperoxide, which rapidly lose dilute hydrochloric acid (HCl) and dilute hydrofluoric acid (HFl) in the presence of water. Reactions with alkane compounds such as 1,1,1-trichloroethane (1,1,1-TCA) or chlorofluorohydro-carbons (Freon) also show rapid destruction in field testing.

THEORY

The KVA process is a patented technology* (the C-Sparge™ process) for *in-situ* treatment of VOCs in groundwater and surface water. The technology combines the unit operations of air stripping and oxidative decomposition in a single process which can be catalytically accelerated. Air and ozone are injected directly into groundwater through Spargepoints®, creating microbubbles which have a very high surface area to volume ratio. Extraction of VOCs from groundwater occurs by aqueous to gas partitioning as the bubbles rise. The ozone contained within the bubble reacts to decompose the chlorinated ethene molecule as it enters the bubble in an extremely rapid gas/liquid phase reaction. The end products are carbon dioxide, very dilute hydrochloric acid, and water. Ozone content in the bubble controls the rate of oxidation. The process is effective on halogenated volatile organic compounds (HVOCs) such as perchloroethene (PCE), trichloroethene (TCE), the dichloroethenes (DCEs: 1,1-dichloroethene, and 1,2-dichloroethene), and vinyl chloride (VC).

Gas entering a small bubble of volume $4\pi r^3$ increases until reaching an asymptotic value of saturation. If we consider the surface of the bubble to be a membrane, a first order equation can be written for the monomolecular reaction:

$$\frac{dx}{dt} = k(Q-x) \tag{1}$$

Where x   = the time varying concentration of the substance in the bubble
     dx/dt = the change in concentration within the bubble

*U.S. patent #5,855,775; other U.S. and foreign patents pending

Q   = the external concentration of the substance
k   = the absorption constant

If at $t = 0$, $x = 0$, then:

$$x = Q(1-e^{-kl}) \tag{2}$$

The constant k is found to be:

$$k = \frac{dx/dt}{Q-x} \tag{3}$$

By multiplying both numerator and denominator by V, the volume of the bubble, we obtain:

$$k = \frac{Vdx/dt}{V(Q-x)} \tag{4}$$

which is the ratio between the amount of substance entering the given volume per unit time and quantity $V(Q-x)$ needed to reach the asymptotic value. By analyzing the concentration change within the fine bubbles sent through a saturated (water-filled) porous matrix interacting with a catalytic matrix (iron silicate), the kinetic rates of reaction can be characterized.

The rate at which the substance quantity k_1QV flows in one unit of time from aqueous solution into the bubble is proportional to Henry's Constant. The second rate of decomposition within the bubble can be considered as k_2, a second rate of reaction ($-k_2x$), where:

$$\frac{dx}{dt} = k_1Q-k_2x \tag{5}$$

At equilibirum, as $dx/dt = 0$:

$$x = \frac{k_1}{k_2} Q \tag{6}$$

However, if the reaction to decompose is very rapid, so $-k_2x$ goes to zero, the rate of reaction would maximize k_1Q, *i.e.*, be proportional to Henry's Constant and maximize the rate of extraction since VOC saturation is not occurring within the bubbles.

PROCESS DESCRIPTION

The concentration of HVOC expected in the bubble is a consequence of rate of invasion and rate of removal. In practice, the ozone concentration is adjusted to yield 0 concentration at the time of arrival at the surface or exit from the reactor:

$$r_{VOC} = -K_{La_{VOC}}(C-C_s) \tag{7}$$

Where r_{VOC} = rate of VOC mass transfer, $\mu g/ft^3 \bullet h$ ($\mu g/m^3 \bullet h$)
      $K_{La_{VOC}}$ = overall VOC mass transfer coefficient, L/h
      C   = concentration of VOC in liquid
      C_s = saturation concentration of VOC in liquid $\mu g/ft^3$ ($\mu g/m^3$)

The saturation concentration of a VOC in wastewater is a function of the partial pressure of the VOC in the atmosphere in contact with the wastewater:

$$\frac{C_g}{C_s} = H_c \tag{8}$$

Therefore: $C_g = H_c \bullet C_s \tag{9}$

Where C_g = concentration of VOC in gas phase $\mu g/ft^3$ ($\mu g/m^3$)
      C_s = saturation concentration of VOC in liquid $\mu g/ft^3$ ($\mu g/m^3$)
      H_c = Henry's Constant

The rate of decomposition is now adjusted to equal the total HVOC entering the bubble.

SET: $[H_c \bullet C_s] = K_{O_3}{}_c[O_3][C_g] \tag{10}$

Therefore surface concentration or exit condition = 0.

The critical factors for speed of reaction become ozone concentration, HVOC concentration, partial pressure, bubble size, and adequate catalyst presence. The reaction is presented as:

$$H_2O + HC_2Cl_3 + O_3 = 2CO_2 + 3HCl \tag{11}$$

An assumed ozone injection rate of 48 grams per hour (1.0 moles/hour) yields the following:

$$1.0 \text{ mol/hr } H_2O + 1.0 \text{ mol/hr } HC_2Cl_3 + 1.0 \text{ mol/hr } O_3 =$$

$$2.38 \text{ mol/hr } CO_2 + 2.98 \text{ mol/hr } HCl \tag{12}$$

Dowideit and von Sonntag (1998) recently confirmed the product yield ratios at 2.38 moles CO_2 and 2.87 moles HCl, very close to the expected values.

The rate of decomposition was measured as millseconds, with intermediate ozonides with a lifetime <2ms, as shown by stopped-flow conductometry.

Since a total HVOC content of 1000μg/L in a flow of 100 gallons per minute contains only .32 mol/hr, the capacity to effectily reduce HVOC content to drinking water levels is considerable at a low energy usage.

The mechanism of ozone attack (Figure 1) suggested by Criegee involves attacking the central double bond. Ozone first forms a π complex that collapses into a zwitterionic α complex. This primary zwitterion closes to the primary ozonide.   The primary ozonide is an unstable trioxide. The primary ozonide cleaves an O-O bond heterolytically. Then the secondary zwitterion decomposes into a carbonyl compound and the tertiary zwitterion. In the presence of water, the compounds decompose to CO_2 and dilute HCl (Masten, 1986).

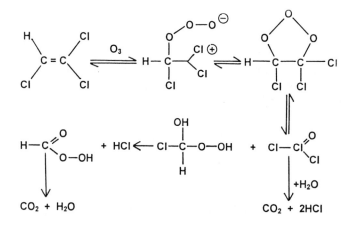

FIGURE 1. Mechanism of attack of TCE by Criegee oxidation (Dowideit and von Sonntag, 1998; Masten, 1986).

Recent laboratory testing at the Max Planck Institute of Aquatic Chemistry has confirmed that the Criegee oxidation pathway dominates over competing mechanisms in aqueous solution (Dowideit and von Sonntag, 1998). Stopped-flow reactions in aqueous solution showed the olefins (ethene, trichloroethene, cis-1,2 dichloroethene, trans-1,1 dichloroethene, and vinyl chloride) nearly exclusively follow the Criegee mechanism and cleave into a carbonyl compound and a hydroxyhydroperoxide, which in the case of a chlorine substituent rapidly loses HCl. In the cases of chlorinated olefins, no HCl-releasing intermediate with a lifetime longer than 1 millisecond was detected.

No organic by-products were found.   No newly-formed volatile organochlorine compounds have been observed in laboratory reactions (Kusakabe, et al., 1991; Dowideit and von Sonntag, 1998). Field analyses have verified independent decomposition of sister compounds PCE, TCE, DCE, without conversion of one form to another (Kerfoot, 1997a, 1998). Field testing has also failed to find mono- or dichloroamines, epoxides, or formaldehyde compounds during treatment (Kerfoot, 1997b).

BATCH TESTING

Five batch tests were conducted. During each test, 250 mL of a given "spiked" source water was placed into a 500 mL Erlenmeyer flask equipped with a magnetic stirring bar. Air and/or ozone were introduced into the solution through a sparger placed in the bottom of the flask. Treatment generally consisted of sparging an air and ozone mixture (containing approximately 150 to 200 parts per million [ppm] into the flask at 5 pounds per square inch (psi) and at a rate of 5 cubic centimeters per second (cc/sec).

The effect of the contact medium, a coarse sand material (iron silicate) of differing size distribution, was evaluated during the testing. For consistency, glass beads were used as a control to bring the total weight of added solids to 10 grams (g) per test.

Samples analyzed by a laboratory and an on-site gas chromatograph (GC) were collected before treatment (time zero) and after 6 minutes of treatment. In addition, for most test runs, samples collected after 3 minutes of treatment were analyzed using the on-site GC.

The results of the batch testing portion of the bench-scale stdies are shown in Table 1. In general, the 6-minute duration of the flask tests was not sufficient to reduce the spiked concentrations of the VOCs (as much as 100-400 µg/L each) to below the method detection limits. Based on the bench-scale results, the best percentage removals were achieved for PCE and the cis- and trans- 1,2-DCE, followed by TCE, and then 1,1,1-TCE. Methylene chloride, carbon tetrachloride, and chloroform had a significantly lower percent removal.

TABLE 1. Bench-scale testing for chlorinated volatile organic compound (HVOC) removal rate (volume 500 cc).

Compound	Concentration (ppb)		% Removal
	0 min.	6 min.	
Perchloroethene (PCE)	500	2	99.6
Trichloroethene (TCE)	470	50	89.4
Cis-1,2 dichloroethene	470	2	99.6
Trans-1,2 dichloroethene	480	7	98.5
1,1,1-TCA	470	82	82.6
Carbon tetrachloride	450	140	68.8
Methylene chloride	450	120	73.3
Chloroform	540	190	64.8

FIELD RESULTS

The following results are from a site occupied by a commercial electronics supplier which had spills from circuitboard manufacture. The predominant soils are a silty fine sand. A single recirculation well was used to inject ozone into the water stream flowing to a low volume (<5gpm) containment well. The injection spargewell was located 97 ft (30 m) from the recovery well.

A variety of haologenated alkenes and alkanes occurred as contaminants at the site. These included PCE, TCE, DCE (1,1- and cis-1,2-) vinyl chloride, 1,1,1-TCA, 1,1-DCA, TCTFE (Freon), and lesser quantities of chloroform (CF) and carbon tetrachloride(CTC).

The observed rate of reduction in alkanes 1,1-DCA, 1,1,1-TCA, and TCTFE were greater than the alkene compounds (PCE, TCE, and DCE). Precise rates for vinyl chloride (VC), CF and CTC were not calculated since the end point was non-detect (Figure 2). (Table 2).

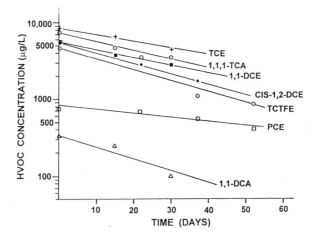

FIGURE 2. Removal rates of HVOCs during the injection pilot test.

A multisite comparison of chlorinated solvent remediation using innovative technology has independently concluded that the Criegee oxidation procedure showed the greatest rate of removal at lowest operating cost (Dreiling, et al., 1998). The use of a 120 volt, 20 amp power supply for the ozone sparger provided a low electrical consumption during the pilot test.

RESULTS AND CONCLUSIONS

The use of engineered microbubble systems with ozone has proved a powerful means of targeting and effectively decomposing chlorinated and fluorinated ethenes and ethanes to harmless by-products of dilute hydrochloric acid (HCl), dilute hydrofluoric acid (HFI), and carbon dioxide (CO_2). The reaction requires a low molar ratio of ozone to attack the central bond of the ethene molecule when ethenes are targeted, resulting in the decomposition of the chlorinated fragments clearing into a carbonyl compound and a hydroxy hydroperoxide, which rapidly lost HCl and HFL in the presence of water.

TABLE 2. Results from ozone pilot test, Hawthorne, California.

Compound	Concentration (PPB) Start	End*	Decay Coefficient (k)	Half Life (Days)
Tetrachloroethene	750	430	-.012	58
Trichloroethene	8200	4500	-.020	35
Cis-1,1 dichloroethene	5600	2800	-.023	30
Cis-1,2 dichloroethene	5600	1700	-.032	22
Vinyl chloride	6	ND	---	<20
1,1,1-trichloroethane	6700	3500	-.023	30
1,1-dichloroethane	340	100	-.035	20
Chloroform	9	ND	---	<20
Carbon tetrachloride	2	ND	---	<20
Trichlorotrifluoroethane	4800	820	-.044	16

*Start and end times vary with sampling. See Figure 2 for plotted results

Independent testing of the reaction has verified that the Criegee mechanism dominates in gaseous and aqueous reactions. The formation of major products can be explained on the sole basis of the Criegee mechanism with ethylene, PCE, TCE, DCE and cinyl chloride. No detrimental products of partial cleavage have been detected in field pilot tests or remedial efforts.

The effectiveness of reaction is not limited to ethenes, but extends to alkane derivatives as well. Chloroform, methylene chloride, and carbon tetrachloride are selectively and rapidly removed due to high Henry's constants, even though cleavage is not as rapid as in the case of ethenes. Reactions in field situations show rapid destruction of chloro-fluoro-hydrocarbons, such as Freon. Kinetic rates of reaction observed in field trials are given for a variety of compounds.

Field use over the past three years has demonstrated economical operation, effectiveness and cleanliness which may qualify the process for use of removing chlorinated ethenes in drinking water aquifers or water streams. The rates of reaction are shown in various groundwater conditions, ranging from acidic sands to basic carbonate limestones. The use of microbubble injection by microporous materials (C-Sparging™) has repeatedly shown the capacity to lower chlorinated solvent spill aqueous plume regions below drinking water MCLs. Normally, a pH change of only .5 units towards acidity is observed. An increase in dissolved oxygen occurs when ozone is supplied in quantities greater than the decomposition reaction. Third party economic analysis has shown operating costs far below standard air sparging.

REFERENCES

Dowideit, P. and C. von Sonntag. 1998. "Reaction of ozone with ethene and its methyl- and chlorine-substituted derivatives in aqueous solution. *Envi. Sci & Tech.* 32(8):1112-1119.

Dreiling, D. N. L. G. Henning, R. D. Jurgens, and D. L Ballard. 1998. "Multi-Site Comparison of Chlorinated Solvent Remediation Using Innovative

Technology." In G. B. Wickramanayake and R. R. Hinchee (Eds.), The Proceedings of the: *Physical, Chemical and Thermal Technologies, Remediation of Chlorinated and Recalcitrant Compounds* , Vol. C 1-5, pp. 247-252. Battelle Press, Columbus, Ohio.

Kerfoot, W. W. 1997a. "Extremely Rapid PCE Removal from Groundwater with a Dual-gas Microporous Treatment System. In P. T. Kostecki, E. J. Calabrese, and M. Bonazountas (Eds.), *Contaminated Soils, pp. 275-284.* ASP Publishers, Amherst, Massachusetts.

Kerfoot W. B. 1997b. "PCE Removal from Groundwater with Dual-gas Microporous Treatment System." *Soil and Groundwater Cleanup,* pp. 38-41.

Kerfoot, W. B., C. J. J. M. Schouten, and V. C. M. van Engen-Beukeboom. 1998. "Kinetic Analysis of Pilot Test Results of the C-sparge™ Process." In G. Wickramanayake and R. E. Hinchee (Eds.), The Proceedings of the: *Physical, Chemical, and Thermal Technologies, Remediation of Chlorinated and Recalcitrant Compounds,* Vol C 1-5, pp. 271-277. Battelle Press, Columbus, Ohio.

Kusakabe, K., S. Aso, T. Wada, Hayashi, S. Morooka. 1991. "Destruction Rate of Volatile Organochlorine Compounds in Water by Ozonation and Ultraviolet Light." *Water Research Watrag,* 25(10):1199-1203.

Masten, S. J. 1986. "Mechanisms and Kinetics of Ozone and Hydroxyl Radical Reactions with Model Aliphatic and Olefinic Compounds." Ph.D. Thesis, Harvard University, Cambridge, Massachusetts.

McCormack, C. E. 1995. "Kinetics and Mechanism of the Reaction between Ozone and Chlorinated Alkenes in the Aqueous Phase." Master's Thesis, Michigan State University, Houghton, Michigan.

FIELD SCALE BIO-SPARGING OF PENTACHLOROPHENOL

R. Donald Burnett and Malcolm K. Man
Morrow Environmental Consultants Inc., Burnaby, B.C., Canada

ABSTRACT: An abandoned dip tank at a former sawmill was the major source of a dissolved pentachlorophenol (PCP) plume which extended 150 m to the ocean. Geochemical assessment indicated that observed attenuation was due to oxygen-limited bio-degradation, however, PCP concentrations were above the applicable criterion at discharge. As bio-degradation appeared to be oxygen limited, a pilot-scale air sparge curtain was implemented as a plume control measure. The pilot test showed that a bio-sparge curtain was an effective control for dissolved-phase PCP contamination. The remedial approach adopted consisted of partial excavation of source area soil and expansion of the sparge curtain across the full width of the plume. The remaining contamination, which is largely below the water table, is being allowed to leach to the groundwater. The resulting PCP plume is being remediated by the sparge curtain. Monitoring has confirmed that dissolved PCP concentrations up-gradient of the curtain decreased rapidly following partial source removal, and that the full scale curtain is successfully controlling the plume. It is expected that the sparge curtain will be operated for approximately two years, after which natural attenuation will be sufficient to degrade the remaining plume.

INTRODUCTION

The growth of fungus on green lumber during shipment results in unsightly discolouration referred to as 'sapstain'. From the 1950's to the early 1990's, pentachlorophenol (PCP) was commonly used as an anti-sapstain fungicide. Lumber was dipped in an alkaline solution of sodium pentachlorophenate (the ionised form of pentachlorophenol) in water and allowed to drain and air dry. Dipping was accomplished using either elevator dip tanks or larger drive-through dip tanks. In the 1980's, spray booths became the accepted application method in order to limit the loss of PCP to the environment.

A former sawmill site located on the shore of Burrard Inlet in south-western British Columbia, Canada, was known to have had a drive-through dip tank and a spray booth for lumber treatment prior to export. The spray booth and the drive-through dip tank were in use until 1984, and had been removed during site redevelopment in the late 1980's. Some PCP soil contamination remained in the area of the former dip tank. During delineation of the dissolved PCP plume down-gradient of the drive-through dip tank, a second, much higher concentration PCP plume was identified. The second plume was traced back to an abandoned elevator dip tank that was used between the early 1950's and 1962. This tank remained in place beneath the paved site.

Objectives of Remediation. The first objective of the remedial program was to control the dissolved PCP plume and prevent further potential discharge to the marine receiving environment. The second objective was to remediate source area soil contamination and the dissolved groundwater plume so that the site could be redeveloped.

SITE HYDROGEOLOGY AND GEOCHEMISTRY

Site Description. The site is located on the margin of a high-energy delta built out into Burrard Inlet, which serves as Vancouver's harbour. The elevator dip tank was located about 150 m north of the current shoreline (see Figure 1). The former drive-through dip tank was located approximately 15 m north of a dredged barge basin, which was infilled in approximately 1989 as part of the redevelopment of the sawmill site. Surrounding land usage is industrial. The shallow aquifer northwest of the site is known to have been impacted by hydrocarbons due to historic activities.

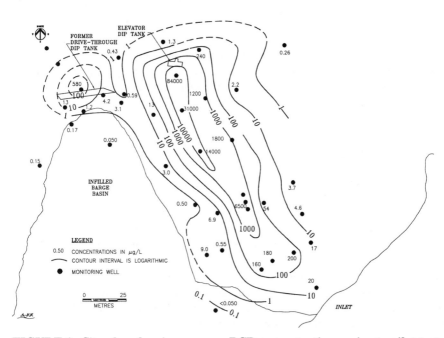

FIGURE 1. Site plan showing average PCP concentrations prior to pilot test.

Soils beneath the site consist of approximately 1 m to 2 m of granular fill and 1 m of mudflat sandy silt overlying coarse-grained sand, gravel and cobble soils of fluvial-deltaic origin. The coarse-grained layers overly an interbedded sequence of marine silty sand and silt alternating with fluvial-deltaic sand and gravel. Locally, the shallow sand, gravel and cobble layers form a surficial aquifer extending to a basal layer of marine silty sand at approximately 9 m depth.

The former barge basin was dredged to approximately the top of the marine silty sand, then later filled with relatively low permeability glacial till excavation spoil.

Site Hydrogeology. Water levels beneath the site vary tidally between 1.4 m and 3.0 m depth beneath ground surface. Daily tidal variation in wells near the shoreline is approximately 1.5 m, or 30% of the tidal variation in the adjacent inlet. Tidal response in the vicinity of the elevator dip tank is approximately 0.5 m. The shallow sand and gravel aquifer varies between confined or semi-confined at high tide and unconfined at low tide.

The shallow aquifer has a tide-averaged horizontal gradient of 0.0006 m/m southwards towards the inlet. Based on pump test analysis, the aquifer has an average horizontal hydraulic conductivity of approximately 3×10^{-3} m/s. This results in a net average linear groundwater velocity of approximately 0.5 m/day or 180 m/year.

Nature and Extent of Contamination. Soil analysis indicated the presence of PCP contamination of soil near the east end of the former drive-through dip tank and for a 15 m radius surrounding the elevator dip tank. Maximum concentrations of 560 mg/kg PCP were measured between 2 m and 3 m depth, coincident with the zone of tidal water table fluctuation.

Dissolved PCP concentrations in groundwater near the elevator dip tank ranged from 30,000 µg/L to 170,000 µg/L and averaged 90,000 µg/L. Source area concentrations near the former drive-through dip tank were much lower, averaging 450 µg/L. The two PCP plumes merged and migrated southwards (see Figure 1). The average PCP concentration 40 m north of the inlet was approximately 200 µg/L. Due to site access limitations, wells could not be installed in the last 40 m of the plume up-gradient of the discharge point to the inlet.

PCP Migration and Natural Attenuation. The dissolved PCP plume had been migrating along its current flow path since filling of the barge basin in 1989. Plume delineation indicated that centreline concentrations decreased under existing conditions from 90,000 µg/L to 200 µg/L over a travel distance of 130 m. The applicable criterion for point of discharge PCP concentration was 1 µg/L. Due to access limitations, point of discharge concentrations could not be measured, therefore, an assessment of natural attenuation rates was made in order to estimate discharge concentrations.

Rates of natural attenuation were assessed using the computer program "BIOSCREEN" (Newell et al., 1996.) Based on site-specific pH and organic carbon measurements, a retardation rate of 1.26 was estimated for PCP. The best fit pseudo-first order degradation rate for site PCP measurements was 0.011 /day, or a half-life of 62 days. Based upon this degradation rate, the potential discharge concentration was estimated to be 50 µg/L, well in excess of the applicable criterion. It was noted, though, that the apparent first-order degradation rate was increasing down the length of the plume, indicating that the estimate of PCP discharge concentration may have been high.

Assessment of geochemical indicators showed that redox conditions up-gradient of the elevator dip tank were mildly reducing, with low dissolved oxygen and elevated dissolved iron concentrations of 14,000 µg/L to 25,000 µg/L. Within the plume, dissolved oxygen concentrations remained generally low (600 µg/L to 1,300 µg/L), but both dissolved iron and PCP concentrations decreased. PCP is known to readily degrade under both anaerobic and aerobic conditions (McAllister et al., 1996). No daughter products of reductive dehalogenation were detected, so it was postulated that aerobic degradation was occurring under oxygen-limited conditions. Enhanced oxygenation due to tidal fluctuation within an unconfined aquifer has been documented at another site (Hardy et al., 1999).

PILOT SCALE BIO-SPARGE TEST

Materials and Methods. A pilot-scale sparge test was carried out to test the effect of increased oxygenation on the degradation of PCP. Three sparge wells were installed at 4 m spacing across the centreline of the plume 80 m down-gradient of the elevator dip tank (see Figure 2). Each sparge well consisted of 50 mm diameter PVC pipe with a 0.6 m length of 0.25 mm slotted screen set near the base of the shallow aquifer (7.9 m to 8.5 m below ground surface.)

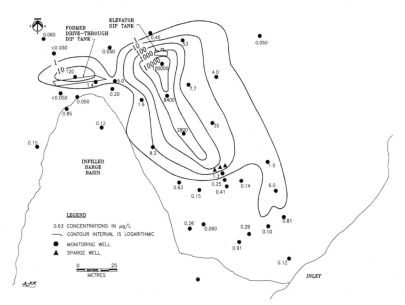

FIGURE 2. Average PCP concentrations during eight month pilot test.

In order to evaluate the effectiveness of the pilot sparge system, shallow wells (screened from the water table to approximately 4 m below ground surface) were installed 4 m and 8 m down-gradient of the center sparge well. Nested shallow and deep wells were located 12 m down-gradient. Testing during commissioning of the system consisted of two short-term (half-hour) sparge tests and one helium injection test using the center sparge well. Monitoring included water level, dissolved oxygen (DO) and helium measurements in the adjacent monitoring wells.

Initial testing indicated that a groundwater mound built up for an 8 m radius around the sparge well during the first ten minutes of sparging. The mound collapsed and returned to static conditions over the next twenty minutes. Bubbles and increased DO were measured only at the 4 m radius monitoring well. During the helium injection test, helium was detected at the 4 m well after less than five minutes, the 8 m well after 20 minutes and at the 12 m well after 35 minutes.

Based on the initial testing, it was decided to operate the pilot sparge system on a cycled basis, with air injected to all three wells on a fifteen minute on, fifteen minute off basis. Sparge rates were 0.17 m³/min (6 cfm) per well. It was hoped that the groundwater circulation induced by the cyclic sparging would increase contact between sparge air and the groundwater to increase oxygenation.

The pilot test ran from April to December 1998. Seven wells were selected for monitoring of the pilot test. Four wells were shallow wells along the plume centreline 29 m up-gradient and 4 m, 8 m and 12 m down-gradient of the pilot sparge curtain. A deep (base of aquifer, 7 m to 8.5 m below ground surface) well 12 m down-gradient was monitored to evaluate vertical effects. Shallow wells located down-gradient of the sparge wells and 17 m west and 10 m east of the plume centreline were also monitored to evaluate possible horizontal plume shifts.

Field and laboratory monitoring data were collected from the shallow and deep wells 12 m down-gradient on a weekly basis, and from the remaining wells on a bi-weekly basis. Field data was collected using a peristaltic pump and flow-through cell, and included DO, pH, and redox potential. Lab analysis included major anions, nitrate, dissolved iron and manganese, and chlorophenols.

Results. PCP and redox indicator species data collected at the shallow well located 12 m down-gradient of the sparge curtain are summarised on Table 1. The PCP plume is shown in plan on Figure 2. The PCP concentrations along the plume centreline up-gradient and down-gradient of the sparge curtain are shown on Figure 3.

Four weeks into the test, groundwater conditions down-gradient of the sparge curtain had changed from iron-reducing to aerobic. PCP concentrations had decreased from approximately 4,000 µg/L to less than 1 µg/L. PCP concentrations less than the discharge criteria of 1 µg/L were maintained down-gradient of the pilot sparge curtain for the duration of the pilot test.

Monitoring indicated that the radius of influence of the sparge system was much greater than anticipated. The PCP concentrations wells located 10 m east of the plume centreline and 17 m west of the plume centerline decreased from 54 µg/L and 7 µg/L respectively pre-test to less than 0.3 µg/L throughout the pilot test. Eight weeks into the test the central sparge well was shut down, leaving two sparge wells at 8 m spacing. This had no apparent effect on down-gradient redox conditions or PCP concentrations.

Deep groundwater down-gradient of the sparge curtain remained anoxic and nitrate-reducing throughout the pilot test. PCP concentrations increased from approximately 16 μg/L pre-test to between 40 μg/L and 70 μg/L eight to 16 weeks into the test, then decreased to approximately 2 μg/L by the end of the pilot test.

TABLE 1: PCP and geochemical indicator concentrations in the shallow monitoring well located 12 m down-gradient of the bio-sparge curtain.

Date	PCP (μg/L)	DO (mg/L)	Nitrate (mg/L)	Fe²⁺ (mg/L)	Mn²⁺ (mg/L)	Sulfate (mg/L)	Alkalinity (mg/L)
1998 03 23	3,800	0.1	<0.25	8.67	4.40	113	185
1998 03 30	Pilot Scale System Activated						
1998 04 06	2,100	3.9	<0.05	0.08	4.96	215	118
1998 04 13	0.52	2.3	0.47	<0.03	0.71	138	88.7
1998 04 20	1,680	8.3	0.09	<0.03	2.36	145	84.0
1998 04 27	0.05	9.0	0.64	0.04	0.28	143	73.2
1998 05 05	0.05	10.8	0.54	<0.03	1.30	153	56.4
1998 05 11	0.19	9.5	0.72	<0.03	0.27	141	68.4
1998 05 19	0.97	1.6	0.85	<0.03	0.44	141	72.1
1998 05 25	0.31	6.4	0.87	<0.03	0.17	144	71.4
1998 06 08	0.21	5.7	-	0.14	0.14	-	-
1998 07 07	0.10	6.6	-	<0.03	0.39	-	-
1998 08 04	0.56	6.8	-	<0.03	0.43	-	-
1998 09 02	0.07	11.4	-	<0.03	0.20	-	-
1998 09 28	0.10	5.5	-	<0.03	0.076	-	-
1998 10 26	2.5	1.1	-	<0.03	1.49	-	-
1998 12 14	Pilot Scale System Deactivated for Full Scale System Installation						
1999 01 07	15	0.5	-	0.05	0.050	-	-
1999 01 18	Full Scale System Activated						
1999 03 04	0.08	5.7	0.57	0.08	0.006	225	123
1999 03 30	0.05	7.2	-	-	-	-	-

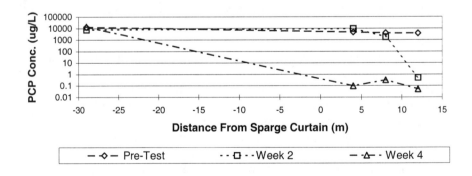

FIGURE 3. Plot of PCP concentrations along the plume centerline during the early portion of the pilot bio-sparge test.

Discussion. The results indicate that the indigenous aerobic bacterial population was attenuating the dissolved PCP plume under oxygen-limited conditions prior to the pilot bio-sparge test. The bacterial population was able to acclimate to the increased dissolved oxygen availability due to sparging within four weeks, and was then able to degrade the PCP from 4,000 µg/L to less than 1 µg/L.

The radius of influence of cyclic sparge wells in this aquifer was greater than 13 m. Bubbles were noted on occasion in the up-gradient control well, over 30 m from the sparge wells. The large radius of influence is likely due to the tidal nature of the aquifer, which becomes confined under high tide conditions. Air bubbles trapped beneath the confining layer migrate laterally, increasing the radius of influence.

Cyclic sparging may have caused some vertical migration of PCP within the aquifer. Dissolved oxygen did not migrate to the base of the aquifer. The increased PCP concentration at the base of the aquifer appeared to be a transient phenomena, as concentrations decreased to below pre-test levels by the end of the eight month pilot test.

Sparging resulted in the precipitation of approximately 13,000 µg/L of dissolved iron and manganese. No increased injection pressures or operation problems resulted from metal precipitates over the eight month pilot test.

FULL SCALE REMEDIAL SYSTEM

The pilot scale bio-sparge curtain was successful in degrading over 99% of the PCP within the dissolved groundwater plume to below the applicable discharge criterion. However, PCP concentrations exceeding the criterion of 1 µg/L were bypassing the pilot system to the east and at the base of the aquifer. In addition, the site owner wished to undertake source removal in order to proceed with site redevelopment and to limit the length of time the sparge curtain would have to be operated.

Partial excavation of source area soil contamination occurred during September 1998. Soil above the water table was excavated to PCP concentrations of less than 5 mg/kg and 14 mg/kg at the drive-through and elevator dip tank source areas, respectively. Soil beneath the water table was excavated in the wet to a PCP concentration of less than 100 mg/kg. Approximately 7,000 tonnes of soil containing an estimated 1,050 kg of PCP were excavated and remediated or landfilled off site.

A full-scale bio-sparge curtain consisting of eleven wells in two rows at 9 m spacing was commissioned in January 1999. The curtain was widened to capture the entire plume. It was postulated that increased PCP concentrations near the base of the aquifer were a result of vertical migration due to circulation or "turn over" of groundwater induced by the sparge curtain. Therefore, a second row of wells was added in an attempt to turn over the groundwater a second time and/or to better aerate the deeper groundwater. In order to accommodate site redevelopment, all wells and piping were installed below grade. The full scale sparge curtain is shown on Figure 4.

Monitoring data indicates that the full scale sparge curtain has been successful in degrading the full width of the PCP plume (see Figure 4). Insufficient monitoring data has been collected to determine the effect of the bio-sparge curtain on PCP concentrations in deeper groundwater.

Dissolved PCP concentrations 25 m down-gradient of the elevator dip tank decreased from approximately 10,000 μg/L to 540 μg/L within seven months after partial source removal. Based upon equilibrium partitioning between soil and groundwater and source depletion estimates, it is anticipated that the sparge curtain may be required for a two year period, after which natural attenuation will degrade the remaining plume to below the discharge criterion. This will have to be confirmed by monitoring.

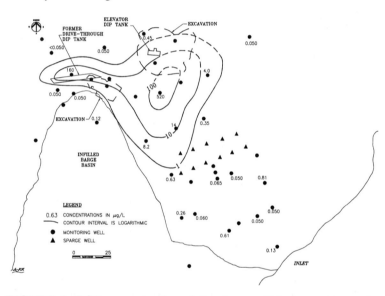

FIGURE 4. PCP concentrations following partial source removal and commissioning of full scale bio-sparge curtain.

REFERENCES

Hardy, L.I., L.A. Launen, M.K. Man, M.M. Moore, and I.D. Thompson, 1999. "Bioattenuation of an LNAPL Plume at a Tidally influenced Site." In B.C. Alleman and A. Leeson (Eds.), *In Situ and On-Site Bioremediation Symposium*, 5(1), 147-152. Battelle Press, Columbus OH.

McAllister, K.A., H. Lee and J.T. Trevors, 1996. "Microbial Degradation of Pentachlorophenol." *Biodegradation* 7:1-40.

Newell, C.J., J. Gonzales and R.K. McLeod, 1996. *BIOSCREEN Natural Attenuation Decision Support System*. US E.P.A., Center for Subsurface Modeling Support, Ada, OK. EPA/600/R-96/087.

FIELD EVALUATION OF DIPOLE METHOD TO MEASURE AQUIFER ANISOTROPY

Michelle Simon, United States Environmental Protection Agency, Cincinnati, OH
Benjamin Hough, Tetra Tech EM Inc., San Diego, CA
Tong Li, Tetra Tech EM Inc., Seattle, WA
Zbigniew J. Kabala, Duke University, Durham, NC

ABSTRACT

The ultimate size of a three-dimensional groundwater circulation cell surrounding a vertical circulation well (VCW) is a strong function of the aquifer hydraulic anisotropy, the ratio of the hydraulic conductivity in the horizontal direction to that in the vertical direction. In designing a VCW, a given aquifer's anisotropy was either assumed or estimated by matching pump test curves. This paper presents results from a field-scale project designed to directly measure the aquifer's anisotropy. A VCW well was installed at the Naval Air Station North Island (NASNI) near San Diego, California, in a sandy aquifer contaminated with chlorinated compounds. An extensive aquifer hydraulic test, including a dipole test, was conducted in August 1998. The interpretation of the dipole test data had to account for the tidal effects from San Diego Bay on the aquifer. The dipole extraction/injection rates were 1.89, 3.79, 5.68, 7.57, and 9.46 x 10^{-2} cubic meters per minute [5, 10, 15, 20, and 25 gallons per minute]. Drawdown and drawup were monitored in the extraction and injection chambers of the VCW and in the surrounding monitoring wells for each step. However, the recovery data of the dipole test proved to be the most consistent and reliable for estimating the aquifer anisotropy. The anisotropic ratio for this aquifer is estimated to be 5:1.

INTRODUCTION

Kabala [1993] proposed a dipole-flow test (DFT), a new approach for aquifer characterization that should, in principle, yield not only the horizontal hydraulic conductivity, K_r, and the specific storage, S_s, but also the vertical hydraulic conductivity, K_z, and consequently the hydraulic conductivity anisotropic ratio, (a^2 = K_r/K_z). The anisotropic ratio typically varies between 1 and 25 or even higher [Weeks, 1969]. Essentially, the DFT isolates two chambers in the well and pumps water at a constant rate, Q, from the aquifer to the extraction chamber, transfers it within the well to the injection chamber, and subsequently allows it to return to the aquifer. Pressure transducers installed in the two chambers measure the transient drawdown and drawup response during the test. These measurements can then be used in inverse modeling to calculate the hydraulic anisotropic ratio. The dipole-flow test develops a strong vertical-flow component near the well, whereas other existing measurement methodologies develop a horizontally dominated flow pattern. The test is, therefore, sensitive to the value of the vertical hydraulic conductivity. The Objective of this study was to apply the method of Kabala to an aquifer in a field setting. The details of this test are covered in Kabala et al. (2000).

The test was performed at Site 9 at the Naval Air Station North Island. The groundwater occurs at approximately 2.44 m below ground surface (bgs) [8 ft bgs]. The upper 33.5 m [110 ft] of the saturated zone contains an unconfined aquifer with a thin 1.5 to 6 m [5 to 20 ft] discontinuous fresh water lens, a brackish mixing zone [9-30.5 m, 30-100 ft] and a seawater wedge intruding inland. The hydraulic gradient various from 0.006, the transmissivity is 235 m²/day [2,526 ft²/day] and the conductivity is 1.0×10^{-2} cm/sec [12 feet/day]

The Navy selected an in-well stripping system to evaluate its effectiveness in remediating the aquifer. The schematic of this well is presented in Figure 1. As The North-South cross-section (Figure 1) shows the VCW, surrounded by two piezometers and seven observation wells discretely screened at various depths. There were three more observation wells in the East-West direction. Nine observation wells were used in aquifer hydraulic tests conducted on the VCW in the summer of 1998. The tests included a step drawdown test, a 32-hour constant rate pumping test, and an injection test in the upper screened interval of the VCW and a step drawdown test in the lower screened interval.

MATERIALS AND METHODS

Under the U.S. Environmental Protection Agency Superfund Innovative Technology Evaluation Program, Tetra Tech EM Inc. conducted a five-step DFT on the VCW at Installation Restoration Site 9 at NASNI on August 3, 1998. Following the notation of Kabala [1993], the thickness of the aquifer around the vertical circulation well is $b = 26.8$ m; the well radius is $r_w = 10.16$ cm; the depth from the water table to the top of the upper screen is $d = 7.9$ m and to the bottom of the lower screen is $l = 18.6$ m; the thickness of the upper screen is $2\Delta_u = 1.22$ m; and the lower screen is $2\Delta_l = 1.83$ m. The pumping/injection rates were 1.89, 3.79, 5.68, 7.57, and 9.46×10^{-2} cubic meters per minute. Aquifer drawdown and drawup were monitored in the extraction and injection chambers of the VCW during the step test. In addition, the water table was monitored in the surrounding observation wells before, during, and after the DFT. The drawup transient response in the upper (injection) chamber is presented in Figure 2. It is significantly affected by ocean tide, as evident from Figure 3, which presents the record of the water table fluctuations in the VCW and the nearby observation wells. The tidal influence has to be removed in order to complete the DFT data analysis.

Removal of Tidal Influence. Although most observation wells exhibit tidal fluctuations of different amplitude - the further from the ocean the smaller the fluctuation - the tidal fluctuations in MW51 observation well are nearly identical to those in the vertical-circulation well. Therefore, via the least square algorithm, we fit the portion of MW51 tidal response involving its first three peaks into the following function:

$$f(t) = \alpha + \beta \sin[2\pi(t-p)/T_1] - \gamma \sin[2\pi(t-p)/T_2] \tag{1}$$

where t is time and the other variables are fitting constants.

Data Analysis of the step data proved to be problematic (Figure 2). Increasing the flow rate via the suction pump caused uneven flow rates into the chamber and confounded the analyses for anisotropy. It was therefor decided to analyze the transient portion of the data.

Kabala [1993] assumed that there was no wellbore storage. The type curves relate the dimensionless drawdown

$$s_D = \frac{s(t)}{s(\infty)} \tag{2}$$

to the dimensionless time

$$\tau = \frac{vt}{r_w^2} \tag{3}$$

where $v = K_r b/S$ is the hydraulic diffusivity. It follows from these type curves that the larger the aquifer hydraulic anisotrophic ratio, the more dimensionless time is required to achieve the steady state.

RESULTS AND DISCUSSION

Figure 4 also presents the normalized DFT recovery response (circles) with time specified in seconds and the normalized DFT recovery response with time scaled by a factor $A = 0.0056$ (dashed thick line) to partially match the type curve for unrealistic value of the hydraulic anisotrophic $a^2 = 0.0001$. No single derived curve closely matches the aquifer data. The same is true for the normalized DFT step-4 response presented in Figure 5. The two figures illustrate that the measured DFT response is initially much faster than the Kabala [1993] model allows. This is most likely a manifestation of the unaccounted wellbore storage effects that are apparently important at the site.

Kabala [1998] formulated a model of the dipole-flow test that accounts for possibly different wellbore storage effects in the extraction and injection chambers and for possibly different screen size in the two chambers. The dimensionless wellbore storage parameters for the upper and lower chambers are specified as

$$C_{DU} = \frac{1 - (r_i/r_w)^2}{4S} \tag{4}$$

and

$$C_{DL} = \frac{(r_i/r_w)^2}{4S} \tag{5}$$

where $r_i = 5.08$ cm is the inner diameter of the eductor pipe, and S is the storativity or specific yield $(S = S_s b$, that is, specific storage times the aquifer thickness). With the well-aquifer geometry of the NASNI VCW, a set of DFT "wellbore storage" type curves is generated for $S = 1$, 0.1, 0.01, and 0.001. The normalized recovery response is plotted in Figure 6 with the "$S = 0.1$" type curves in the background. It is clear that none of the type curves can be matched. The normalized recovery response cannot fully match any of the "$S = 0.1$" type curves presented in Figure 4 either. However, it is similar in shape to the "$S = 0.01$" type curve presented in Figure 5. Indeed, with time scaled by A = 30 s^{-1}, the recovery response fits well between the two type curves of $a^2 = 3$ and $a^2 = 10$. It follows then from Figure 6, that

$$\frac{vt}{r_w^2} = At \qquad (6)$$

and hence

$$K_r = \frac{Ar_w^2 S}{b} = \frac{(30)(0.1016)^2(0.01)}{(26.82)} = 0.000,115 m/s \qquad (7)$$

This value is in the lower range ($0.000,109 < K_r \leq 0.000,259$ m/s) obtained independently from pumping tests conducted at the site on observation wells. With the horizontal conductivity given by (7) and with the average maximum drawup from the step DFT, $s_{max} = 1.05$ m, we obtain the value $4\pi K_r b s_{max} = 130.0$. This value leads, via the relation plotted in Figure 7 and derived by *Kabala* [1993, eq. 8], to a reasonable $a^2 = 4.93$, which confirms the appropriateness of the typecurve hydraulic anisotrophic estimate of $3 \leq a^2 \leq 10$.

Since the aquifer is unconfined, it was possible to use the observed delayed yield effects and estimate the aquifer hydraulic anisotropy from the Neuman [1975] delayed yield model *($\beta=r^2 K_z/b^2 K_r$)*. Our estimate is very close to the average anisotrophic ratio obtained for the average of eight observation wells at the site, that is, $a^2 = 5.88$. It would not be possible to use the Neuman model with data from analogous tests conducted in a confined aquifer.

CONCLUSIONS

A number of general conclusions can be reached from this preliminary analysis.

1.   The wellbore storage in the VCW may be important for the DFT in aquifers such as that of NASNI.
2.   The shallow aquifer parameters estimated from the DFT, that is, $0.001 \leq S \leq 0.01$, $K_r = 0.000,115$ m/s, and $a^2 = 4.93$, are consistent with the corresponding average parameters obtained from constant discharge pumping tests conducted on eight observation wells.
3.   The recovery response is much less noisy than the pumped DFT and contains the best information for DFT interpretation.
4.   In conducting a step dipole-flow test, it is imperative to maintain a pumping rate as steady as possible. The pumping rate should be adjusted once for each step and the change in pumping rate should be measured after the steady state is reached.

5.   To minimize the nonlinear effects in unconfined aquifer DFT, the step pumping rates should be selected judiciously so that the total drawdown/drawup should represent a small fraction of the aquifer thickness.

ACKNOWLEDGMENTS

The authors appreciate the support of the United States Navy, United States Environmental Protection Agency, Tetra Tech EM Inc., and Duke University in the performance of this study. Although the research described in this paper has been funded in part by the U.S. EPA, it has not been subjected to Agency review. Therefore, it does not reflect the official views of the Agency.

REFERENCES

Kabala, Z. J., M. A. Simon, B. L. Hough, T. Li. 2000. "Dipole-Flow Test on a Vertical-Circulation Well at Naval Air Station North Island, San Diego, California." Unpublished manuscript. 20 pages.

Kabala, Z. J.. 1998. "The dipole-flow test: a new type-curve interpretation technique accounting for wellbore storage." Unpublished manuscript. 10 pages.

Kabala, Z. J., "The dipole-flow test: a new single-borehole test for aquifer characterization." *Water Resour. Res., 29*(1), 99-107, 1993.

Nueman, S. P. 1975. "Analysis of pumping test data from anisotropic unconfined aquifers considering delayed gravity response." *Water Resour. Res., 11* 329-342.

Weeks, E. P. 1969. "Determining the ratio of horizontal to vertical permeability by aquifer-test analysis." *Water Resour. Res., 5,* 196-214.

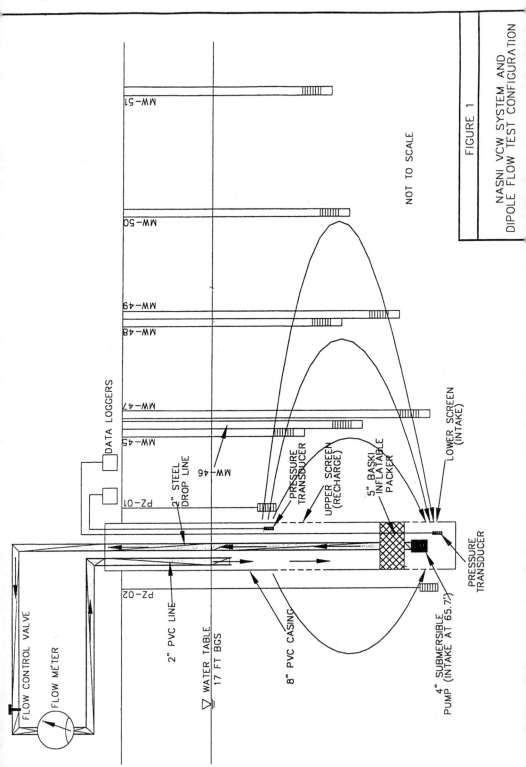

FIGURE 1

NASNI VCW SYSTEM AND
DIPOLE FLOW TEST CONFIGURATION

Figure 2

Figure 3

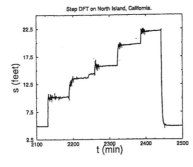

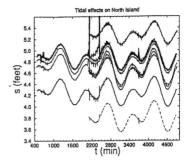

Figure 4

Figure 5

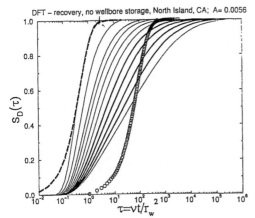

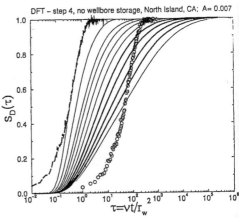

Figure 6

Figure 7

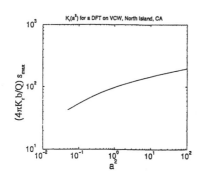

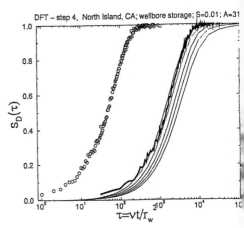

NUMERICAL SIMULATION AND PILOT TESTING OF A RECIRCULATION WELL

Sarah R. Forman (Dames & Moore, Linthicum, Maryland)
Tim Llewellyn and Scott Morgan (Dames & Moore, Linthicum, Maryland)
Don Green and Kimberly Gates (U.S. Army Garrison, APG, Maryland)
George DeLong (Lockheed Martin Energy Systems, Oak Ridge, Tennessee)

Abstract: A field scale Pilot Test has been in operation at the Graces Quarters peninsula of Aberdeen Proving Ground (APG), Maryland since September 1999. The recirculation well technology is used to deliver an optimal mixture of vitamin B_{12}, titanium citrate, and glucose to remediate contaminant source areas of the site. The recirculation well and surrounding monitoring network were designed and located based on a numerical simulation developed specifically for this project/site. Hydraulic testing was conducted to evaluate the effectiveness of the recirculation well technology on site. This paper discusses the hydraulic effectiveness of the recirculation well technology, the numerical flow model used to design the well, and the model validation based on the actual recirculation cell.

INTRODUCTION AND PURPOSE

Recirculation well technology was used to deliver an optimal mixture of vitamin B_{12}, titanium citrate, and glucose to degrade Chlorinated Volatile Organic Compounds (VOCs) at a National Priorities List (NPL) site in eastern Maryland. As part of system design, a computer simulation was developed using the United States Geological Survey (USGS) Modular Flow Model (referred to as MODFLOW (McDonald & Harbaugh, 1988; 1996) to aid in the understanding of groundwater flow within the recirculation cell. MODPATH was used to aid in the understanding of the time required for advective recirculation from the upper screen to the lower screen of the recirculation well along various flow paths. Based on the results of this modeling, the design the recirculation well and the monitoring network were optimized. Following the installation of the recirculation well and the monitoring network, hydraulic testing of the recirculation well was conducted. The computer simulation was compared to the hydraulic testing results and refined as appropriate. The Pilot Test has been operational since September 1999. This project was developed and implemented under contract and oversight by AIMTech, a U.S. Department of Energy program managed by Lockheed Martin Energy Systems, Inc., in conjunction with the APG Installation Restoration Program. Distribution restriction statement approved for public release, distribution is unlimited, OPSEC Number 3373-A-4.

The purpose of this paper is to discuss the development of the computer simulation used to design the recirculation well and monitoring network and to discuss the comparison between the computer simulation and the results of the hydraulic testing.

SITE BACKGROUND

The investigation site is located at Graces Quarters, a part of the Edgewood Area portion of the Aberdeen Proving Ground (APG) in Eastern Maryland. Graces Quarters is located on the Gunpowder Neck Peninsula as shown in Figure 1. The Primary Test Area (the site of the recirculation well in the surficial aquifer Pilot Test) is approximately 22 acres (0.089 square kilometers (km^2)) in size and is underlain by a surficial aquifer. This aquifer is underlain by a silty clay confining unit, which is in turn underlain by a confined aquifer. Based on previous data, the confining unit is absent at various locations at the Primary Test Area. In areas missing the confining layer, the surficial and confined aquifers are in contact, thus forming a single hydrologic unit.

Groundwater exists under unconfined conditions within the surficial aquifer. The surficial aquifer is recharged mainly through the infiltration of precipitation and flows predominantly south-southwest, away from the topographic clay uplands toward marshes to the south. Groundwater from the surficial aquifer also recharges the confined aquifer through pathways in the confining unit. Therefore, recharge conditions have formed a radial flow pattern in the confined aquifer, which is superimposed on an overall flow field to the south in that aquifer.

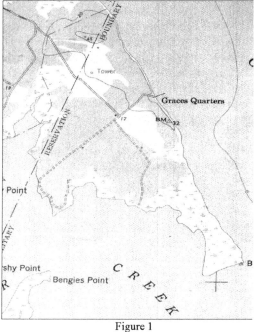

Figure 1
Site Location Map

The contaminants of concern at Graces Quarters are chlorinated VOCs. The most frequently detected chlorinated VOCs include 1,1,2,2-tetrachloroethane, trichloroethene, carbon tetrachloride, tetrachloroethene, and chloroform. The highest VOC detection was 1,1,2,2-tetrachloroethane with a concentration of 181,000 micrograms per liter (μg/L). This was a single unrepeated peak detection. Consistent and reproducible concentrations are on the order of 7,000 μg/L (total VOCs). It is believed that these contaminants were introduced to the surficial aquifer, migrated south with the flow of groundwater in the surficial aquifer, entered the confined aquifer through holes in the confining layer, and spread out in an overall southward flowing groundwater system in the confined aquifer.

DESCRIPTION OF TECHNOLOGY

Recirculation wells aid in the reduction and removal of contamination without above-ground handling or disposal of fluids. The USEPA does not require fees or permits for this technology. This technology involves setting-up a hydraulic gradient stimulating circular flow of groundwater into the well from the formation and out of the well back into the aquifer. Recirculation wells are typically used in conjunction with various treatment technologies to reduce the concentration of contaminants.

The recirculation well and its monitoring network (9 nested piezometer locations, a total of 24 individual piezometers) at Graces Quarters were installed and hydraulically tested, prior to the initiation of the surficial aquifer treatment system (Mowder et. al., 2000)). Hydraulic testing was conducted to assess the ability of the site to develop a recirculation cell at that location and whether the piezometer network would adequately monitor that cell.

Recirculation wells have two screens which are separated by a packer thus sealing off the upper screened interval from the lower screened interval within the casing. The screen separation and the geology on site (i.e., vertical and horizontal conductivity) and prevailing groundwater flow patterns, determine the shape and extent of the recirculation cell (Herrling, B., J. Stamm, W. Buermann, 1991). Pumping and injection rates determine the time frame for the development of the recirculation cell. For example, pumping rate influences travel time within the cell. A recirculation cell will eventually reach a static point in its outward growth, or effective radius of influence. This may take several years. Within the maximum radius of influence the groundwater would be constantly recirculating at velocities decreasing outward from the recirculation well itself.

DEVELOPMENT OF COMPUTER SIMULATION

The recirculation well technology was chosen for evaluation in the surficial aquifer at Graces Quarters because of the predominance of clean, fine, well sorted sand with no laterally continuous clay layers. Therefore, initial geologic conditions indicated that a "sand-box" conceptual model could be used to develop a groundwater flow model. For this flow model, the recirculation well was placed in the center of a square model domain. The base of the domain was assumed to be horizontal and was based on the depth of the confining unit measured from geoprobe samples and monitoring wells located at the "hot-spot" in the Primary Test Area. Therefore, the recirculation well was to be installed solely in the surficial aquifer to a depth of 34 feet (10.36 meters(m)). The lateral extent of the model domain were subjectively estimated so that they were of sufficient distance from the recirculation well and would be beyond its area of influence. Pumping test data provided guidance in estimating the area of influence (APG, 1996). Therefore, a model domain of 300 feet (91.44 m) from the recirculation well was chosen for the simulation.

Once the model domain was chosen, boundary conditions were evaluated. Initial conditions and boundary conditions were simplified to allow for the visualization of an ideal recirculation well and the placement of monitoring points. It was assumed that the majority of lateral groundwater flow enters the model from the northeast and flows out to the southwest. Vertical flow through the silty clay (model base) was assumed to be insignificant. Based on these simplifying assumptions, constant (specified) head boundaries were used to control water entering and leaving the model while the base of the model is, by design, a no flow condition. As data becomes available the specifies head boundaries will be fit to measured water levels to simulate field conditions. For initial model runs, the starting heads and specified heads were set to average elevations measured at the site.

A fine grid spacing was needed to accurately simulate groundwater flow within the recirculation cell. Based on the size of the grid and geologic data, a constant cell spacing of 5 feet (1.52 m) in the x and y directions was chosen. This fine spacing was deemed adequate to see hydraulic gradient changes caused by the simulated recirculation well and was smaller than the estimated formation dispersivity, which would be useful for future solute transport uses of the model.

A vertical (z direction) grid spacing of 2.5 feet (0.762 m) was chosen for ease of data manipulation. Thus the model was divided into 10 layers, each 2.5 feet (0.762 m) thick, for a total of 25 feet (7.62 m) (the estimated saturated thickness of the surficial aquifer). This vertical spacing allowed the hypothetical injection and withdrawal screens to be broken into three 5 foot (1.52 m) by 5 foot (1.52 m) by 2.5 foot (0.762 m) cells.

Aquifer constants used within the model were based upon field studies and pumping test (July 1995 (APG, 1996)) conducted at the Primary Test Area. These initial values are listed in Table 1.

TABLE 1.   Aquifer Parameters

Variables	Initial	Final
Vertical K	1.4 feet per day (ft/day) 0.427 meters per day (m/day)	2.1 ft./day (0.640 m/day)
Horizontal K	14 ft/day	21 ft./day (6.40 m/day)
Porosity	0.20	0.20
Leakance		0.84

To verify the model, it was calibrated to both steady state and transient conditions. Due to the use of constant head boundaries, the steady state calibration did little to confirm operation of the model. However, transient model runs were compared to drawdown data collected from a 72 hour pumping test conducted at well Q14 in 1995. Using the drawdown measured in Q14, as well as observation wells OBS1 and OBS2 , the input values were adjusted until the model error was minimized. Final error estimates are provided in Table 2.

TABLE 2.   Final Error Estimates

Parameter	Error
Mean Error	0.37
Mean Absolute Error	0.37
Root Mean Squared Error	0.47

The model indicated that a maximum effective treatment radius of 32 feet (9.75 m) at 2 gallons per minute (gpm) (0.126 liters per second (L/sec)) would be reached at the end of six months. Analytical and numerical design of the recirculation well at Graces Quarters indicates that it would take approximately 2.5 years for the radius of influence to reach 80 feet (24.38 m) (at 2 gpm (0.126 L/sec). It would take approximately 5.5 years to attain a maximum radius of 113 feet (34.44 m). Based on this model, a complete pore flush at a radius of 40 to 45 feet (12.19 to 13.72 m), with multiple pore flushes within that area, would occur by the end of the six-month Pilot Test.

An analytical model (adapted from Herrling and Buermann (1990)) was also used to estimate capture radii of recirculation wells at the Primary Test Area for the purpose of "hot-spot" remediation. Based on this model, it was estimated that the capture radius would be approximately 100 feet (30.48 m). However, this model was found to be limited in its usage and did not demonstrate the flow of a recirculation well, mounding, or seepage velocities, such as those predicted by the numerical model.

USE OF MODPATH TO DETERMINE TIME FOR RECIRCULATION AND EXTENT

MODPATH was used to simulate the particle pathways (Figure 2) which were then used to determine the locations of chemical and hydraulic monitoring locations (i.e., nested piezometer locations). This simulation was also used to determine the seepage velocities and travel time of groundwater in the reciculation cell and to determine the groundwater sampling schedule (associated

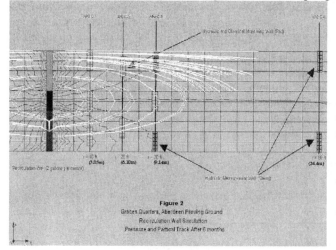

with the chemical portion of the Pilot Test (Mowder et. al, 2000)) of the Pilot Test.

MODPATH indicated that the effective treatment radius would reach approximately 32 feet (9.75 m) by the end of 180 days (6 months—the duration of the Pilot

Figure 2
Graces Quarters, Aberdeen Proving Ground
Recirculation Well Simulation
Pressure and Particle Track After 6 months

Test). Therefore, it was determined that a total of nine nested piezometer locations would be placed at distances of 10 to 15 feet (3.05 to 4.57 m), 30 to 35 feet (9.14 to 10.67), and 65 to 80 feet (19.81 to 24.38 m) for chemical and/or hydraulic monitoring. Figure 3 presents a plan view of the recirculation well and nested piezometer locations. These piezometers were designed to intercept specific locations or distances and depths along the particle pathways so that the degradation of contaminants could be monitored. The screened intervals of the peizometers were designed to be placed at depths of 12 to 14 feet (3.66 to 4.27 m) bgs, 21 to 23 feet (6.40 to 7.01 m) bgs, and 29 to 31 feet (8.84 to 9.45 m) bgs, assuming that the recirculation well would extend to a maximum depth of 34 feet (10.36 m) bgs. The sampling schedule was based on the particle tracking so that sampling events were scheduled to intercept specific particle pathways at various distances along each pathway.

INSTALLATION OF RECIRCULATION WELL AND MONITORING NETWORK

The recirculation well was installed using an air rotary drilling rig with a roller bit to advance the borehole. The air rotary method was selected to minimize potential skin effects around the screens and to allow the greatest probable well efficiency. The casing was advanced by direct-push and air percusion using the air hammer.

The 10-inch (25.4 centimeters (cm)) inner diameter (ID) recirculation well was installed to an actual total depth of 36 feet (10.97 m) below ground surface (bgs) (placed at the bottom of the surficial aquifer on top of the clay confining layer). It is constructed of 0.010-inch (0.0254 cm) ID, wire-wrapped stainless steel screens and PVC risers. The two screens are each 8 feet (2.44 m) in length and were placed from 36 to 28 feet (10.97 to 8.53 m) bgs and from 18 to 10 feet (5.49 to 3.05 m) bgs. These screens are separated by schedule 80 PVC riser which is placed from 28 to 18 feet (8.53 to 5.49 m) bgs. A packer set in the riser between them separated the two screens. Additionally, a submersible pump was set mid-screen in the lower portion of the recirculation well.

A total of 24 individual (9 nested locations) 2-inch (5.08 cm) ID piezometers were installed by hollow stem augers and a truck-mounted rig at each location. As presented in Figure 3, the piezometers were placed along three lines projected outward from the recirculation well. These spokes were placed at angles of

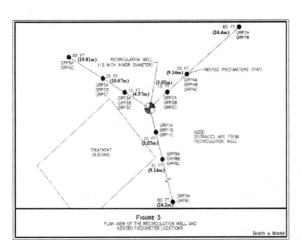

FIGURE 3
PLAN VIEW OF THE RECIRCULATION WELL AND
NESTED PIEZOMETER LOCATIONS

greater than 90 degrees apart to monitor for anisotropic conditions in addition to chemical and hydraulic changes.

Each piezometer is constructed of Schedule 40 PVC riser with 2 foot (0.610 m) long, 0.010-inch, wire-wrapped PVC screens. Shallow, intermediate, and deep piezometers ranged in depth from 15 to 17 feet (4.57 to 5.18 m) bgs, 25 feet (7.62 m) bgs, and from 31 to 33 feet (8.84 to 10.06 m) bgs (bottom of surficial aquifer), respectively.

HYDRAULIC TESTING OF THE RECIRCULATION WELL AND ANALYSIS OF DATA

Following installation and development activities, hydraulic testing of the recirculation well was conducted. Hydraulic testing consisted of two 300-minute step tests (one per each screened interval) and a 72-hour recirculation test. Step test data was collected to assess the reduction in specific capacity with increasing yields. In addition to assessing screen efficiency and estimating sustainable yield, the step tests were conducted to provide a baseline for screen efficiency in case of biofouling. These activities were conducted in August 1999.

Prior to hydraulic testing activities, background water levels were measured over in well Q14 and piezometer QRP3C. Pressure transducers were placed in these wells for a 48 hour period.

Changes in water levels were monitored during the step tests using pressure transducers with data loggers and electronic water level indicators (to manually measure water levels). During the lower screen step test, a total drawdown of 3.48 feet (1.06 m) was observed. During the upper screen step test, a total drawdown of 6.66 feet (2.03 m) was observed. Based on the drawdowns observed in the lower screened interval step test, it was determined that the maximum withdrawal rate of 5 gpm (0.315 L/sec) would not stress the lower zone of the surficial aquifer. However, based on the drawdowns observed in the upper screened interval step test, it was determined that a maximum groundwater withdrawal rate of 4.5 gpm (0.284 L/sec) could be maintained although the water level was drawn into the upper screen. The maximum drawdown in the upper screened interval was used to estimate the potential rise in groundwater level during the 72 hour recirculation test. Therefore, it was estimated that the groundwater level would rise to 4 feet (1.22 m) below top of casing (TOC) during the 72 hour recirculation test. Based on these results, a sustainable recirculation rate was estimated to be 5 to 6 gpm (0.315 to 0.379 L/sec).

Changes in water levels were monitored during the 72 hour recirculation test using pressure transducers with data loggers and electronic water level indicators (to manually measure water levels at predetermined intervals). The 72 hour test consisted of withdrawing the groundwater from the lower screened interval of the recirculation well at sequential rates of 2 gpm (0.126 L/sec), 4 gpm (0.252 L/sec), and 5 gpm (0.315 L/sec), while simultaneously injecting groundwater through the upper screened interval into the aquifer. The groundwater withdrawal rate was not increased until the hydraulics stabilized as evidenced by depth to water measurements. The recovery period following completion of the recirculation test was monitored for 48 hours using transducers and data loggers.

During the course of the 72 hour recirculation test, the depth to water rose in the upper screened interval of the recirculation well to 4.0 feet (1.22 m) below TOC. Thus, the maximum draw-up/rise in the groundwater level recorded in the upper-screened interval of the recirculation well, caused by the 72 hour recirculation test, was 5.47 feet (1.67 m). This rise in groundwater elevation (5.47 feet) dropped-off rapidly (i.e., within a 10 foot (3.05 m) distance from the recirculation well) along all three spokes.

Based on a flow net analysis (Figure 4) of the water level measurements at the end of the 72 hour recirculation test, the recirculation cell generally functioned as predicted. Groundwater levels measured in the piezometers along the southern spoke (i.e., QRP2A/B/C, QRP4A/B/C, and QRP7A/C) and the eastern spoke (i.e., QRP3A/B/C, QRP5A/B/C, and QRP6A/C) all decreased during the 72 hour

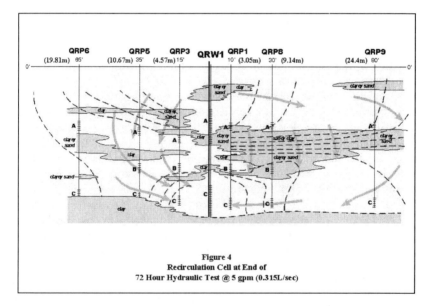

Figure 4
Recirculation Cell at End of
72 Hour Hydraulic Test @ 5 gpm (0.315L/sec)

recirculation test. However, groundwater levels in piezometers QRP1A and QRP8A, which are screened in the upper zone of the surficial aquifer along the western spoke, rose 0.17 feet (0.052 m) and 0.05 feet (0.015 m), respectively. All other piezometers along the western spoke decreased in groundwater elevation during the 72 hour test. Thus, a small amount of mounding was observed in the aforementioned piezometers, which correlates to a distance of 30 feet (9.14 m) from the recirculation well. As presented in Figure 4, clay layers in the upper zone of the surficial aquifer caused variable radii of influence ranging from 40 to 80 feet (12.19 to 24.38 m).

COMPARISON OF COMPUTER SIMULATION TO THE DATA
GENERATED BY THE HYDRAULIC TESTING

The actual recirculation cell setup was largely as predicted by the numerical model. Aquifer heterogeneities resulted in some asymmetry in the areas of influence and layer assignations will be made in this and future models to account for this important variable.

REFERENCES

Aberdeen Proving Ground (APG). 1996. *Graces Quarters, Aberdeen Proving Ground, Maryland, Focused Feasibility Study Groundwater Data Report.*

Mowder, C., T. Llewellyn, S. Forman, S. Lesage, S. Brown, K. Millar, D. Green, K. Gates, G. DeLong, F. Tenbus, 2000. "Field Demonstration of *In Situ* Vitamin B_{12}-/Catalyzed Reductive Dechlorination." In proceedings for the *Second International Conference on Remediation of Chlorinated and Recalcitrant Compounds.*

Herrling, B., J. Stamm, W. Buermann, 1991. "Hydraulic Circulation System for In Situ Bioremediation and/or In Situ Remediation of Strippable Contamination." In proceedings for *First International In Situ and On-Site Bioreclamation Symposia.*

IMPROVING AIR SPARGING SYSTEMS BY PULSED OPERATION

Gorm Heron (SteamTech Environmental Services, Bakersfield, CA)
Tom Heron (NIRAS, Aarhus, Denmark)
Carl G. Enfield (US EPA, Cincinnati, OH)

ABSTRACT: Air sparging was evaluated for remediation of tetrachloroethylene (PCE) present as dense non-aqueous phase liquid (DNAPL) in aquifers. A 2-dimensional laboratory tank with a 50-cm high DNAPL zone was created, with a PCE pool accumulating on an aquitard. During air injection, the vadose zone DNAPL was removed within a few days, and the recovery in the extracted soil vapors decreased to low values. Pulsing the air injection led to improved mass recovery, as the pulsing induced water circulation and increased the DNAPL dissolution rate. The induced circulation of water led to limited spreading of the dissolved contaminant, but accelerated mass removal by 40 to 600%, depending on the aggressiveness of the pulsing. For field applications, pulsing with a daily or diurnal cycling time may increase mass removal, thus reducing the treatment time and saving in the order of 40 to 80% of the energy cost used to run the blowers. PCE DNAPL located below the sparge point was unaffected by the air sparging and cannot be removed using this technology.

INTRODUCTION

Several studies have shown the effectiveness of air sparging for dissolved plumes (Bass & Brown 1995; Johnston et al.1998). However, air sparging at sites contaminated with DNAPLs may potentially be much less efficient (Bausmith et al. 1996). One major difference is that the presence of DNAPL leads to a multiphase-system in the porous medium, where air and water may have a tendency to flow around DNAPL areas, rather than through them. Also, air channeling during air sparging has been shown to be highly heterogeneous and the channel distribution is unpredictable (Hein et al. 1997; Ji et al. 1993).

Pulsed air injection was proposed for improved performance of air sparging systems (Rutherford & Johnson 1996, Johnson et al. 1997). Pulsed injection leads to both vertical and horizontal groundwater flow, potentially increasing the mixing between treated water (which has been in direct contact with the air channels) and untreated water, and increasing the radius of influence of each sparging well. The poor understanding of field-scale air sparging phenomena leads to the design of field systems without good guidelines for selecting air injection rates, pulsing frequency, and pulse duration. This field has been dominated by trial-and-error applications, and only semi-empirical rules have been established (Bass & Brown 1995; Johnson et al. 1997).

In this study, the effect of pulsed air injection on removal of tetrachloroethylene (PCE) DNAPL was studied in a laboratory soil tank. Detailed analysis led to a description of the processes and recommendations for pulsing.

LABORATORY AIR SPARGING SIMULATION

Air sparging was studied in the tank shown in Figure 1. The soil-filled volume was 4-cm thick, 116-cm long, and 56-cm high. All materials contacting the soil were either stainless steel, aluminum, teflon (PTFE), glass or viton. A total of 7.24 kg of a low-permeable silty soil was packed into the bottom of the tank. Then the tank was filled to capacity with medium sand. The back aluminum sheet contained a total of 48 water sampling ports. The box was flushed with 10 pore volumes of CO_2 and wetted slowly through three bottom ports, allowing the box to saturate within 7 days. Then the water table was lowered to an elevation of 36 cm, 20-cm below the top of the box. A PCE plume was then created by the rapid release of 100-mL liquid PCE dyed red with 0.2 g/L Sudan(IV). After 48 hours, a stable PCE distribution was observed. The water table was lowered to the top of the silt layer and raised back up, resulting in limited horizontal and significant vertical smearing of the PCE. Another 14 days of equilibration was allowed prior to air sparging from a stainless steel filter inserted 12-cm above the silt-sand interface in the left side of the box (Figure 1). Air injection rates of 100, 200, 400 and 800 mL/min were used, with inlet pressures ranging from 6 to 50 cm of water. Vacuum extraction was done from the top central port at extraction rates equal to 1.5 times the sparge rate, with passive air inlets in the corners at the top of the box. Effluent PCE and water vapor concentrations were sampled at the outlet line using a gas-tight syringe. Between and during air sparging pulses, water samples were taken using a 1000 uL gas-tight syringe through the sampling ports in the back plate of the box, and analyzed by gas chromatography. After air sparging was finalized, soil samples were collected by a cork bore through the back holes after removal of the sampling ports and analyzed for residual PCE.

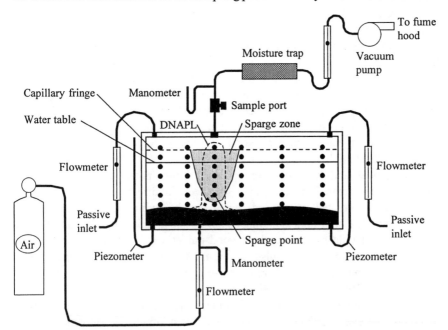

Figure 1. Setup for the 2-dimensional air sparging simulation. The soil tank was 4-cm thick, 116-cm wide and 56-cm high. Sampling points are indicated as dots.

RESULTS AND DISCUSSION

Air injection through the sparge point caused a cone-shaped sparge zone (Figure 1). At the glass surface, an irregular pattern was observed, with several air channels and preferentially contacted areas. Dyed PCE disappeared from the soil above the water table within 4 days of air sparging, and 25% of the PCE mass was removed (Figure 2). After 9 days of flow stoppage, sparging was resumed on day 13, and the sparge rate was doubled at day 15. Both these changes resulted in increased mass fluxes in the effluent vapor, and removal of another 25% of the PCE mass. At this time PCE vapor concentrations were below 1 mg/L, corresponding to a removal of 1.7 gram of PCE per day. Only small individual droplets and ganglia remained in the soil above the sparge point. This situation represents a typical field case, where low recovery of contaminants is seen in the extracted vapors, but concentrations rebound after a down-period, due to the slow dissolution of NAPL into the water, in agreement with the conceptual model proposed by Johnson et al. (1993) and Rabidean et al. (1998).

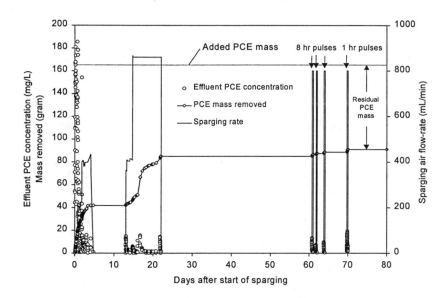

Figure 2. Effluent PCE concentrations, accumulated PCE mass removal, and air injection rate during the air sparging.

Following the constant air injection tests, cyclic air injection using 8-hr air injection and 16-hr equilibration was tested (3 pulses total; right side of Figure 3). Effluent concentrations peaked at 15 to 30 times the concentrations obtained by continuous sparging after 15 to 30 minutes after the onset of sparging, followed by an exponential decrease with time. Asymptotically low recovery rates were achieved after 5 to 8 hrs of sparging, indicating that shorter pulses might be more efficient for the mass removal. The mass of PCE removed by the 8-hr pulses depended strongly on the equilibration time allowed prior to each pulse. Waiting

30 days before a pulse (pulse 1, Figure 3) resulted in removal of 1.6 g of PCE during a single pulse, which is double the amount removed by a pulse after 16 hours downtime (Pulse 2, Figure 3).

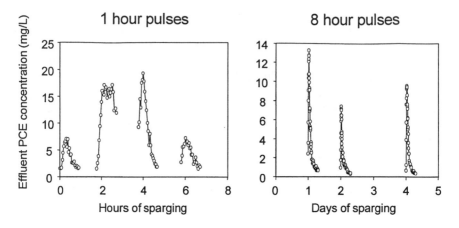

Figure 3. PCE effluent concentration during the sparging pulses at 800 mL/min.

Ten minutes after a pulse, concentrations were lowered in the sparge zone to levels less than 10% of the aqueous solubility. This is explained by the stripping of PCE by the air channels, which apparently occurs faster than the dissolution of PCE into the aqueous phase (Johnson 1998). After 16 hours equilibration time, PCE concentrations were close to the solubility limit in the entire area containing DNAPL droplets or ganglia.

More aggressive pulsing was performed with a 1-hr injection period followed by 1-hr equilibration (left side of Figure 3). The peak PCE concentrations in the effluent vapor were in the same range as for the 8-hr pulses, but the disturbance led to higher PCE fluxes during pulse 2 and 3. The increased fluxes may be caused by new pathways for the injected air caused by the more aggressive pulsing. After four pulses, the PCE mass removed per pulse dropped to approximately 0.3 g, indicating that the DNAPL in direct contact with the newly formed channels had been removed.

Dissolved PCE concentrations were reduced very significantly within the sparge zone, and by up to 24 mg/L in the immediate vicinity of the DNAPL area. At further radial distance from the sparge point, concentrations increased by up to 43 mg/L, as a result of water migration. This observation suggests that air sparging in a NAPL zone potentially can lead to spreading of the contamination, and that the induced water movement is significant.

Soil concentrations often determine the success of the remedial action. The final PCE distribution in the soil remained extremely high in the area below the sparge point, where a DNAPL pool was present (above 5,000 mg/kg dry soil). It also appears that dissolved PCE is spread from the DNAPL area to surrounding soil volumes by the induced water movement, and that these volumes may not be affected by the sparge air. A significant fraction of DNAPL tends to accumulate on top of clay and silt layers, resting on the capillary barrier (Hunt et al. 1988). Air sparging filters are typically more than 10-cm long, and the air has been observed to preferentially migrate out of the top of a screen (Johnson, 1997). It is likely that the majority of the air will be injected above a DNAPL pool resting on a low-permeable layer. These problems are inherent to application of air sparging in the presence of DNAPL in aquifers (Nyers & Suthersan 1993; Bass & Brown 1995).

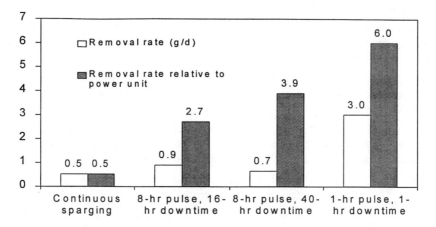

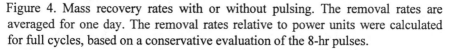

Figure 4. Mass recovery rates with or without pulsing. The removal rates are averaged for one day. The removal rates relative to power units were calculated for full cycles, based on a conservative evaluation of the 8-hr pulses.

Air sparging is often used for the remediation of dissolved plumes, where the presence of a DNAPL phase has not been realized, due to the difficulties in finding such phases. In such cases, unexplainable rebound and slow mass recovery may be due to DNAPL presence, and thus the mass-transfer limitations during this application become important.

For typical in-situ air sparging systems, the clean-up period is months or years, and the overall cost may be dominated by the energy consumption of the blowers and the off-gas treatment system. Our results indicate that improved mass removal, and thus reduced treatment times, may be achieved by pulsing. Figure 4 shows the PCE mass removal rate obtained by continuous operation, and three different pulsing scenarios. By aggressive pulsing (1-hr pulses, 1-hr downtime), a six-fold increase in the daily PCE removal rate was obtained, spending half the energy for air injection. In other words, for every dollar spent on running the blower, 12 times more PCE was removed from the aquifer system.

For field applications, longer pulsing frequencies are required, since water table fluctuations induced by the pulsing are less rapid. Typical mounding (after air sparging onset) and collapse (after air sparging cessation) last in the range of hours in real systems, whereas they lasted in the order of minutes in our laboratory system. The more often the system is disturbed (by pulsing), the higher the overall recovery rate. An optimal scenario for a field scale application would be a daily or diurnal cycling. Overall, this could potentially reduce the treatment time and cut the energy cost of the system at least in half.

REFERENCES

Bass, S.D., and R.A. Brown. 1995. Performance of air sparging systems- a review of case studies. In Proceedings of the Petroleum Hydrocarbons and Organic Chemicals in Ground Water Conference: Prevention, Detection and Restoration, 621-636, by National Ground Water Association. Dublin, Ohio: NGWA.

Bausmith, D. S., D. J. Campbell, and R. D. Vidic. 1996. In Situ Air Stripping: Using Air Sparging and Other In Situ Methods Calls for Critical Judgments. Water Environment and Technology 8(2): 45-51.

Hein, G. L., J. S. Gierke, N. J. Hutzler, and R.W. Falta 1997. Three-dimensional experimental testing of a two-phase flow-modeling approach to air sparging, Ground Water Monitoring and Remediation 17, 222-230.

Heron, G., J.S. Gierke, B. Faulkner, S. Mravik., A.L. Wood, and C.G. Enfield. 1999. Potential for Downward Spreading of Dense Nonaqueous Phase Liquids during Air Sparging. Submitted to Ground Water Monitoring and Remediation.

Hunt, J.R., N. Sitar, and K.S. Udell. 1988. Nonaqueous phase liquid transport and cleanup 1. Analysis of mechanisms. Water Resources Research 24(8): 1247-1258.

Ji, W., A. Dahmani, D.P. Ahlfeld, J. Ding Lin, and E. Hill III. 1993. Laboratory study of air sparging: air flow visualization. Ground Water Monitoring and Remediation Fall: 115-126.

Johnson, P.C., A. Das, R.L. Johnson, A. Leeson, D. McWorther, and R.E. Hinchee. 1997. Effects of IAS process changes on the removal of immiscible-phase hydrocarbons. In In Situ and On-Site Bioremediation. Vol. 1, Proceedings of the Fourth International In-Situ and On-Site Bioremediation Symposium, New Orleans, Apr. 28-May 1, pp 135-140.

Johnson, P.C. 1998. Assessment of the contribution of volatilization and biodegradation to in situ air sparging performance. Environmental Science and Technology 32: 276-281.

Johnson, R.L., J.F. Pankow. 1992. Dissolution of dense chlorinated solvents into groundwater. 2. Source functions for pools of solvent. Environmental Science and Technology 26: 896-901.

Johnson, R.L., P.C. Johnson, D.B. McWorther, R. Hinchee, and L. Goodman. 1993. An overview of in situ air sparging. Ground Water Monitoring and Remediation 13, no. 4: 127-135.

Johnston, C. D., J. L. Rayner, B. M. Patterson and G. B. Davis. 1998. Volatilization and Biodegradation During Air Sparging of a Petroleum Hydrocarbon-contaminated Sand Aquifer. Proceedings of Groundwater Quality '98: Remediation and Protection. 250, 125-131, September 20-25, 1998, Tuebingen, Germany.

Nyer, E.K, and S.S. Suthersan. 1993. Air sparging: savior of ground water remediations or just blowing bubbles in the bath tub? Ground Water Monitoring and Remediation Fall: 87-91.

Rutherford, K.W., and P.C. Johnson. 1996. Effects of process control changes on aquifer oxygenation rates during in situ air sparging in homogeneous aquifers. Ground Water Monitoring and Remediation Fall: 132-141.

COMETABOLIC AIR SPARGING FIELD DEMONSTRATION WITH PROPANE TO REMEDIATE A CHLOROETHENE AND CHLOROETHANE CO-CONTAMINATED AQUIFER

Adisorn Tovanabootr, Mark E. Dolan, and Lewis Semprini
(Oregon State University, Corvallis, Oregon)
Victor S. Magar and Andrea Leeson (Battelle, Columbus, Ohio)
Alison Lightner (Air Force Research Laboratory, Tyndall AFB, Florida)

ABSTRACT: Cometabolic air sparging (CAS) is an innovative form of conventional air sparging, and is designed to degrade or remove chlorinated aliphatic hydrocarbon compounds (CAHs) in groundwater and to reduce off-gas CAH emissions during air sparging. This CAS demonstration was conducted at McClellan AFB, California, for removal of chloroethenes (TCE, *cis*-DCE, and 1,1-DCE) and dichloroethane (1,1-DCA) from groundwater. Propane was used as the cometabolic cosubstrate and was selected based on microcosm studies conducted by Oregon State University (OSU). The demonstration system had two test zones, a propane active zone that was sparged with a propane and air mixture and a control zone that was sparged only with air. The sparge systems were compared for their performance for CAH removal with and without propane. An initial lag period of four to six weeks occurred before in situ propane utilization was observed. A similar lag period was observed in laboratory microcosms using unacclimated soils and groundwater from McClellan AFB. After approximately six weeks of repeated addition of propane to groundwater, TCE, *cis*-DCE, and DO levels decreased in proportion with propane usage. TCE, *cis*-DCE, and other CAH concentrations continuously decreased with repeated sparging of propane and air to the groundwater. Little propane utilization or CAH transformation was observed in the vadose zone during the first 188 days of operation. On day 188, ammonia was added to the active zone sparge gas to provide bioavailable nitrogen to the treatment zone. CAH concentrations in the soil gas monitoring points closest to the groundwater began to decrease, as did propane concentrations at some points as the test ended on day 220.

INTRODUCTION

The U.S. Air Force and Navy have been developing in-situ remediation technologies, such as air sparging, which have the potential to remediate sites much less expensively and more effectively than conventional ex-situ technologies. In-situ air sparging remediates groundwater through a combination of volatilization and enhanced biodegradation of contaminants (Marley and Bruell, 1995). Cometabolic air sparging (CAS) is an innovative form of conventional air sparging designed to remediate CAH-contaminated groundwater and to reduce off-gas CAH emissions. As with traditional air sparging, CAS also involves air injection directly into an aquifer. However, CAS is unique in that it also includes the addition into an aquifer and the overlying vadose zone of a gaseous cometabolic growth substrate such as methane (Travis and Rosenberg, 1997), or propane, or butane to promote the in-situ cometabolic degradation of CAH compounds. Previous microcosm studies performed with groundwater and aquifer solids showed indigenous propane-utilizers were effective at transforming CAH mixtures of trichloroethylene (TCE), chloroform (CF), and 1,1,1-trichloroethane (1,1,1-TCA) using saturated soils from McClellan Air Force Base (MAFB) (Tovanabootr and Semprini, 1998). Thus, a CAS demonstration with propane as a cometabolic substrate was conducted in the MAFB subsurface.

Objective. The objective of this demonstration was to evaluate the effectiveness of CAS for remediating CAH contamination in groundwater and the vadose zone using indigenous propane-utilizing microorganisms. After air and the co-substrate (propane) were delivered to the subsurface, oxygen and propane distribution in the groundwater and vadose zone was determined and propane and oxygen uptake rates and CAH transformation rates were measured. The effectiveness of sparging with and without propane addition was determined to compare cometabolic CAH biodegradation versus the CAH air stripping from the groundwater.

SITE BACKGROUND AND SITE CHARACTERIZATION

McClellan AFB is located approximately 7 miles north of Sacramento, CA. Operable Unit A (OU A) was selected as the treatment site, based on previous microcosm studies conducted at Oregon State University (OSU) (Dolan and Semprini, 1999). OU A was logistically favorable, being located underneath a parking lot currently used for vehicle storage. OU A is located at the southern end of the Base and in an area with relatively high CAH concentrations in groundwater, with trichloroethylene (TCE) and 1,2-cis-DCE (*cis*-DCE) concentrations exceeding 500 μg/L. Figure 1 shows the site plan view of McClellan AFB and the cometabolic air sparging field demonstration with multi-level (groundwater and the soil-gas) monitoring wells, the soil vapor extraction wells (SVE), and the sparge wells at OUA site.

Two sparge wells were installed to compare air sparging with and without propane. One sparge well received air only, and one received a mixture of propane and air. The two sparge wells were located approximately 100-ft apart. Propane gas was fed to one sparge well, and an air compressor was used to deliver air to both wells. The SVE blower draws air from the SVE wells, which were plumed to a two stage GAC treatment unit. The SVE system was never used in experiments conducted to date. Six multi-level monitoring points for soil-gas and groundwater sampling surrounded each sparge well. The distance from the sparge well to the inner and outer multilevel monitoring wells was 7.5 ft. and 15 ft., respectively. Four additional groundwater monitoring wells MW-1 and MW-2, and MW-3 and MW-4 were located 50-ft from each sparge well, one upgradient and one downgradient of the anticipated sparged radius of influence.

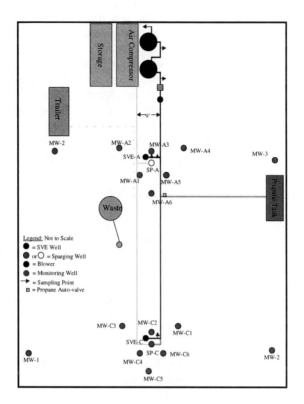

FIGURE 1. Site plan view of the McClellan AFB cometabolic air sparging field demonstration including multi-level (groundwater & soil-gas) monitoring wells, soil vapor extraction (SVE), and sparge wells at OU A site.

APPROACH AND METHODS

The following nomenclature was used to identify the sparge system and monitoring assembly. The cometabolic or active propane zone (C-zone, MW-C1 through C6) received propane and air and the air zone (A-zone, MW-A1 through A6) received air only; otherwise, both zones were operated equally in relation to the flow rate and time of sparging. GW1 and GW2 represent groundwater monitoring locations at 113-ft and 117-ft below ground surface (bgs), respectively. SG1 through SG4 represent soil gas sampling points in the vadose zone at 35-, 65-, 95-, and 105-ft bgs.

A mixture of contamination was present in both the groundwater and vadose zone of the test site, including TCE, *cis*-DCE, 1,1-dichloroethene (1,1-DCE), chloroform (CF), and 1,1-dichloroethane (1,1-DCA). A deep vadose zone exists at the site, with the contaminated groundwater level being approximately 108-ft bgs. TCE and *cis*-DCE were the major contaminants in the groundwater and the vadose zone.

Propane and air were pulse injected into the saturated zone once or twice a week for 5 to 10 hr durations. This pulse frequency was initiated after propane utilization was observed. A mixture of air containing propane (2%) was injected at an airflow rate of 5 standard cubic foot per minute (scfm). A separate control zone (A-zone) was subjected to the same treatment; however, only air was injected. Soil vapor extraction (SVE) was not operated during the test.

The sparging resulted in propane and air being added to both the saturated and the vadose zones. The tests were performed in a series of steps as follows.

- Day 0-36   Background concentration measurements were collected
- Day 36   Propane injection initiated at a concentration of approximately 2% in air.
- Day 36-76   Propane concentrations in the saturated and vadose zone were monitored without air sparging during this time.
- Day 80-120   Weekly injection of propane and air
- Day 120-160   Frequency of propane and air sparging was increased to twice a week in an attempt to accelerate the bioremediation of TCE and c-DCE.
- Day 160-188   No propane or air was added so that propane oxidation rates could be verified and compared to earlier rates and to determine if chlorinated solvent transformation would continue without additional propane.
- Day 188-220   Ammonia was added with propane and air at a concentration of about 0.01 %, after it was determined that nitrogen was limiting in the treatment zone.

RESULTS AND DISCUSSION

Background data collected during the first 40 days of testing showed higher groundwater concentrations of TCE and *cis*-DCE in the C-zone ranging from 100 to 1,000 µg/L, compared to the A-zone where concentrations ranged from 60 to 400 µg/L. TCE and *cis*-DCE concentrations also were higher in the C vadose zone (500 to 600 µg/L) than in the A vadose zone (300 to 400 µg/L).

After sparging with propane on day 36, a fairly long lag period of 22 to 59 days occurred before propane utilization was evident in the saturated zone. Table 1 shows initial propane and DO concentrations achieved in the groundwater C-zone monitoring wells after the initial sparge and lag times before propane utilization. Three out of six monitoring wells located at both 113- and 117-ft levels showed propane and DO were successfully delivered, with propane concentrations ranging from 0.3 to 1.5 mg/L and DO concentrations ranging from 6.0 to 7.5 mg/L. The results show a non-uniform distribution of propane and air in the saturated zone. The dissolved propane concentrations in the saturated zone varied greatly, and were correlated with the dissolved oxygen concentrations, indicating some areas were more effectively sparged than others. The rates of propane utilization also varied at the different monitoring locations. After successive propane additions, the rate of propane utilization increased as the microbial population was stimulated.

TABLE 1. Propane and DO concentration achieved in C-zone groundwater monitoring wells after the initial sparging and the time period before a 50% reduction in propane concentration was observed

GW Monitoring Wells	Well Location Depth (ft.)					
	113 ft. (GW1)			117 ft. (GW2)		
	Propane Conc. (mg/L)	DO Conc. (mg/L)	Time Period before propane concentration decreased by 50% (days)	Propane Conc. (mg/L)	DO Conc. (mg/L)	Time Period before propane concentration decreased by 50% (days)
MW-C1	0.06	2.4	ND	0.04	1.6	ND
MW-C2	0.4	6.5	25	1.5	7.7	33
MW-C3	0.3	6.6	59	0.4	7.3	44
MW-C4	1.1	6.0	22	0.8	6.9	46
MW-C5	0.02	2.3	ND	0.3	5.7	ND
MW-C6	0.02	2.0	ND	0.0	3.4	ND

ND = not detected

Effective removal of TCE, *cis*-DCE, and other CAHs present at low concentrations (CF, 1,1-DCE, and 1,1-DCA), was observed in the saturated zone where effective propane and DO delivery occurred. Little or no removal of the CAHs was observed at monitoring locations where the propane and air were unsuccessfully delivered, and at several of these locations CAH concentrations increased during the test. Figure 2 shows the decrease in TCE and *cis*-DCE concentrations in the active zone compared to that of the control zone. In the biostimulated C-zone, three out of six monitoring wells located at both the 113- and 117-ft levels that received propane and air showed significant TCE and *cis*-DCE removal. Concentrations of *cis*-DCE decreased more rapidly than TCE, which is consistent with the cometabolic transformation observed in our microcosm tests. At the 117-ft level, all the contaminants were successfully removed in the zone of active propane utilization. TCE concentrations decreased below 5 µg/L and *cis*-DCE concentrations were below 2 µg/L at two monitoring locations.

At the 113-ft monitoring level, higher initial TCE and *cis*-DCE concentrations were observed. Three out of six monitoring wells showed a gradual decrease in TCE and *cis*-DCE concentrations in response to propane and DO consumption over the first 160 days of the demonstration. In comparison, *cis*-DCE and TCE were less effectively removed from the A-zone. Relative rates of *cis*-DCE removal were similar to that achieved for TCE indicating physical removal by stripping in the A-zone.

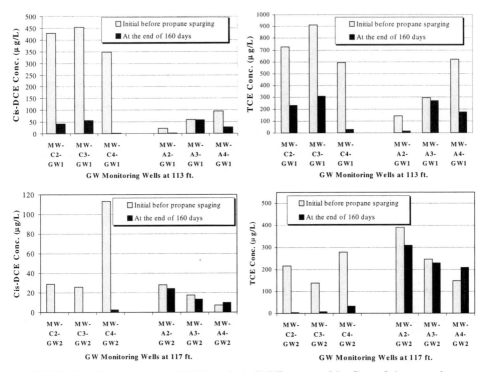

FIGURE 2. Comparison of TCE and *cis*-DCE removal in C- and A-zones in both GW monitoring levels at 113 ft. and 117 ft.

After 160 days of operation, the rates of propane uptake decreased at several locations that received high propane doses. Nitrate was found to be limiting in the groundwater during this period and effective removal of TCE and *cis*-DCE ceased as indicated by an increase in TCE and *cis*-DCE concentrations. To supply nitrogen, ammonia gas was added to the sparge gases (air and propane) in the C-zone after 190 days of operation. Upon adding ammonia, the propane uptake rate increased and TCE and *cis*-DCE removal resumed.

The study was less successful in demonstrating the stimulation of propane-utilizers in the vadose zone. Overall propane concentrations tended to increase during the demonstration as more propane was sparged into the zone. Several locations showed decreases in propane concentration after ammonia was added at 190 days, suggesting that nitrogen may have been limiting in the vadose zone. Figure 3 shows TCE and *cis*-DCE removals in the C- and A-zones in two soil gas-monitoring wells at 95-ft bgs, approximately 13-ft above the water table.

At vadose zone monitoring well MW-C2-SG3, propane concentrations gradually increased over time toward the sparge gas concentrations. Decreases in propane concentrations were observed early on (75 to 90 days) when propane addition was suspended and at the end of the test after ammonia was added. TCE and *cis*-DCE concentrations decreased from 500 to 600 µg/L to about 120 µg/L at the end of the test. Similar observations were made at vadose zone monitoring well MW-C4-SG3. TCE and *cis*-DCE dropped from 400 to 700 µg/L to about 250 µg/L over the course of the demonstration. Decreases in TCE and *cis*-DCE appeared to be coincident with propane utilization at these locations. Minimal or no loss of TCE and *cis*-DCE was observed in the control zone. The results provide some evidence of biostimulation in the vadose zone and TCE and *cis*-DCE transformation, but further studies are needed.

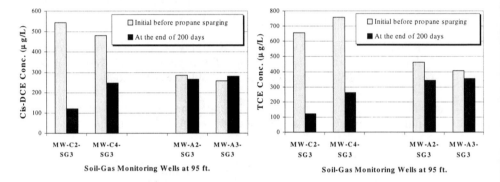

FIGURE 3. Comparison of TCE and *cis*-DCE removal in C- and A-zones at the 95-ft soil-gas monitoring wells

SUMMARY

- Cometabolic sparging resulted in a non-uniform distribution of propane and air, an indication of heterogeneous gas flow in the saturated zone.
- Stimulation of propane-utilizing microorganisms in contaminated groundwater occurred after lag periods of 22 to 59 days.
- With pulsed additions of air and propane for over 200 days, effective TCE and *cis*-DCE removal was observed at monitoring locations in the saturated zone where effective propane delivery and uptake occurred.
- The stimulation of indigenous propane utilizing microorganisms resulted in transformation of CAH mixtures (TCE, *cis*-DCE, 1,1-DCE, and 1,1-DCA) in the C-zone resulting in concentrations below MCL levels at some locations.
- At locations where propane and air were not successfully delivered, due to aquifer heterogeneities, little removal of TCE and *cis*-DCE was observed.
- Minimal loss of CAHs due to physical air stripping was observed in the control A-zone monitoring wells located at 117-ft bgs. However, TCE

and *cis*-DCE concentrations decreased significantly at the 113-ft level, indicating that physical removal by sparging was a significant factor.

ACKNOWLEDGMENTS

The field demonstration was funded by the Environmental Security Technology Certification Program (ESTCP) under the project title *Use of Cometabolic Air Sparging to Remediate Chloroethene-Contaminated Groundwater Aquifers.* The authors wish to thank Dennis M. DeBacker (Battelle), Chris Coonfare (Battelle), and Derek Rogers (OSU) for their invaluable assistance in this effort.

REFERENCES

Dolan. M.E. and L. Semprini. 1999. "Aerobic Cometabolic TCE Transformation in Propane-fed Aquifer Microcosms" 4th International Symposium on Subsurface Microbiology, Vail, CO, August 22-27.

Marley, M. and C.J. Bruell. 1995. *In Situ Air Sparging: evaluation of Petroleum Industry Sites and Considerations for Applicability, Design and Operation.* American Petroleum Institute Publication No. 4609.

Tovanabootr, A. and L. Semprini. 1998."Comparison of TCE transformation abilities by methane and propane-utilizers." *Bioremediation Journal* 2:105-124.

Travis, B.J. and N.D. Rosenberg. 1997. "Modeling *In Situ* Bioremediation of TCE at Savannah River: Effects of Product Toxicity and Microbial Interactions on TCE Degradation." *Environ. Sci. Technol.* 31:3093-3102.

DEEP AIR SPARGING OF A CHLORINATED
SOLVENT SOURCE AREA

Scott A. Glass (Naval Facilities Engineering Command,
North Charleston, South Carolina)

ABSTRACT: A full-scale air sparging/soil vapor extraction (AS/SVE) system with off-gas treatment has been in operation at the Aircraft Intermediate Maintenance Department (AIMD) Seepage Pit (Site 16) at the former Naval Air Station, Cecil Field in Jacksonville, Florida since June 1999. The goal of the project is to reduce the concentrations of Volatile Organic Compounds (VOCs), primarily Trichloroethene (TCE), to less than 1000 ug/l in the source area. The remediation system was designed to treat elevated VOC concentrations at a depth up to 100 ft (30.5 m) below ground surface (bgs). Performance data collected during the first several months of operation suggests that effective removal of VOCs from the aquifer, based on significant decreases in groundwater contaminant concentrations and total VOC concentrations measured in the pre-treated off-gas.

INTRODUCTION

The Base Realignment and Closure Cleanup Team (BCT) for the former Naval Air Station, Cecil Field consists of representatives from the Navy, U.S. Environmental Protection Agency Region IV and the Florida Department of Environmental Protection. The BCT is tasked with the environmental cleanup of the former Naval Air Station to support transfer of the facility to the local community. The BCT determined that implementing AS/SVE as a source remediation system held the greatest promise for effective remediation of the site in the shortest time frame and in the most cost effective manner.

The source remediation system is currently being operated and is reducing the concentrations of chlorinated solvents in the groundwater. Data has been collected to measure the effectiveness of the treatment system. Factors supporting system design and the results of the first several months of system operation are presented.

Objective. The objective of this project is to reduce total VOC concentrations in groundwater, at the source area. TCE is the most widespread of the chlorinated solvents at the site and exists at the highest concentrations, therefore reductions of TCE in groundwater to less than 1000 ug/l was selected as the cleanup goal. High concentrations of TCE at depths as great as 100 ft (30.5 m) bgs, have been identified, therefore it was necessary to design a system that would effectively treat source contamination at these depths.

Site Description. High concentrations of VOCs including TCE, Trichloroethane (TCA), Dichloroethane (DCA), Dichloroethene (DCE), and Vinyl chloride are present in the unconfined aquifer, as a result of past disposal practices at the AIMD Seepage Pit. The seepage pit was used for the disposal of liquid wastes from 1959 to 1980. A source removal action was conducted in May 1994 to remove the seepage pit and approximately 1,500 yd^3 (1,147 m^3) of contaminated soil.

The majority of the site at the source area is flat and covered with grass that is well maintained. The former seepage pit was located in this grassy area. Areas adjacent to the source area are paved with asphalt or concrete. Underlying the site is an unconfined surficial aquifer consisting primarily of fine-grained sands. Groundwater at the site is at approximately 6 ft (1.8 m) bgs. Groundwater flows to the southeast towards the adjacent flightline. The depth to the confining unit is approximately 100 ft (30.5 m) bgs. The highest VOC concentration identified at the site, prior to start-up of the source remediation system, was 978,000 ug/l TCE, in a monitoring well (45I) screened from 25 ft to 32 ft (7.6 m to 9.8 m) bgs, located in the center of the source area.

MATERIALS AND METHODS

A pilot-scale test was conducted in September 1998 to determine the effective radius of influence for air injection and vapor extraction wells. Samples of extracted vapors were analyzed and it was determined that off-gas treatment would be necessary. Bench-scale treatability tests identified a site specific TCE solubility concentration ranging from 1,230,000 ug/l to 1,600,000 ug/l.

Vapor extraction tests evaluated the radius of influence of 2 wells located approximately 40 ft (12.2 m) apart with 1 ft (0.30 m) of screen below the water table (Figure 1). The tests identified a limiting vacuum pressure of 45 inches of water (3.3 in Hg). Above this vacuum pressure significant extraction of groundwater occurred. The corresponding vapor extraction flow rate at this limiting vacuum pressure ranged from 10 ft^3/min to 15 ft^3/min (4,720 cm^3/sec to 7,080 cm^3/sec). A conservative design extraction flow rate of 10 ft^3/min (4,720 cm^3/sec) per well was adopted, providing an estimated radius of influence of 15 ft (4.6 m) based on measured vacuum in the piezometers located radially outward from the extraction wells.

Air injection tests evaluated the radius of influence of a well pair screened at a depth of 58 ft to 60 ft (17.7 m to 18.3 m) and 95 ft to 100 ft (29.0 m to 30.5 m) bgs (Figure 1). These depths were selected to ensure air would be delivered below the highest VOC concentrations observed in the aquifer. Air injection tests measured dissolved oxygen, oxidation reduction potential, and organic vapor concentrations in piezometers and monitoring wells located radially outward at various distances from the air injection wells. By evaluating these parameters measured at the wells, along with physically observing air bubbles in the wells, an estimated radius of influence of 30 ft (9.1 m) for each pair of air injection wells was adopted. These results correspond to a design injection flow rate of 10 ft^3/min (4,720 cm^3/sec) per well.

Based on pilot-scale test results, 3 pair of air injection wells and 19 vapor extraction wells were installed within the source area. A duel vessel (operated in series) gas-phase granular activated carbon (GAC) adsorption system was selected to reduce VOC concentrations in the off-gas to acceptable levels. Figure 2 shows the layout of the AS/SVE wells and Figure 3 shows the AS/SVE system equipment configuration.

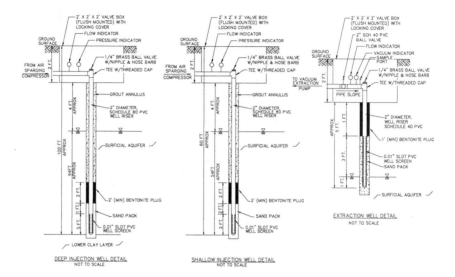

FIGURE 1. Air injection and soil vapor extraction well construction details.

RESULTS AND DISCUSSION

System Operation. Start-up of the full-scale remediation system was conducted in June 1999. Routine system maintenance has been ongoing since start-up. No major operational difficulties have been encountered. The system is occasionally shut down for scheduled maintenance activities and during groundwater sampling events.

The average vapor extraction flow rate measured at the well head is 7.8 ft^3/min (3,682 cm^3/sec) compared to the design flow rate of 10 ft^3/min (4,720 cm^3/sec) per well. The average air injection flow rate measured at the well head is 18.6 ft^3/min (8,779 cm^3/sec) compared to the design flow rate of 10 ft^3/min (4,720 cm^3/sec) per well. Piezometers within the treatment area are monitored for pressure/vacuum on a monthly basis and if a positive pressure is detected, the system is adjusted until a vacuum, or zero pressure is measured.

Off-Gas Treatment Analysis. Since the fresh air bleeder line does not contain a flow meter, the total airflow through the blower is estimated using the measured vacuum at the blower inlet compared to the vacuum pressure blower curves provided by the manufacturer. Using this method, a conservative average discharge airflow rate has been estimated at 220 ft^3/min (103,840 cm^3/sec). Based on this average flow rate, total VOC kg/day loading and emissions rates are calculated using analytical results from the influent and effluent to and from the

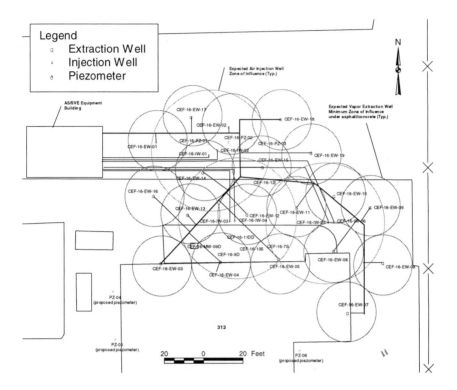

FIGURE 2. Air sparging/soil vapor extraction system well configuration.

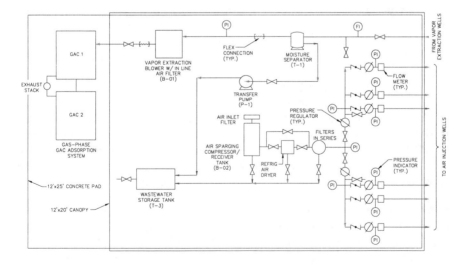

FIGURE 3. Air sparging/soil vapor extraction system equipment layout.

GAC treatment system. Average total VOC removal efficiencies are exceeding 99%. For the first week of full-scale operations, loading rates exceeded 10 kg/day. After 8 months of operation, loading rates have dropped below 0.5 kg/day (Figure 4). Based on these loading rates, an estimated 544 kg of total VOCs have been removed from the aquifer during the first 8 months of operation.

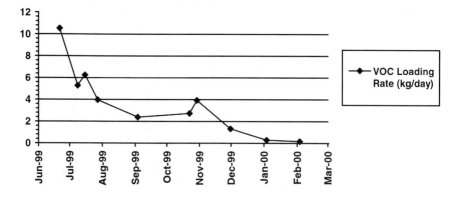

FIGURE 4. VOC loading rate of off-gas treatment system

Groundwater Monitoring Results. Quarterly groundwater sampling of the source area wells has shown significant reductions of total VOCs in groundwater. Some increases in VOC concentrations were encountered during the sampling event following start-up of the AS/SVE system, but in all cases, these concentrations have decreased to below or at the concentrations prior to startup. All VOC concentrations in the source area from the February 2000 sampling round are below 1000 ug/l. The highest concentration remaining is 845 ug/l TCE at well 10S.

After the November sampling round, monitoring well 45I was removed from the sampling program. The well experienced excessive pressure that blew sand out of the riser, and eventually filled the well, rendering it non-productive. The TCE concentration at well 45I was 140 ug/l when it was last sampled in November. Well 45I will be replaced to allow assessment of groundwater contamination concentrations in the intermediate zone within the source area to resume. Table 1 shows VOC concentrations in groundwater detected at the source area.

CONCLUSIONS

Total VOC concentrations in the source area have significantly decreased since start-up of the AS/SVE source remediation system. Direct evidence of substantial mass removal from the aquifer is provided by the estimated 544 kg of total VOCs measured in the off-gas. It is expected that the initial increase of VOC concentrations experienced after startup of the system was due to the redistribution of contaminants after being agitated by the injected air. The subsequent decrease of concentrations in all wells, along with the total VOC mass measured in the off-gas air stream suggests that the AS/SVE system is effectively removing VOC contamination from the aquifer rather than simply moving the contamination out of the treatment zone.

Table 1. VOC concentrations in groundwater at the source area.

Monitoring Well	VOC	Sampling Event (concentrations in ug/l)			
		Apr-99#	Aug-99	Nov-99	Feb-00
10S	TCA	673	1,380	3.9	ND
	DCA	802	<400	19.4	4.9
	cis-DCE	11,900	6,330	838	163
	TCE	836	13,800	2770	845
	VC	1,550	<400	12.2	1.2
45I	TCA	8,030	<500	<1	NS*
	DCA	2,710	<500	<1	NS*
	cis-DCE	1,130	<500	<1	NS*
	TCE	978,000	27,300	140	NS*
	VC	<200	<500	<1	NS*
9D	TCA	<1	<1	<1	ND
	DCA	<1	<1	<1	ND
	cis-DCE	0.57	<1	7.2	1.1
	TCE	1.2	9.6	497	13.4
	VC	<1	<1	<1	ND
11DD	TCA	ND	NS**	NS**	NS**
	DCA	ND	NS**	NS**	NS**
	cis-DCE	ND	NS**	NS**	NS**
	TCE	ND	NS**	NS**	NS**
	VC	ND	NS**	NS**	NS**

ND = Non-Detect
NS = Not Sampled
# System start-up was June 1999. April 1999 represents pre start-up data.
* Not sampled due to well no longer being productive.
** Not sampled due to initial samples being non-detect.

No groundwater TCE concentrations above 1000 ug/l were detected during the February 2000 groundwater sampling event. Design criteria requires at least two consecutive sampling rounds of less than 1000 ug/l TCE in order to consider that cleanup objectives have been met. In addition, well 45I could not be sampled during the February 2000 groundwater sampling event to determine if TCE concentrations at this well remain below 1000 ug/l.

Off-gas concentrations appear to be reaching an asymptotic level of less than 0.5 kg/day. Contamination remaining in the aquifer within the zone of influence is likely now in a diffusion limited state, indicating that removal efficiencies from the AS/SVE system may also be limited. Based on pre start-up groundwater data, an estimated 410 kg total VOCs is expected to have been present in the groundwater source area. The measured mass removal of 544 kg total VOCs, based on off-gas concentrations, compares favorably to the estimated pre-treatment mass of total VOCs in the aquifer. All these factors combined support the conclusion that the AS/SVE system has effectively treated the source area groundwater for VOC contamination.

Prior to discontinuing active remediation of the source area, several parameters will be evaluated. Well 45I will be replaced and groundwater contamination concentrations in the source area will continue to be assessed to ensure concentrations remain below the cleanup target level. Monitoring of off-gas concentrations will continue to determine if asymptotic levels have been reached indicating limited efficiencies of the active treatment system. When it is determined that the cleanup objectives have been met and the AS/SVE system can be turned off, groundwater sampling will continue to assess any potential rebound effects. If rebounding of groundwater contamination concentrations occur, the BCT will determine if restarting the AS/SVE system is warranted.

REFERENCES

Tetra Tech NUS, Inc. 1999. *Amended Record of Decision, Operable Unit 7, Site 16, Naval Air Station Cecil Field, Jacksonville, Florida,* Southern Division, Naval Facilities Engineering Command.

Tetra Tech NUS. 1999. *Groundwater Remedial Design, Operable Unit 7, Site 16, Naval Air Station Cecil Field, Jacksonville, Florida.* Southern Division, Naval Facilities Engineering Command.

CH2Mhill Constructors, Inc. *Quarterly Operations and Maintenance Report, Air Sparging System at Operable Unit 7, Site 16, 2nd Quarter of Operation, Naval Air Station Cecil Field, Jacksonville Florida.* Southern Division, Naval Facilities Engineering Command.

PASSIVE SOIL VAPOR EXTRACTION OF CHLORINATED SOLVENTS USING BOREHOLES

A.G. Christensen, E.V. Fischer, H.H. Nielsen & T. Nygaard
NIRAS, Consulting Engineers and Planners, Allerød, Denmark
H. Østergaard
County of Frederiksborg, Hillerød, Denmark
S.R. Lenschow
County of Ribe, Ribe, Denmark
H. Sørensen
County of Storstrøms Amt, Nykøbing F., Denmark
I.A. Fuglsang
Miljøstyrelsen (Danish EPA), Copenhagen, Denmark
T.H. Larsen
Hedeselskabet, Roskilde, Denmark

ABSTRACT: Passive soil vapor extraction (PSVE) relies on the naturally occurring variations in barometric pressure (Ellerd et al. 1999, Rossabi 1998). In Denmark, these oscillations are primarily resulting from passage of weather fronts. As the barometric pressure rises, a gradient is imposed on the soil gas, which drives fresh air into the soil. When the barometric pressure drops, soil gas will ascend into the atmosphere. When low permeability layers are present, they tend to damp and delay the pressure signal in the deeper vadose zone, resulting in a net varying vertical gradient. Boreholes screened in this vadose zone will act as a preferred flow path, and a significant flow of VOC contaminated soil gas out of the borehole will be the result of the pressure gradient. Different ways to enhance and control the flow out of boreholes have been investigated. These include one-way valves allowing outflow only, GAC-canisters to adsorb the VOCs and different depths to the screened portion of the boreholes.

INTRODUCTION

PSVE has been tested and implemented at 4 former dry cleaners, where the existing buildings and low permeability clay in the upper 5-10 m exclude the use of ordinary remediation by active soil vapor extraction (SVE) or excavation. The contaminated clay at the sites acts as a long-term source for contamination of the deeper vadose zone found directly under the clay. At the bottom of this zone the groundwater table is found, and caused by infiltration and gas phase diffusion in the vadose zone, the VOCs are entering the capillary fringe and hence the groundwater.

The goal of this project is to develop, implement and document PSVE as a low-cost in situ containment and extraction technology for sites where the VOC-contamination resides in a confined vadose zone. By continuously removing mass from the vadose zone, it is the expectation that the net mass transport to the groundwater will be significantly reduced, and that the groundwater concentration will decrease over time.

The project is sponsored by the 3 countys and by the Danish EPA under the "Technology Demonstration and Evaluation Program", and will be carried out over a 2 year period. The initial development and testing of the GAC has been documented (Christensen et al., in press). The monitoring will be published in a series of progress reports and a final report will be prepared in year 2002.

RESULTS

At the 4 sites, a total of 23 boreholes have been installed. At each site, a datalogger records the barometric pressure, airflow and temperature out of one of the wells as well as the differential pressure (driving force). In Figure 1, data from November 1999 from one site shows 3 periods (day 1-2, 3-4 and 7-8) with a positive differential pressure, caused by the decreasing barometric pressure in these periods. Conversely, increasing barometric pressure generates a negative differential pressure, but due to the one-way valves, no airflow into the well is generated, Figure 2.

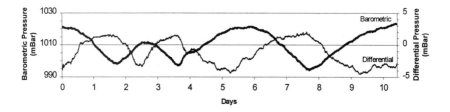

FIGURE 1. Barometric pressure and differential pressure between the vadoze zone and the atmosphere.

In agreement with the above 3 periods of positive differential pressure, there can be seen 3 periods with airflow from the recorded data, Figure 2. The peak flows of 5.5 m³/h occur at a time of maximum differential pressure. The concentration in the outflowing air shows initial rapid rise and then reaches a steady level around 200 mg/m³. During the 10 days, a total air volume of 300 m³ and a total mass of 45 g chlorinated solvents have been passively extracted.

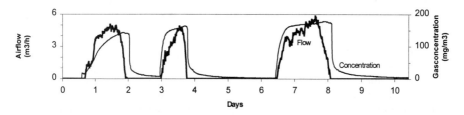

FIGURE 2. Airflow and gasconcentration of chlorinated solvents (PCE+TCE) recorded in well PV-2.

Based upon the initial data from the 4 sites, a yearly extraction rate of 1-25 kg chlorinated solvents per site can be expected. The maximum flow rate recorded is 15 m³/h, and the maximum concentration of chlorinated solvents (PCE+TCE) is 500 mg/m³.

REFERENCES

Christensen A.G. & Husum H., 1999. Teknologiudviklingsprojekt. "Passiv Ventilation (PV). Design dimensionering og afprøvning af anlægskomponenter." *København: Miljøstyrelsen.* (www.mst.dk)

Ellerd M.G., Schwaegler D.P., Massman J.W. & Rohay V.J, 1999. "Enhancements for passive vapor extraction": *The Hanford Study. Groundwater* 37(3): 427-437.

Rossabi. J., 1998. "Passive vapor extraction for interim remediation at the Savannah River site. In physical, chemical and thermal technologies: Remediation of chlorinated and recalcitrant compounds." Vol 1(5):161-167. *Batelle Press.*

DIESEL FUEL RECOVERY USING BIOSLURPING – A CASE STUDY

Barry Christian, Bruce Clarke, P.E., Nizar Hindi, Michael Wolf C.P.G., and
Rhonda Gibson E.I., Earth Tech, Grand Rapids, Michigan, USA

ABSTRACT: Decades of locomotive fueling at a Pontiac, Michigan rail yard resulted in the release of diesel fuel oil no. 2 (oil) to surface soil. Oil infiltrated the sand soil to the water table. The oil plume covered 2 acres (0.81 ha) and was measured up to 4 ft (1.22 m) thick in wells. A bioslurping system consisting of 27 bioslurping wells, an air-liquid separator, an oil-water separator and a rotary lobe blower was installed. To date the system has operated for 12,627 hours and has recovered: 21,695 gal (82,124 L) of oil (averaging 41 gal/24 hrs or 155 L/24 hrs); 2,440,000 gal (9,240,000 L) of groundwater which is discharged back to the plume (averaging 3.2 gpm [0.2 L/s]); and 126,000,000 cu ft (3570000 cu m) of air (averaging 166 scfm [0.08 cu m/s]). The area and thickness of the oil plume have decreased. Design and operating information are provided in this paper.

INTRODUCTION

Site Description. Diesel fuel oil no. 2 (oil) was released to surface soil over several decades and infiltrated to the water table at a railroad locomotive refueling operation in Pontiac, Michigan. Surface soils are railroad ballast. Subsurface soils are sand to a depth of approximately 20 ft (6.1 m) where a clay aquaclude is encountered. The water table averages about 10 ft (3.05 m) below grade. The oil plume covers about 2 acres (0.81 ha) with an average observed thickness of 1 ft (0.35 m) and a maximum thickness of 4 ft (1.22 m). The volume of recoverable oil was estimated to be 21,500 gal (81,400 L). The leading edge of the plume migrated onto adjacent residential property. Groundwater is not impacted by dissolved constituents. A site map is presented in Figure 1.

Project History and Objectives. An interceptor trench was installed in late 1996 to limit off-site migration. A feasibility study was performed in early 1997 to evaluate conventional technologies and multi-phase extraction, or bioslurping. Bioslurping was determined to be aggressive and cost-effective based on net present value. A two-week bioslurping pilot study was performed on three wells in June 1997 to evaluate oil recovery rates in response to various vacuum levels and water flow rates. Results were promising and authorization to design a bioslurping system came in August 1997. The system was operating in March 1998. Earth Tech operates the system in about 16 to 20 hours per week. Project objectives are to prevent the further oil migration and to remove all recoverable oil in four years or less. Operating objectives include minimizing water and emulsion production.

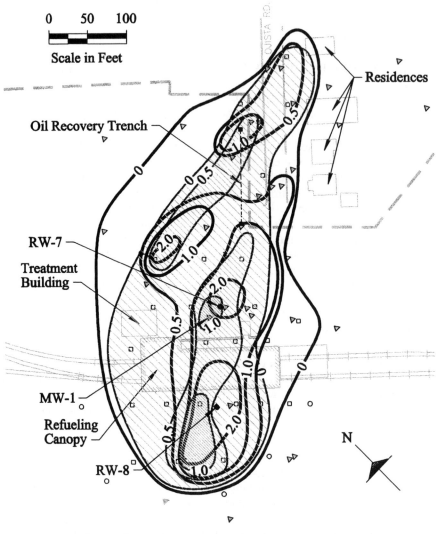

FIGURE 1. Bioslurping Layout and 1998 and 1999 Oil Thickness Contour Site Map

SYSTEM DESIGN

Process Description. Figure 2 is a process flow diagram. The system includes: 27, 6-in (15 cm) bioslurping wells; a 400-gal (1514 L) air-liquid separator (ALS) with a 1.5-hp (1.12 kW) progressive cavity discharge pump; two 1,800-lb (816 kg) vapor-phase carbon (GAC) vessels in series; a 75-hp (56 kW) blower; heat exchangers; a 7.5-hp (5.6 kW) rotary lobe, air injection blower; 2 air injection wells; a 3,000-gal (11356 L) oil-water separator (OWS) with a 0.75-hp (0.56 kW) oil discharge pump; a secondary, 50-gpm (3.15 L/s) inclined plate OWS; a 100-gal (378.5 L) oil transfer tank with an 0.5-hp (0.37 kW) discharge pump; an oil storage tank; a 100-gal (378.5 L) water transfer tank with a 1-hp (0.75 kW) discharge pump; 6, 6-in (15 cm) water injection wells; a programmable logic controller; an auto-dialer; and a 22- by 30-ft (6.7- by 9.1-m) treatment building. Oil, water and soil vapors are sucked into the ALS. Soil vapor exits the top of the ALS, passes through GAC and is ultimately reinjected. Water and oil are pumped to the OWS system where they are separated. Oil is disposed off site. Groundwater is injected within the capture zone to reduce water treatment costs.

While the bioslurping system recycles groundwater in a closed loop, two 6-in (15 cm) groundwater recovery wells (RW-7 and RW-8) with submersible pumps located in the heart of the oil plume maintain hydraulic control of the plume and prevent its expansion by discharging to a storm sewer that flows off site. Five-foot screens in these wells were placed at the bottom of the water table aquifer just above a relatively impermeable clay layer (20 ft [6.1 m] below grade). Groundwater from RW-7 and RW-8 is pumped through 2, 100-lb (45.4 kg) GAC vessels and then to the storm sewer.

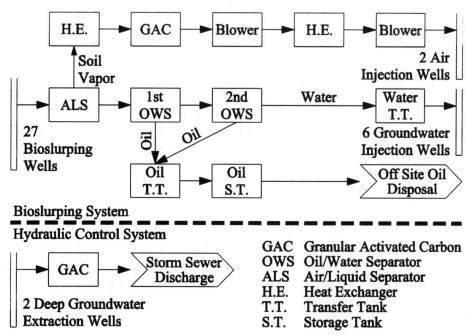

FIGURE 2. Process Flow Diagram

Blower Selection. Others (Battelle, 1995) recommend liquid-ring pumps (LRPs) for bioslurping applications, in part because an LRP is inherently explosion proof relative to the process fluid. Our experience is that LRPs have high O&M costs and are prone to problems. LRP problems often stem from using process water for sealing and cooling. Alternatively, supplying clean water can be expensive. We selected a Roots DVJ-412 positive displacement, rotary-lobe, high-vacuum blower (Roots) over a comparably sized LRP (Squire-Cogswell 150A02TM, or equivalent).

The Roots and LRP have similar capital costs and high vacuum capability (i.e., approximately 27 in [68.6 cm] of mercury [Hg]). Normal operation is at much lower vacuum levels, (10 to 16 in [30.5 to 40.6 cm] of Hg). In this range the Roots blower requires less horsepower. Unlike the direct drive LRPs, the Roots blower can be inexpensively re-shieved to yield flexibility in flow rate and operating power. Roots blowers have proven to be robust, low-maintenance, long-life blowers. Critical to not using an LRP was the ability to prevent explosive conditions in the process air stream. Michigan regulations require treatment to reinject air. The recovered oil flash point is at or above 130 deg F (54.4 deg C). There is no possibility for the oil to reach this temperature and the lower explosive limit (LEL) of 0.7%, or 7,000 ppmv, should never be reached. Potential for the LEL to be exceeded at the blower is virtually eliminated by GAC. An in-line LEL meter set to trigger system shut-down provides added safety.

Bioslurping from a Trench. Early in the project, an oil recovery trench was installed along approximately 100-ft (30.5-m) of the property boundary with an orientation that is approximately perpendicular to the direction of groundwater flow. This 36-in (91 cm) deep trench was constructed with three overlapping sections of an ADS® 1.5-in (3.8 cm) wide by 15-in (38 cm) tall perforated drain tile, wrapped with a non-woven, geotextile fabric. The trench terminates at a 12-in (30.5 cm) diameter sump at each end. The trench bottom is several inches below the observed water table. This design accommodates large water table fluctuations and intersects the full oil bearing zone. Prior to start-up of the bioslurping system, approximately 3,000 gal (11356 L) of oil were recovered from this trench. The two trench sumps are used as bioslurping wells. This strategy has been quite successful and the header that the trench is connected to has produced the largest volume of oil on the occasions when individual header production has been evaluated.

RESULTS

Cumulative Oil Recovery. In 22 months of operation, beginning in March 1998, 21,695 gal (82124 L) of oil have been recovered. Oil recovery data are presented in Figure 3. While the recovery rate may be decreasing slightly, large amounts are still recovered on a weekly basis. The data indicate that higher rates occur in late summer and fall. Higher recovery rates during this period may result from a lower water table exposing more oil to the vacuum. In January 1999 1,500 gal (5678 L) of a 50% oil-in-water emulsion was removed from the primary OWS. In

December 1999, a large oil volume was recovered when the system was restarted after being shut down for several weeks.

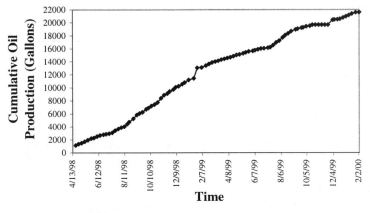

FIGURE 3. Cumulative oil recovered

Bioslurping Results. From April 1, 1998 through February 7, 2000 the system has operated for 12,627 hours and has removed the following: 21,695 gal (82,124 L) of oil for an average rate of 41 gal/24-hour day (155 L/24 hrs); 2,440,000 gal (9,240,000 L) of groundwater which is discharged back to the plume (average rate of 3.2 gpm [0.3 L/s]); and 126,000,000 cu ft (3,570,000 cu m) of air extracted from the vadose zone (average rate of 166 scfm [0.08 cu m/s]). The ratio of oil:water:air by volume is 1:116:45,000. Vacuum levels at the blower inlet mainly ranged from 8 to 10 in (20 to 25 cm) of Hg, occasionally dipping to 7 in (17.8 cm) or rising to 12 in (30.5 cm) of Hg vacuum. Vacuum levels at the well heads have ranged from 0 to 50 in (0 to 127 cm) of water column. Based on the output from an in-line turbine air flow meter and an assumption of equal flow distribution, it is concluded that air velocity in the 1-in (2.54 cm) slurp tubes ranges from 2,000 to 3,000 ft/min (10.2 to 15.2 m/s). It ranges from 600 to 1,200 ft/min (3.0 to 6.1 m/s) in the horizontal transmission pipe which consists of 1.5- to 4-in (3.8 to 10.2 cm) PVC pipe.

Hydraulic Control Results. Expansion of the oil plume is prevented by extracting groundwater from the bioslurping wells and from RW-7 and RW-8 at continuous average rates of 3.3 and 0.5 gpm (0.21 and 0.03 L/s), respectively. The groundwater extraction produces a depression of 1.5 to 2.0 ft (0.46 to 0.61 m) in the water table below the oil plume. Groundwater modeling using Winflow® indicates that the re-injected groundwater is captured by the bioslurping and recovery wells. Groundwater captured from RW-7 and RW-8 contains only trace concentrations of dissolved hydrocarbons (HCs) and requires only minimal treatment prior to discharge. Groundwater flows average 0.5 gpm (0.031 L/s) and the carbon vessels have been changed twice to meet the discharge limit of 20 µg/L total BTEX. The groundwater quality withdrawn from RW-7 and RW-8 has remained relatively constant even though the injected water oil and grease concentration is several hundred thousand µg/L.

Plume Size Changes. Figure 1 shows the maximum measured oil thickness for May to November 1998 and for May to November 1999. These data were used in the oil volume calculations. Not only has the areal extent of measurable oil decreased from 1998 to 1999, the oil thickness is significantly reduced in the center of the plume. In many wells the oil thickness has been reduced by 50% or greater.

DISCUSSION

Oil Thickness and Volume Estimates. Based on observations at this and other similar sites, oil thickness measurements can fluctuate over time. Designing a remediation system based on limited measurements could lead to a potentially inefficient system. If oil thickness measurements are biased to low or high water table periods, misleading information can result. Significant water level and associated oil thickness fluctuations can be less than one to two days, and may be caused by a large precipitation event, an unreported spill, cycling of nearby pumping wells, or locomotives moving on nearby tracks.

The computer program SPILLVOL$^{®}$ was used to estimate oil volumes at the site. In December 1996, Earth Tech estimated 21,500 gal (81,400 gal) of recoverable oil based on 16 observation points. By spring 1998, a total of 66 monitoring and bioslurping wells were used to measure water levels and oil thickness. To account for water table and oil thickness fluctuation, revised estimates have been based on maximum observed oil thicknesses at each well during a six-month period (five events – May through November 1998 and four events – May through November 1999). Using this approach, the calculated total oil volume at the start of bioslurping operation in 1998 was approximately 108,000 gal (408,800 L), with 95,000 gal (359,600 L) recoverable. The calculated total oil volume during the last half of 1999 is approximately 40,500 gal (153,300 L), with a calculated recoverable oil volume of approximately 29,500 gal (111,700 L). Since only 21,000 gal (79,500 L) of oil have been recovered from the site, there is a discrepancy of 44,500 gal (168,400 L). The magnitude of this discrepancy illustrates the level of difficulty in assessing the volume of oil present.

Oil is often said to float on groundwater. Oil in a well accumulates to a greater thickness than exists in the surrounding aquifer, a result of density and surface tension differences. These two useful concepts, are of questionable accuracy (Schwille, F. 1984). Observed oil thickness and overall oil recovery rates at this site are impacted by seasonal water table variations (see Figure 4). When the water table is near its seasonal high, observed oil thickness and recovery rates tend to be at a minimum. This project did not research this phenomenon, but knowledge of the subject suggests that oil does not float up and down with seasonal water table variations, but rather shares void space with water and air within the seasonally fluctuating, zone of capillary rise.

Emulsions. Oil and water, when vigorously mixed, can form an emulsion that is troublesome and costly to manage. Emulsions may lack sufficient net heat value to be sold as fuel, thus resulting in disposal costs. Conversely, recovered oil can be directly reused or sold as a fuel. Emulsions can be viscous reducing the efficiency of oil-water separation and handling equipment.

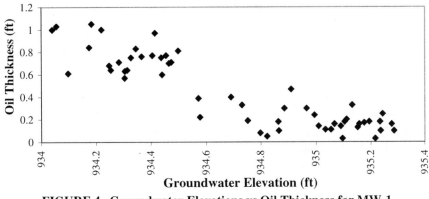

FIGURE 4. Groundwater Elevations vs Oil Thickness for MW-1

Several design elements help minimize emulsion production. Sharp turns and valves were minimized in process pipe. The ALS is upstream of the blower, so no oil passes through the blower. A low shear, progressive cavity pump provides the high inlet vacuum required to move liquids from the ALS to the primary OWS, but minimizes mixing energy. The primary OWS is a simple baffle design. It has a residence time of 10 to 12 hours allowing time for complete separation of the oil and emulsion from the water that flows through the OWS under gravity. Oil with no emulsion is pumped off the top allowing many months of residence time for accumulated emulsions to separate. A secondary, corrugated plate, inclined plane separator provides polishing of the water. The net result is that the primary separator has only been cleaned once.

Project Difficulties. Design, construction and operation difficulties have been experienced. The project schedule was very aggressive, requiring that construction contracts and equipment purchase orders be signed prior to completion of the design. The site is intersected by numerous active railroad tracks and roadways (not shown on Figure 1). The tracks are closely spaced allowing little room for construction equipment and requiring track lock-outs to protect construction crews from accidents with moving trains. Wintertime excavation of trenches under railroad tracks required extra effort in backfilling and compaction to prevent eventual slumping of the track under the weight of locomotives. There are numerous underground obstructions related to approximately 100 years of rail yard use.

The client required that there be nothing sticking up and that there be no trip hazards. This required that all well heads be located in manholes with traffic-rated covers. Moving heavy covers and entering the manholes is frequently required to inspect and adjust slurp tubes and vacuum levels since the area experiences water table fluctuations. Working in and around these manholes located so close to active tracks is difficult for the operator and ultimately limits our ability to monitor and optimize the performance of individual bioslurping wells.

Pipes connecting manholes in series had to be located below frost line and had to maintain the proper slope. Manhole bottoms are often quite close to the relatively shallow capillary fringe in which the oil is located. The closeness of the capillary fringe to the ground surface presented opportunities for airflow short-circuiting through excavated soils. Plastic sheeting was installed in the floor of the manhole excavations to minimize this potential problem.

KEY OBSERVATIONS
Key observations from our experience with bioslurping are as follows:
- Bioslurping is aggressive and can recover oil at a high rate.
- Fluctuating water tables require frequent slurp tube adjustments.
- Rotary lobe blowers can be safely and successfully used with diesel oil.
- Lower water tables yield larger oil thickness and oil recovery rates.
- Emulsion production can be minimized and successfully managed.
- Reinjecting water back to the oil plume can eliminate treatment costs.
- Oil estimating models are subject to difficulties in accurately measuring oil thickness.
- Slurp tube air velocities of 2,000 to 3,000 ft/min. (10.16 to 15.24 m/s) work well for oil recovery.

REFERENCES
Battelle, 1995. *Test Plan and Technical Protocol for Bioslurping.* Prepared for Air Force Center for Environmental Excellence by Battelle, 505 King Avenue, Columbus, Ohio, 43201.

Schwille, F., 1984. Migration of Organic Fluids Immiscible with Water in the Unsaturated Zone. In: Pollutants in Porous Media, The Unsaturated Zone Between Soil Surface and Groundwater (Yaron, B. et al., Eds.) Berlin, Springer-Verlag.

LESSONS LEARNED FROM A DUAL-PHASE
EXTRACTION FIELD APPLICATION

Barry N. Rice
Roy F. Weston, Inc.*
U.S. Department of Energy
Grand Junction Office
Grand Junction, Colorado

ABSTRACT: A field-scale Dual-Phase Extraction (DPE) system was installed and operated at the Pinellas Science, Technology, and Research (STAR) Center, formerly the U.S. Department of Energy (DOE) Pinellas Plant, in Largo, Florida, from August 1997 through September 1999. The goal of applying the DPE system was to enhance the pump-and-treat remediation of volatile organic compounds (VOCs) (primarily vinyl chloride, toluene, trichloroethene, and 1,2-dichloroethene) from a shallow surficial aquifer at the Pinellas STAR Center 4.5 Acre Site. Initial operating data for the DPE system demonstrated an aggressive groundwater recovery rate; however, influent groundwater contaminant concentrations were less than those experienced with the pump-and-treat system. In addition, multiple issues complicated initial operations. Following numerous system and operational changes, the system responded with consistent on-line time and daily operation; although contaminant recovery rates did not increase.

INTRODUCTION

DPE Technology. DPE is a remediation method that uses vacuum to extract groundwater and soil vapor from a recovery well. Some other common terms used for DPE are Multi-Phase Extraction, Two-Phase Extraction, and Vacuum-Enhanced Recovery (EPA, 1999). Typically, an extraction tube (also referred to as a drop tube or stinger tube) is inserted in a sealed recovery well to the desired depth of recovery at or below the water table. A vacuum pump creates negative pressure in the extraction tube and groundwater is lifted up the tube. The water table is then drawn down to the intake of the extraction tube, at which time vadose-zone vapors are drawn into the extraction tube.

DPE has a number of advantages over conventional pump-and-treat systems. One advantage is that it uses both gravity and pressure differential to move groundwater. Another advantage is its ability to move air and water through formations previously inaccessible. Also, DPE uses two carriers (liquid and air) for contaminant recovery (Nyer et al., 1996).

---

*Work performed under DOE Contract No. DE–AC13–96GJ8335 for the U.S. Department of Energy.

Some disadvantages of DPE technology are the associated equipment and maintenance costs, potentially lengthy start-up periods, and depth/vacuum lift limitations (EPA, 1999).

Site Description. The former DOE Pinellas Plant, now the Pinellas STAR Center, is owned by the Pinellas County government and occupies approximately 100 acres (40.5 hectares) in Pinellas County, Florida. The Pinellas Plant operated from 1956 to 1994, manufacturing components for nuclear weapons under contract to DOE. The site where the field application took place, the 4.5 Acre Site, was previously a waste resin and solvent disposal area. In 1984, DOE began to identify potential environmental problems at the STAR Center, including the 4.5 Acre Site. A source removal activity in June 1985 at the site removed 303 tons (275 metric tons) of waste, including 83 drums, solidified drum contents, and 5,000 ft^3 (141.5 m^3) of contaminated soil (S&ME, 1987).

An Interim Remedial Action consisting of groundwater extraction and treatment by air stripping began in May 1990. Operation of the pump-and-treat system continued through July 1997 when, because of reduced contaminant recovery rates, a DPE system was installed to enhance recovery. The site contractor at that time, Lockheed Martin Specialty Components, Inc., designed and installed the DPE system. Following start-up of the system in August 1997, DOE transferred responsibility for environmental restoration activities of the STAR Center to the DOE Grand Junction, Colorado, Office and MACTEC Environmental Restoration Services (MACTEC). Roy F. Weston, Inc., is the technical remediation contractor for MACTEC.

Subsurface Conditions. The 4.5 Acre Site consists of a shallow surficial aquifer contaminated with VOCs, including trichloroethene, *cis*-1,2-dichloroethene, toluene, and vinyl chloride. The water table is generally 3 to 5 ft (0.9 to 1.5 m) below ground surface (bgs) (LMSC, 1997). The surficial aquifer ranges in thickness from 24 to 32 ft (7.3-9.8 m) bgs and is composed primarily of fine sand. Hydraulic conductivity is approximately 2.0×10^{-4} cm/s with a variable presence of silt and clay (S&ME, 1987). The Hawthorn Group, composed primarily of clay, underlies the surficial aquifer.

REMEDIATION METHODS

4.5 Acre Site Groundwater Recovery and Treatment System. The Pinellas 4.5 Acre Site pump-and-treat system began operation in May 1990. The pump-and-treat system consisted of seven recovery wells containing pumps that transferred groundwater to an adjacent treatment system. The treatment system consisted of an equalization tank, a pretreatment phase that settled and collected naturally occurring metals (e.g., iron and calcium) that had fouled the air stripper during initial system start-up, and a contaminant treatment phase in which the groundwater was air stripped to remove VOCs. Effluent was transferred to an industrial wastewater facility that adjusted the pH of the water and discharged it to the county sewer system. This pump-and-treat system, with some performance modifications, operated through July 1997.

4.5 Acre Site Dual-Phase Extraction System. The DPE system was proposed as a modification to the pump-and-treat system to provide a more aggressive means of contaminant recovery. A network of 22 DPE wells was installed during May and June 1997. In 11 DPE wells, organic vapors were detected during well installation, or VOCs were reported in well development water. Concentrations were generally higher in the deeper portions of the wells (MACTEC, 1997).

The extraction well system was divided into three legs to improve operational flexibility. The DPE wells were fully screened through the saturated thickness of the surficial aquifer (from approximately 5 ft [1.5 m] bgs to 30–32 ft [9.1–9.8 m] bgs). In each well a vacuum extraction tube was installed to approximately 22 ft (6.7 m) bgs.

On July 11, 1997, the existing pump-and-treat system was shut down and the seven recovery wells were capped. The DPE system was installed adjacent to the existing groundwater treatment system. The three extraction well legs were merged into a manifold that was connected to a large, phase-separation tank in a pit. To supply the needed vacuum, a 60-horsepower (45-kW) liquid-ring vacuum pump was installed at the ground surface and connected to the top of the phase-separation tank in the pit. Vapors recovered by the pump were routed to the blower intake of the existing air-stripper tower. Groundwater from the bottom of the phase-separation tank was used to maintain the minimum 23 gallons per minute (gpm) (87 L/min) flow necessary for the liquid-ring vacuum pump seal. All influent groundwater was eventually transferred to the influent tank of the existing groundwater treatment system. The groundwater treatment system (air stripper and pretreatment phase) processed the water recovered by the DPE system in the same manner as it had with groundwater recovered by the previous seven recovery wells.

By 1999, it appeared that significant modifications would be necessary to maximize the effectiveness of the DPE system. In September 1999, the DPE system was shut down. A new remediation system using biosparging through horizontal wells was installed at the 4.5 Acre Site and became operational in November 1999.

RESULTS AND DISCUSSION

DPE System Operation. The DPE system was started in August 1997 and continued operation through September 1999. During the initial months of operation, multiple problems complicated the daily operations, thwarted attempts to enhance contaminant mass recovery, and resulted in low system on-line time. Following numerous system and operational changes to address these problems, the DPE system responded with relatively high on-line time (>90 percent) and minimal daily operations complications. Some of the problems and responses to those problems are as follows:

- Groundwater recovery rate following start-up exceeded the treatment system capacity of about 40–45 gpm (151–170 L/min) with all 22 extraction wells operating (MACTEC, 1997). This resulted in the DPE system cycling on and off to avoid overloading the treatment system. Eight of the extraction wells were turned off, resulting in an influent flow of 30–35 gpm (114–132 L/min) and cycling reduction.

- In contrast, as recharge from rainfall decreased and adequate influent flow was not available, use of domestic water was necessary to supplement the minimum 23-gpm (87 L/min) seal water supply. In addition, a larger sheave was installed on the vacuum pump in August 1998 that decreased the minimum seal water supply to 17 gpm (64 L/min).

- In June and July 1998, the piping supplying compressed air to the 4.5 Acre Site began to rupture due to internal degradation of the 10-year-old HDPE pipe. Replacement of the pipe prevented system operation during August and September 1998.

- From May through July 1999, fluctuations in the domestic water supply pressure periodically interrupted the domestic water supplement to the vacuum pump. Fluctuations in domestic water pressure ceased after July 1999.

DPE System Performance. During operation of the DPE system, concentrations of total VOCs in the groundwater influent were relatively low in comparison to previous pump-and-treat influent concentrations. From January through July 1997, the pump-and-treat system's average concentration of total VOCs in the influent was 2,170 µg/L per month. From August 1997 through August 1999, the DPE system's average influent concentration was 416 µg/L per month. Continuous operation of the DPE system did not result in increased influent concentrations (Figure 1).

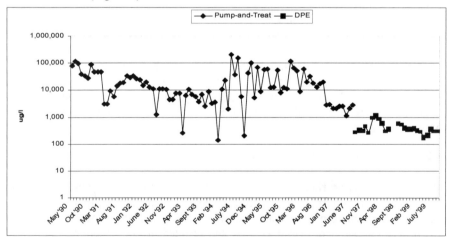

FIGURE 1. 4.5 Acre Site influent VOC concentrations

In contrast, monitoring well M044, which is probably the best groundwater well to monitor effectiveness of cleanup at the site, showed declining total VOC levels (MACTEC, 1999). VOC concentrations in well M044 declined from 1,160 µg/L in May 1997 to 3.1 µg/L in October 1999. Sampling of extraction wells during the DPE system's operational time period showed declining total VOC levels. However, some of the extraction wells with the highest levels of contamination still had concentrations in excess of 10,000 µg/L after two years of DPE system operation. Extraction well E012, for example, had total VOC concentrations in excess of 40,000 µg/L in early 1998; prior to shutting down the DPE system, well E012 had 15,800 µg/L total VOCs.

Concentrations of total VOCs in the influent vapor were also low. Vapors monitored before the vacuum pump during this period ranged from 0 to approximately 100 µg/L (HSW, 1998). Subsequent vapor sampling did not reveal significant increases in contaminant concentrations.

In contrast with influent groundwater contaminant concentrations, the influent groundwater volume increased significantly in comparison to the previous 12–18 months of pump-and-treat operations. From January 1996 through July 1997, the influent volume from the pump-and-treat system averaged 317,835 gallons (1.20×10^6 L) per month. From August 1997 through August 1999, the DPE system's average influent volume was 671,291 gallons (2.54×10^6 L) per month (Figure 2).

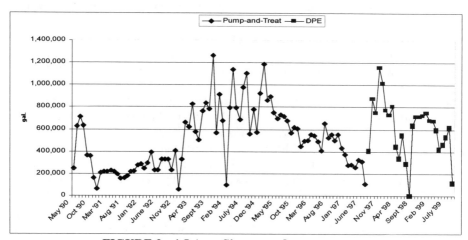

FIGURE 2. 4.5 Acre Site groundwater recovery

VOC mass recovery with the DPE system (both vapor and water phases) was approximately the same or slightly more than with the previous year's pump-and-treat operations. In comparison to historical contaminant mass recoveries, the DPE system was recovering less mass. Continuous DPE system operation through 1998 and 1999 did not show a significant increase in mass recovery over initial operations (Figure 3).

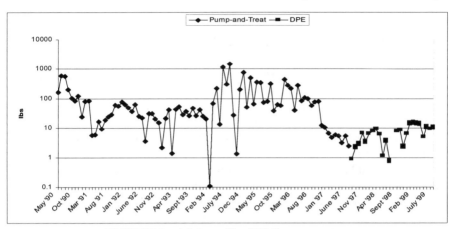

FIGURE 3. 4.5 Acre Site VOC mass recovery

Multiple adjustments were made to the system in an attempt to increase the recovery. The primary focus was to increase the groundwater table drawdown and recover the deeper contamination indicated during installation of the extraction wells. Some of the adjustments and the system's responses are as follows:

- The intakes of the slurp tubes were extended to the bottom of the extraction wells (about 28–31 ft [8.5–9.4 m] bgs). However, the water table was only drawn down to approximately 22 ft bgs (6.7 m) and influent concentrations did not increase.
- Individual legs of the DPE system were isolated by turning off the flow from the other operating legs. The resulting drawdown of the water table was not significantly greater than during previous operations.
- In an attempt to produce airlift pumping, stinger tubes were placed down the slurp tubes of the extraction wells, and a valve on the aboveground end of the stinger tube was opened slightly to bleed air into the slurp tube. The water level in a nearby monitoring well rose approximately one-tenth of a foot, and operation of the stinger tubes was ceased.
- Due to the presence of naturally occurring iron in the site groundwater, precipitation of iron in the extraction well piping and valves caused a significant reduction in groundwater flow and required continual maintenance.
- During initial DPE operations, contaminants were continually detected in the air-stripper tower's effluent. The DPE effluent vapor piping was rerouted up the side of the air-stripper tower and contaminant detections in the effluent ceased.

Costs.

Construction*	$206,000.00
Operations and Maintenance*	$400,000.00
Water Consumption and Disposal	$ 53,727.91
Electrical Power	$ 28,559.40
Total	$688,287.31

*Indicates an estimated cost.

CONCLUSIONS

Lessons Learned. The following are lessons learned during operation of the 4.5 Acre Site DPE system:

- A DPE system should be designed to operate efficiently at the optimum state of recovering both vapor and water. Biasing the system design to move large quantities of water may limit the ability to move large quantities of air.
- Precipitation of naturally occurring iron in the site groundwater can cause a significant reduction in groundwater flow and require continual maintenance.
- A consistent, uninterrupted supply of utility services is essential for a DPE system to reach and maintain the optimum state of recovering both vapor and water.
- The ability to isolate and operate only specific extraction wells attached to a DPE system allows the focus of remediation efforts in specific areas.
- Special attention should be paid to how the vacuum pump seal is maintained. If the pump requires a liquid-ring seal, management of the water supply can greatly affect operation of the DPE system.
- Limitations of a DPE system's lift capability may limit contaminant recovery to shallow depths (<20 ft bgs, 6 m bgs).
- Introducing recovered contaminant vapors into the intake of an air-stripper blower may partition contaminants from the vapor phase to the liquid phase, resulting in detections in the air-stripper effluent discharge.

Summary. The DPE system recovered 177 pounds (80 kg) of VOCs from the 4.5 Acre Site during approximately 23 months of operation. Groundwater recovery totaled approximately 15,726,000 gallons (59.5 × 10^6 L). During that period, costs for installing and operating the DPE system were approximately $688,000.

DPE system operations at the 4.5 Acre Site demonstrated an aggressive groundwater recovery rate and a decrease in contaminant levels in some wells. However, contaminant recovery was not as aggressive as expected and was at times actually less than the previous pump-and-treat system's recovery. Deeper contamination was not effectively addressed. Consistent operations were hindered at times by utility supply interruptions, iron precipitation, and seal water supply demands. Adjustments and alterations to the DPE system to correct these issues did not produce favorable results. Future applications of DPE systems at other remediation sites should include careful consideration of the lessons learned from this application.

ACKNOWLEDGMENTS

Operations and performance modifications to the DPE system were made possible through the support of David Ingle (DOE), the Roy F. Weston/MACTEC-ERS Grand Junction and Pinellas Office, and ConsuTec, Inc.

REFERENCES

HSW. January 15, 1998. *Status Report on the Dual-Phase Extraction System Vapor Monitoring.* HSW Engineering, Inc., Tampa, FL.

LMSC. May 1997. *Pinellas Plant Environmental Restoration Program Soil Investigation at the 4.5 Acre Site.* Lockheed Martin Specialty Components. Largo, FL.

MACTEC Environmental Restoration Services. July 1997. *4.5 Acre Site Interim Remedial Action Quarterly Progress Report.* Grand Junction, CO.

MACTEC Environmental Restoration Services. April 1999. *4.5 Acre Site Interim Remedial Action Quarterly Progress Report.* Grand Junction, CO.

Nyer, E.K., S. Fam, D.F. Kidd, F.J. Johns II, G. Boettcher, P.L. Palmer, S.S. Suthersan, and T.L. Crossman. 1996. *In Situ Treatment Technology.* CRC Press, Inc., Boca Raton, FL.

S&ME. 1987. *Interim Remedial Action Plan, Department of Energy 4.55 Acre Site.* Prepared by Haztech, Inc./S&ME, Inc. for General Electric Company, Neutron Devices Department, Largo, FL.

U.S. Environmental Protection Agency. June 1999. *Multi-Phase Extraction: State-of-the-Practice.* EPA 542-R-99-004.

DESIGNING PIPING NETWORKS FOR MULTI-PHASE EXTRACTION SYSTEMS

Matthew P. Peramaki, P.E.
Brad A. Granley, E.I.T.
Rebecca A. Carlson, E.I.T.
Mark D. Nelson, P.E.
Leggette, Brashears & Graham, Inc., St. Paul, Minnesota

ABSTRACT: A lack of readily available standard design methods for calculating the pressure drop in multi-phase flow (both soil vapor and ground water) applications has left remediation engineers without essential tools required for the proper design of multi-phase extraction (MPE) systems. The lack of a simple way to quantify pressure drop for multi-phase flow in a closed conduit has resulted in design methods that are based on the previous experience of the designer or, at worst, guesswork. This paper investigates the applicability of using empirical correlations developed by researchers from the petroleum refining, refrigeration and chemical processing industries to predict pressure losses in two-phase vertical and horizontal flow under high-vacuum conditions. All of the correlations existing in chemical engineering literature were developed at pressures ranging from near atmospheric to well over 100 pounds per square inch (p.s.i.). Vacuum measurements, vapor flow rates, ground-water flow rates and other relevant data were collected from an operating MPE system and compared to the pressure drop predicted by these correlations. It was determined that, if used properly, these correlations are valuable tools to engineers who design piping networks associated with MPE systems.

INTRODUCTION

An MPE system designer must be able to calculate the pressure drop in a two-phase flow system in order to properly size vacuum pumps and minimize pressure losses in the system piping. Minimizing pressure drop is of supreme importance because vacuum pumps are limited in the vacuum that they can produce. Most readily available vacuum pumps cannot produce a vacuum greater than 25 inches of mercury. While designers may be able to get away with a sloppy air sparging system pipe design by over sizing the air compressor, no such luxury exists for the MPE system designer. When an inefficient MPE piping network is constructed, the penalty is a reduced well casing vacuum and reduced water and vapor extraction rates. These reductions can often be so significant that the system fails or must run years longer than anticipated.

This paper presents empirical correlations developed by researchers from the petroleum refining, oil production, chemical processing and refrigeration engineering industries that have been determined to predict pressure drop for two-phase flow systems existing under positive pressures. These correlations account for changes in vapor flow, liquid flow, pipe diameter, pipe length, pipe geometry, fitting losses and pressure drop in a closed conduit. The empirical correlations were compared to data collected from existing remediation systems operating under a strong vacuum and some were found to be suitable for MPE system design.

TWO-PHASE FLOW

The major complications in comparing two-phase flow to single-phase flow are the variety of flow patterns that can be produced in a gas-liquid system. Two-phase flow cannot simply be described as laminar, transitional and turbulent. There are various flow regimes or "patterns" that are common to two-phase flow systems, each having different characteristics and associated pressure loss behavior. The two-phase patterns vary not only with flow rates and fluid properties, but also with pipe diameter and inclination.

In horizontal pipe, the two-phase flow patterns are (in order of lowest gas velocities to highest) bubble flow, plug flow, stratified flow, wavy flow, slug flow, annular flow and spray flow. In vertical pipe there are bubble, slug, churn, ripple, annular flow and mist flow. In the typical MPE application, there are four primary horizontal flow patterns and three vertical flow patterns. The common horizontal flow patterns (occurring in connective piping between the well and the vacuum pump) are stratified, wavy, slug and annular flow. Two-phase flow regimes that are common in vertical piping (i.e. the drop tube within a well or a vertical connective piping run) are described as churn, ripple and annular flow.

Most remediation professionals would recognize the appearance of these patterns from conducting pilot tests and from observations gathered during full-scale system operation. (Note: The term "pattern" refers to visual observations and the flow "regime" applies to flow behavior that can be described by more quantitative expressions.) A description of each of the patterns commonly occurring in MPE systems is given below.

Horizontal Flow Patterns

Visual flow patterns for two-phase flow in horizontal pipe were classified by Alves, 1954. The following patterns are clearly observable and are shown in Figure 1.

Stratified flow: This flow pattern is very common in typical MPE systems. In stratified flow, the liquid flows along the bottom of the pipe and the gas flows at the top of the pipe over a smooth liquid/gas interface. Liquid velocities <0.5 feet/second with gas velocities of 2 to 10 feet per second tend to produce stratified flow.

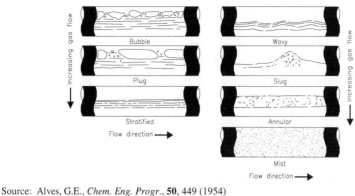

Source: Alves, G.E., *Chem. Eng. Progr.*, **50**, 449 (1954)
Figure 1. Horizontal Flow Patterns

Wavy flow: This flow pattern is very common in typical MPE systems. Similar to stratified flow, the liquid flows along the bottom of the pipe. However, an increased gas velocity above the liquid produces waves on the liquid/gas interface. Liquid velocities of approximately 1 foot/second with gas velocities of approximately 15 feet per second tend to produce wavy flow.

Slug flow: This flow pattern is common in typical MPE systems. When wavy flow wave crests become sufficiently high to bridge the pipe, they form frothy slugs that move much faster than the average liquid velocity. Slug flow can cause severe and sometimes dangerous vibrations in piping and equipment because of the impact of the high-velocity slugs against bends and other fittings.

Annular flow: This flow pattern can occur in a MPE system. In annular flow, the liquid flows as a thin film along the pipe wall and gas flows in the core. Gas velocities >20 feet per second tend to produce annular flow.

Vertical Flow

Visual definitions of flow patterns in vertical flow appear to cause more difficulty than do those in horizontal flow. Vertical flow patterns tend toward radial symmetry, which is not the case for horizontal flow. Figure 2 illustrates the vertical flow patterns and they are described below.

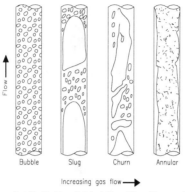

Source: Taitel, Y., D. Bornea and A.E. Dukler, " Modeling Flow Pattern Transitions for Steady Upward Gas-Liquid Flow in Vertical Tubes," *AIChE J.,* **26**, 345 (1980).

Figure 2. Vertical Flow Patterns

Slug flow: This flow pattern is not typical as gas flow rates in MPE systems are too high for slug flow to occur. However, bioslurping systems are designed to operate in this regime. In this pattern gas bubbles coalesce to form larger bullet shaped bubbles. These bubbles are commonly called "Taylor Bubbles". Taylor bubbles move uniformly upward and are separated by slugs of liquid that bridge the pipe and contain small gas bubbles. Surficial gas velocities of 2 to 20 feet/second tend to produce slug flow.

Churn (froth) flow: This flow pattern is very common in typical MPE systems. Churn flow is similar to slug flow, although much more chaotic and disorderly. The Taylor Bubbles narrow and become distorted. The continuity of the liquid in the slug between bubbles is repeatedly destroyed by a high local gas concentration within the slug. As the liquid slug is destroyed, it falls. This liquid accumulates, forms a bridge and is again lifted by the gas.

       Churn flow, at low air to water ratios, has the highest relative pressure drops common to MPE systems. The interaction between the gas and liquid phases increases the pressure drop compared to other flow patterns. This flow pattern should be avoided. Additionally, the slugging produces strong and sometimes damaging vibrations in piping. Surficial gas velocities of approximately 20 to 30 feet/second tend to produce churn flow.

Ripple flow: This flow pattern is very common in typical MPE systems. Ripple flow, which is characterized by an upwards-moving, wavy or ripply layer of liquid on the piping wall, is far less chaotic than that of churn flow. The zone of ripple flow is narrow and actually forms a transition region between churn and annular flow. The pressure drop associated with this transition is low and designers should attempt to produce this flow pattern in vertical pipe runs and drop tubes.

Annular (film) flow: This flow pattern is common in typical MPE systems. The annular flow pattern consists of a annular liquid layer moving on the pipe wall with a gas core moving at much higher velocities. Surficial gas velocities of >30 feet/second tend to produce annular flow.

Flow Pattern Prediction

The above description of two-phase flow patterns lists various superficial fluid velocities that tend to produce certain flow patterns (Perry, 1984). (NOTE: Superficial velocity is defined as the velocity of a phase as if it was flowing alone within the pipe.) While superficial liquid and gas velocities can produce a very rough pattern prediction, empirically or theoretically derived flow pattern maps are more accurate.

Horizontal Flow: The chart presented by Baker, 1954 is the most widely used chart to predict horizontal flow pattern. The chart was developed using 1, 2 and 4-inch diameter pipe and air/water mixtures near atmospheric pressures at a temperature of 68 degrees F. Taitel, Y., and A.E. Dukler, 1976 presents a non-empirical, mechanistically based methodology for predicting horizontal two-phase flow patterns.

Vertical Flow: A chart prepared by Govier et. al., 1957 is the most widely used chart to predict vertical flow pattern. This chart was developed using a 1.025-inch diameter smooth pipe with air and water at 70 degrees F at 36 pounds per square inch actual (p.s.i.a.). Taitel, Y., D. Bornea and A.E. Dukler, 1980 presents a non-empirical, mechanistically based methodology for predicting vertical two-phase flow patterns.

TWO-PHASE PRESSURE DROP FUNDAMENTALS

Pressure drop calculations for two-phase flow are more complex than for single-phase flow. The total pressure drop occurring during two-phase flow in a pipe can be regarded as the sum of the frictional, hydrostatic and accelerational components [Equation (1)].

$$\left(\frac{\Delta P}{\Delta Z}\right)_{TP} = \left(\frac{\Delta P}{\Delta Z}\right)_{frictional} + \left(\frac{\Delta P}{\Delta Z}\right)_{hydrostatic} + \left(\frac{\Delta P}{\Delta Z}\right)_{accelerational} \qquad (1)$$

Where:

TP = Total two-phase flow pressure drop
frictional = two-phase flow pressure drop due to friction losses
hydrostatic = two-phase flow pressure drop due to hydrostatic or
        gravitational losses (i.e. losses associated with vacuum lift)
accelerational = two-phase flow pressure drop due to accelerational losses
        caused by a change in phase from liquid to gas

Powers, 1992 states that the actual pressure drop in horizontal construction dewatering pipe containing both water and air can be from 150 to 200 percent of the drop due to water flowing in the pipe alone. The two-phase flow pressure drops in an MPE system can be as much as ten times the sum of the superficial gas and liquid pressure drops.

The introduction of a second phase causes a sharp rise in pressure drop due to the following: 1) the cross-sectional area available for the flow of one phase is reduced by the introduction of the second, 2) the flow boundary is no longer a relatively smooth pipe was, but is a rough, irregular, mobile interface between phases, 3) the introduction of a second phase can cause a very unsteady flow accompanied by intense turbulence.

Accelerational Pressure Drop

In a typical MPE application, soil vapor and water are the two-phase flow components. At the conditions typical of an MPE system, the accelerational pressure drop can be neglected. This is because the accelerational pressure drop can be viewed as negligible if the liquid phase does not boil and become transferred to gas phase. The accelerational pressure drop can be neglected in both horizontal and vertical flow as water does not boil at vacuums and temperatures typical of MPE system operation. (Water boils at 133 degrees Fahrenheit at 25-inches of Mercury vacuum).

However, if a non-aqueous phase liquid (NAPL) is recovered in the MPE fluid stream, acceleration losses may become a significant component of the total pressure drop. Therefore, designers should determine the boiling point of the NAPL to determine if losses due to acceleration can be neglected. This component of pressure drop must also be considered for fluid recovery at steam injection sites.

Horizontal Pipe

For horizontal flows during most MPE applications the hydrostatic pressure drop can be eliminated as most horizontal MPE pipe is flat or has a slight loss in elevation in the direction of flow. Therefore, the horizontal pressure drop is a function of frictional losses at the interface between the phases and at the boundaries of the conduit:

$$\left(\frac{\Delta P}{\Delta Z}\right)_{TP} = \left(\frac{\Delta P}{\Delta Z}\right)_{frictional} \tag{2}$$

Vertical Pipe

Equation (3) describes the components that make up the total pressure drop in vertical piping such as a drop tube. These components not only include the frictional losses but also the hydrostatic losses or pressure drop that is due to "lift."

$$\left(\frac{\Delta P}{\Delta Z}\right)_{TP} = \left(\frac{\Delta P}{\Delta Z}\right)_{frictional} + \left(\frac{\Delta P}{\Delta Z}\right)_{hydrostatic} \tag{3}$$

PRESSURE DROP PREDICTION

Two approaches may be used to solve the problem of predicting pressure drop in a two-phase flow system. One may consider the detailed hydrodynamics of individual flow patterns or groups of flow patterns. A mechanistic approach such as this will usually be more difficult to apply quantitatively. Alternatively, empirical correlations could be used. The benefit of correlations is that they are easy to use. However, they must be applied to problems having conditions similar to those used to obtain the original data. When this criteria is met, the results can be quite satisfactory. However, improper use of the correlations can result in significant prediction error. Therefore, an engineer's first step is to use Baker, 1954, Govier et. al., 1957, or one of the many other published flow pattern maps/methods to determine what type of flow will be encountered. If one uses correlations that have been developed under conditions that are comparable to typical MPE operating conditions (other than pressure) the designer can predict the pressure drop with an accuracy of about 25 to 50 percent, and in many cases much higher accuracy.

Gas-Phase Density Considerations: All published pressure drop correlations are based on positive pressure testing. That is, researchers injected compressed air at the downstream side of the test apparatus to develop a pressure gradient. MPE obviously uses vacuum on the upstream side to produce the needed pressure gradient. The significance of this is the gas density of the empirical correlations is significantly greater than that of the typical MPE application. Fortunately, gas-phase density appears to have little impact on the pressure drop in two-phase extraction systems. The effect of gas density under pressure conditions ranging from 18.0 to 110 p.s.i.a. was investigated by Brown, et. al 1960. This research indicated that the density of the gas phase has no significant effect on the pressure drop in vertical pipe. The results of this current research on MPE systems operating under vacuum conditions up to approximately 15 inches of mercury also appear to indicate that changes in gas phase density has little impact on pressure drop calculations.

Horizontal Flow

Two correlations were compared to horizontal pressure drop field data, using Lockhart and Martinelli, 1949 and Dukler, et. al., 1964.

The Lockhart and Martinelli, 1949 correlation was empirically derived and has been shown to have predictive capability within plus or minus 50 percent, with a great majority of the predictions within 25 percent. The correlation was developed using 1 to 4-inch diameter pipe with air/water mixtures at 18 p.s.i.a. Previous research indicates that the correlation is more precise for certain flow patterns. In general, predictions are high for stratified, wavy and slug flows, and low for annular flow.

Dukler, et. al., 1964 presents a method to calculate pressure drop in a horizontal pipe that includes not only the frictional losses but also the accelerational losses. The correlation was developed in pipes from 1 to 5-inches in diameters for air/water and water/steam mixtures at pressures ranging from 25 to 1400 p.s.i.a. This method generally under predicts for plug and annular flow.

Neither of these two correlations predicts hydrostatic losses, as there is no hydrostatic component without a change in elevation [Equation (2)]. Additionally, the Lockhart and Martinelli, 1949 method does not account for losses due to acceleration.

Vertical Flow

Two vertical flow correlations were also compared to the field data. The Govier et. al., 1957, correlation was developed using a 1.025-inch diameter smooth pipe with air and water at 70 degrees F at 36 p.s.i.a. This work was later expanded to include pipe diameters from 0.63-inch to 2.5 diameter (Govier and Short, 1958). The latter correlation is recommended for pipe diameters of 0.5 to 3.0 inches. However, neither is recommended for annular flow. Both are also restricted to cases where the change in gas density in the pipeline section under evaluation is small enough to use an average value. These papers also provide the equations required to calculate both the frictional and hydrostatic pressure drop in vertical upflow [Equation (3)].

Fitting Losses: Wallis, 1969 states that the commonly used single-phase flow design method of accounting for fittings losses by using equivalent pipe lengths is applicable to two-phase flow. However, he states that the equivalent pipe lengths for two-phase flow tend to be "somewhat longer" in the two-phase flow case.

Chenoweth and Martin, 1955 also determined that two-phase flow fitting losses can be handled by using equivalent pipe lengths. Their research indicates that in pipe fittings that promote mixing (e.g. orifice and globe valve) single phase flow equivalent lengths tend to over predict the actual pressure drop. In pipe fittings that tend to separate the liquid and gas phases due to centrifugal action (i.e. a 180 degree elbow) the equivalent lengths tend to under predict pressure drop. They conclude that almost all of the fitting losses they studied agree with the predicted pressure drop within plus or minus 50 percent of the measured values.

Hsu and Graham, 1976 summarizes several papers that present empirical constants for two-phase flow through pipe fittings.

DATA COLLECTION

The data acquired to validate the correlations for use in MPE applications were collected from an remediation system currently in operation. For vertical flow correlations, vacuum measurements were recorded in the annular space and at the top of the drop tube of a MPE well. Horizontal pipe run vacuum data was collected at both the wellhead and at the system manifold. Vapor flow rates, converted to standard conditions (measured with a thermal anemometer after the vacuum pump), and ground-water flow rates (measured by recording the amount of water entering the vapor/liquid separator over time) were also recorded along with observed flow patterns. Measurements were recorded over a range of vapor and water flow rates in three drop tube pipe diameters in order to validate the correlations over a range of conditions that are likely to exist at a typical remediation site.

The accuracy of the vertical pressure drop data should be quite good as a direct measurement of the pressure drop through a known length of pipe. However, the unitized (i.e. "Hg/100 feet) horizontal pressure drop data is an approximation of the actual pressure drop due to the fact that the exact buried pipe length was unavailable, the number of buried pipe fittings was approximated as was the length of underground vertical pipe runs.

To account for the pressure drop associated with the fittings and vertical pipe, single-phase flow equivalent lengths were added to the horizontal pipe length. Additionally, the pressure drop due to vertical pipe runs between the wellhead and system manifold were subtracted from the total pressure drop. The corrected horizontal pressure drop was then normalized by dividing it by the equivalent pipe length and then multiplied by 100 feet.

RESULTS AND DISCUSSION

The following four tables present the actual pressure drop, water flow, vapor flow and pipe diameter information collected in the field in conjunction with the calculated pressure drop prediction for both vertical and horizontal flows. Tables 1 and 2 present the Govier et. al., 1957 and Govier and Short, 1958 vertical pressure drop prediction data, while tables 3 and 4 presents the Lockhart and Martinelli, 1949 and Dukler et. al., 1964 derived horizontal flow pressure drops.

Table 1.   A comparison of actual vertical pressure drop and predicted vertical pressure drop calculated using the correlation of Govier et. al., 1957.

Dia. of Pipe	Gas Flow Rate	Water Flow Rate	Govier, 1957 Press. Drop	Actual Press. Drop	Press. Drop Difference	Percent Difference	Pipe Dia. Average Prediction
inches	scfm	gpm	"Hg/ft	"Hg/ft	"Hg/ft	%	%
1	18.2	2.2	0.40	0.44	-0.045	-10%	
1	20	1.7	0.33	0.41	-0.078	-19%	1" Diameter
1	22	1.6	0.29	0.35	-0.057	-16%	-15%
1.5	22.3	3.5	0.31	0.27	0.040	15%	
1.5	31	3.1	0.20	0.24	-0.044	-18%	
1.5	18.1	2.9	0.20	0.2	0.003	1%	
1.5	18.5	2.6	0.19	0.22	-0.032	-14%	
1.5	16.3	2.9	0.17	0.17	-0.002	-1%	1.5" Dia.
1.5	20.1	2.8	0.18	0.15	0.035	23%	1%
2	25.7	3.6	0.12	0.26	-0.137	-53%	
2	44	2.9	0.11	0.21	-0.101	-48%	
2	27	3	0.11	0.19	-0.081	-43%	2" Diameter
2	19.9	2.8	0.09	0.19	-0.098	-52%	-49%

Table 2. A comparison of actual vertical pressure drop and predicted vertical pressure drop calculated using the correlation of Govier and Short, 1958.

Dia. of Pipe	Gas Flow Rate	Water Flow Rate	Govier, 1958 Press. Drop	Actual Press. Drop	Press. Drop Difference	Percent Difference	Pipe Dia. Average Prediction	Pattern Flow Average Prediction
inches	scfm	gpm	"Hg/ft	"Hg/ft	"Hg/ft	%	%	%
1	18.2	2.2	0.35	0.44	-0.09	-20%		
1	20	1.7	0.31	0.41	-0.10	-23%	1" Dia.	Annular
1	22	1.6	0.27	0.35	-0.08	-24%	-22%	-24%
1.5	22.3	3.5	0.33	0.27	0.06	22%		
1.5	31	3.1	0.20	0.24	-0.04	-17%		
1.5	18.1	2.9	0.21	0.2	0.01	5%		
1.5	18.5	2.6	0.18	0.22	-0.04	-18%		
1.5	16.3	2.9	0.24	0.17	0.07	42%	1.5" Dia.	
1.5	20.1	2.8	0.24	0.15	0.09	61%	16%	Churn/
2	25.7	3.6	0.19	0.26	-0.07	-26%		Ripple
2	44	2.9	0.19	0.21	-0.02	-11%		5%
2	27	3	0.20	0.19	0.01	4%	2" Dia.	
2	19.9	2.8	0.17	0.19	-0.02	-11%	-11%	

Vertical Pipe

As shown in Table 1, the Govier et. al., 1957 correlation has limited use for MPE applications. The correlation is only predictive for 1 and 1.5-inch diameter pipe, and only when the flow pattern is not annular. The Govier and Short, 1958 and correlation (Table 2) is predictive for MPE drop tubes and piping from 1 to 2-inches (probably from 0.5 to 3-inch) in diameter for churn and ripple flow. As stated by Govier and Short, 1958 it under predicts pressure drop in annular flow patterns (~15 to 25% average). This correlation predicts ripple and churn flow in 1.5 and 2-inch drop tubes to within +/-50 percent (including outliers), with most data within +/-25 percent (5% average)

Table 3. A comparison of approximate actual horizontal pressure drop and predicted horizontal pressure drop calculated using the correlation of Lockhart and Martinelli, 1949.

Diameter of Pipe	Gas Flow Rate	Water Flow Rate	Lockhart/Martinelli Pressure Drop	Actual Press. Drop	Press. Drop Difference	Percent Difference
inches	scfm	gpm	"Hg/100 ft	"Hg/100 ft	"Hg/100 ft	%
2.067	22.3	3.5	1.02	1.18	-0.16	-14%
2.067	25.7	3.6	1.30	0.65	0.64	98%
2.067	22.0	1.6	0.53	0.72	-0.19	-27%
2.067	18.2	2.2	0.55	0.61	-0.06	-11%
2.067	27.0	3.0	1.16	0.99	0.17	17%
2.067	19.9	2.8	0.75	1.07	-0.32	-30%
2.067	18.5	2.6	0.65	1.12	-0.47	-42%
2.067	16.3	2.9	0.61	1.24	-0.63	-51%
2.067	18.1	2.9	0.69	1.34	-0.64	-48%
2.067	20.1	2.8	0.75	1.27	-0.52	-41%
					Average	**-15%**

Table 4. A comparison of approximate actual horizontal pressure drop and predicted horizontal pressure drop calculated using the method of Dukler et. al., 1964.

Diameter of Pipe	Gas Flow Rate	Water Flow Rate	Dukler et. al., 1964 Pressure Drop	Actual Press. Drop	Press. Drop Difference	Percent Difference
inches	scfm	gpm	"Hg/100 feet	"Hg/100 ft	"Hg/100 ft	%
2.067	22.3	3.5	0.95	1.18	-0.23	-19%
2.067	25.7	3.6	1.06	0.65	0.41	63%
2.067	22.0	1.6	0.64	0.72	-0.09	-12%
2.067	18.2	2.2	0.60	0.61	-0.01	-2%
2.067	27.0	3.0	1.11	0.99	0.11	11%
2.067	19.9	2.8	0.74	1.07	-0.32	-30%
2.067	18.5	2.6	0.67	1.12	-0.45	-40%
2.067	16.3	2.9	0.65	1.24	-0.59	-48%
2.067	18.1	2.9	0.71	1.34	-0.63	-47%
2.067	20.1	2.8	0.75	1.27	-0.51	-41%
					Average	**-16%**

Both the Lockhart and Martinelli, 1949 and Dukler, et. al., 1964 methods predicted actual pressure drop to within +/-50 percent.

Additionally, the pressure drops calculated using these two methods were very similar to one another (table 5). This is in spite of the fact that the Dukler, et. al., 1964 method accounts for losses due to acceleration and changes in gas-phase density while Lockhart and Martinelli, 1949 does not. The similarity of the predictions from both methods may indicate that the accuracy of field data may had more to do with the deviation from the predictions than the correlations themselves.

Table 5. A comparison of horizontal pressure drop calculated using the methods of Dukler et. al., 1964 and Lockhart and Martinelli, 1949.

Diameter of Pipe	Gas Flow Rate	Water Flow Rate	Dukler et. al., 1964 Pressure Drop	Lockhart/ Martinelli Press. Drop	Press. Drop Difference	Percent Difference
inches	scfm	gpm	"Hg/100 feet	"Hg/100 ft	"Hg/ft	%
2.067	22.3	3.5	0.95	1.02	-0.07	-7%
2.067	25.7	3.6	1.06	1.30	-0.24	-18%
2.067	22.0	1.6	0.64	0.53	0.11	20%
2.067	18.2	2.2	0.60	0.55	0.05	10%
2.067	27.0	3.0	1.11	1.16	-0.05	-5%
2.067	19.9	2.8	0.74	0.75	-0.01	-1%
2.067	18.5	2.6	0.67	0.65	0.02	3%
2.067	16.3	2.9	0.65	0.61	0.04	6%
2.067	18.1	2.9	0.71	0.69	0.02	2%
2.067	20.1	2.8	0.75	0.75	0.00	0%
					Average	**1%**

CONCLUSIONS

Published empirically-derived two-phase flow pressure drop correlations can be valuable tool for the design of MPE systems.

It was determined that the Govier and Short, 1958 method predicts the pressure drop in vertical MPE pipe (e.g. drop tubes) with acceptable accuracy. For ripple and churn flow in 1.5 and 2-inch drop tubes this method has an accuracy of +/-25% (not including outliers). However, it appears to under predict pressure drop in annular flow patterns.

Both the Lockhart and Martinelli, 1949 and Dukler, et. al., 1964 methods predict actual horizontal pressure drops within approximately +/-50 percent. Both methods appear to under predict actual pressure losses. Lockhart and Martinelli, 1954 is much simpler method than Dulker. et. al., 1964 with little loss in accuracy.

ACKNOWLEDGMENTS

The authors wish to thank Tom Peragin of the Chevron Research and Technology Company and George Mickelson of the Wisconsin Department of Natural Resources for their review and input during the preparation of this paper.

REFERENCES

Alves, G.E., *Chem. Eng. Progr.*, **50**, 449 (1954).

Baker, O., "Simultaneous Flow of Oil and Gas," *Oil Gas J.* **53**, 185 (1954).

Brown, A.S., G.A. Sullivan, and G.W. Govier, "The Upward Vertical Flow of Air Water Mixtures: III. Effect of Gas Density on Flow Pattern, Holdup and Pressure Drop," *Can J. Chem. Eng.*, **38**, 62 (1960).

Chenoweth, J.W and Merritt W. Martin, "Turbulent Two-Phase Flow," *Petroleum Refiner*, October, 1955 pp. 151-155.

Dukler, A.E., M. Wicks III and R.G. Cleveland, "Frictional Pressure Drop in Two-Phase Flow: B. An Approach Through Similarity Analysis," *AIChE J.*,**10**, 44 (1964).

Govier, G.W., B.A. Radford and J.S.C. Dunn, "The Upward Vertical Flow of Air Water Mixtures: I. Effect of Air and Water-Rates on Flow Pattern, Holdup and Pressure Drop," *Can J. Chem. Eng.*, **35**, 58 (1957).

Govier, G.W., and W. L. Short, "The Upward Vertical Flow of Air Water Mixtures: II. Effect of Tubing Diameter on Flow Pattern, Holdup and Pressure Drop," *Can J. Chem. Eng.*, **36**, 195 (1958).

Hsu, Y.Y and R.W. Graham, Transport Processes in Boiling and Two-Phase Systems Including Near-Critical Fluids, McGraw-Hill (1976).

Lockhart, R.W. and R.C. Martinelli, "Proposed Correlation of Data for Isothermal Two-Phase, Two-Component Flow in Pipes," *Chem. Eng. Prog.*, **45**, 39 (1949).

Perry, R.H., D.G Green, and J.O. Maloney, *Perry's Chemical Engineers' Handbook*, 7th Edition (1997)

Perry, R.H., *Perry's Chemical Engineers' Handbook*, 6th Edition (1984).

Powers, J. P., *Construction Dewatering*, John Wiley & Sons, 1992.

Taitel, Y., and A.E. Dukler, "A Model for Predicting Flow Regime Transition in Horizontal and Near-Horizontal Gas-Liquid Flow," *AIChE J.*, **22**, 47 (1976).

Taitel, Y., D. Bornea and A.E. Dukler, " Modeling Flow Pattern Transitions for Steady Upward Gas-Liquid Flow in Vertical Tubes," *AIChE J.*, **26**, 345 (1980).

Wallis, G.B., *One Dimensional Two-Phase Flow*, McGraw-Hill (1969).

CASE-STUDY IN THE CONSTRUCTABILITY TESTING AND OPERATION OF AN EX-SITU SOIL TREATMENT CELL

William A. Plaehn (Parsons Engineering Science, Inc., Denver)
Timonthy C. Shangraw, Michael F. Steiner, and Mark Murphy
(Parsons Engineering Science, Inc., Denver)
Lori T. Tagawa (Waste Management, Denver)
Dennis D. Bollmann (City and County of Denver)

ABSTRACT: The Former Tire Pile Area (FTPA) waste pits are located at the Lowry Landfill Superfund Site in Arapahoe County, Colorado. From the mid-1960s until 1980, the Site was operated as an industrial liquid waste and municipal solid waste landfill. Waste disposed at the site contained hazardous substances such as chlorinated volatile organic compounds (VOCs) and heavy metals. The initial Record of Decision (ROD) specified removal of surface and subsurface drums, associated free liquids, and other visible contamination from three waste pits, followed by off-site disposal of the excavated materials.

Toxicity Characteristic Leaching Procedure (TCLP) and bench-scale treatability testing performed in 1996 determined that only VOCs (chlorinated solvents and petroleum hydrocarbon compounds) failed toxicity criterion and that vapor extraction could be used to reduce the VOC concentrations to acceptable concentrations. An aboveground treatment cell for *ex situ* treatment via vapor extraction from waste material involving both active extraction and injection was designed. To increase the porosity of the excavated material, the soil/sludges were mixed with approximately 10-percent shredded tire chips. Multiple layers of extraction and injection laterals were constructed into the treatment cell. The total volume of bulked material placed in the treatment cell is approximately 21,000 cubic yards at a moisture content of approximately 20 to 25-percent. A significant amount of woody debris encountered in the waste pits was also mixed with the material and placed in the treatment cell.

A pilot-scale test performed in November 1998 verified design parameters such as lateral spacing, flow-rates, vacuums/pressures, and influent air quality. The pilot-scale test also indicated that the treatment cell as constructed behaved as a large biological reactor, similar to a compost pile. Internal temperatures increased from 60 to 90-degrees Fahrenheit prior to active air injection, to approximately 80 to 130-degrees Fahrenheit with the introduction of oxygen.

Subsequent to the pilot-scale test, full-scale characterization of the treatment cell indicated that approximately 33-percent of the solids failed TCLP for VOCs with tetrachloroethene (PCE) being the target compound with the highest levels at 1.0 to 3.0 mg/L. Full-scale treatment commenced in February 2000. Close monitoring of soil gas quality, cell temperatures, and flow-rates are being conducted to evaluate system performance.

The presentation will provide a detailed summary of the constructability testing, design parameters and construction of the treatment cell, and operational data collected to date. The significance of the presentation is the dissemination of data regarding above-ground hazardous waste remediation.

INTRODUCTION
The Lowry Landfill Superfund Site is located approximately 20 miles southeast of downtown Denver, Colorado, in unincorporated Arapahoe County. From the mid-1960s until 1980, the Site was operated as an industrial liquid waste and municipal solid waste landfill. Liquid wastes disposed of at the Site included hazardous substances such industrial liquids and sludges containing chlorinated, solvents, heavy metals, and petroleum hydrocarbons.

On March 10, 1994, the EPA Region VIII and the Colorado Department of Public Health and Environment (CDPHE) issued a Record of Decision (ROD) that formally defined the cleanup plan for the Site. One of the remedies specified for the site was to excavate, remove, and treat surface and subsurface drums, and contaminated soils within the Former Tire Pile Area (FTPA). As part of the overall remedy for landfill solids, the FTPA waste pit material was considered to be a "principal threat" that required removal.

During the remedial investigation for landfill solids, it was estimated that there were approximately 1,350 buried drums containing approximately 1,300 gallons of liquid waste, and that there were approximately 15,000 cubic yards of contaminated soil and debris in the area. Pre-design investigations conducted during 1995 and 1996 confirmed the presence of approximately 1,300 buried drums, as well as approximately 10,400 cubic yards of "visibly" contaminated soil and debris.

During remedial design (RD), the EPA and CDPHE approved an excavation with on-site treatment approach that would meet all applicable or relevant and appropriate requirements (ARARs). The following paragraphs discuss the RD process that was followed to assure compliance with the ARARs.

METHODS, RESULTS, AND DISSCUSION
Three sequential testing phases that were or are being completed to examine the treatability of the excavated material on-site to render the waste non-hazardous are detailed in Table 1.

Bench-Scale Testing. A waste pit characterization program was implemented during the fall of 1995. The field program involved advancement of 109 Geoprobe® borings to determine the lateral and vertical extents of visual contamination, and to characterize the contents of the waste pits.

The Toxicity Characteristic Leaching Procedure (TCLP), and ignitability, corrosivity, and reactivity tests were used to determine if the waste pit material was RCRA hazardous or non-hazardous. Analytical results indicated that some, but not all, of the waste pit materials exhibited hazardous characteristics. These are summarized as follows:

- Toxicity for VOCs were detected above TCLP regulatory limits for 1,1-dichloroethene (1,1-DCE), tetrachloroethene (PCE), benzene, methyl ethyl ketone (MEK), and trichloroethene (TCE).
- Toxicity for semivolatile organic compounds (SVOCs), pesticides and herbicides were not detected above TCLP regulatory limits in any of the samples.
- Toxicity for lead was the only metal detected above the relevant TCLP limit at a concentration of 27.2 milligrams per liter (mg/L) in one sample compared to

TABLE 1. Summary of Treatability Testing Parameters

Parameter	Bench-Scale	Pilot-Scale	Full-Scale
Soil Volume (yd^3)	0.2	4,350	21,000
Moisture Content (%)	23-26	21.9	<25.0
Air Filled Porosity (%) (Est.)	29.0	29.0	29.0
Bulking Agent/Ratio (waste:agent)	wood chips/1:1	tire chips/9:1	tire chips/10:1
Pore Volume Exchange Per Day	64.3	3.2	6.0 (target)
Off-Gas Treatment	carbon	carbon	catalytic oxidation
Duration	2-months	1-week	12-18 months
Objective	treatment efficiency	constructability	closure
Status	completed	completed	on-going

the TCLP limit of 5 mg/L. Metals concentrations in all other samples were below relevant TCLP limits.
- Toxicity for lead was the only metal detected above the relevant TCLP limit at a concentration of 27.2 milligrams per liter (mg/L) in one sample compared to the TCLP limit of 5 mg/L. Metals concentrations in all other samples were below relevant TCLP limits.
- Ignitability at greater than 140 degrees Fahrenheit (^{o}F) was measured in seven of nine individual borehole samples, and two of three composite samples.
- Reactivity and corrosivity values were within acceptable limits (i.e., non-hazardous) for all samples.

A treatability testing program for the waste pit material was conducted during the summer of 1996. Physical drying/controlled aeration and enhanced bioremediation were the two treatment technologies evaluated. The bench-scale apparatus used for the testing is detailed in Figure 1. General excavating equipment was used to obtain the test solids from three separate waste pits. Waste material was placed into a 20 yd^3 rolloff container for blending. Waste material was then conditioned for treatment using wood chips at a 1:1 ratio. The use of a bulking agent reduced the overall moisture content of the excavated material and allowed for better air permeability through the soil. Subsequent test cell construction, time series monitoring of the hazardous characteristics and off- gas were conducted to assess treatment efficiency. The information collected was considered representative of conditions that will likely be encountered during full-scale remediation, and were used extensively during full-scale design.

Specific objectives against which each technology was evaluated during bench-scale testing included:
- Reducing ignitability such that the flash point is greater than 140 °F; and
- Reducing TCLP VOC concentrations to below regulatory limits.

Physical drying/controlled aeration treatment successfully achieved the first two objectives as illustrated in Figure 2. Enhanced bioremediation treatment successfully achieved the first objective, but not the second (Figure 2). Based on these results, physical drying/controlled aeration was recommended for full-scale application.

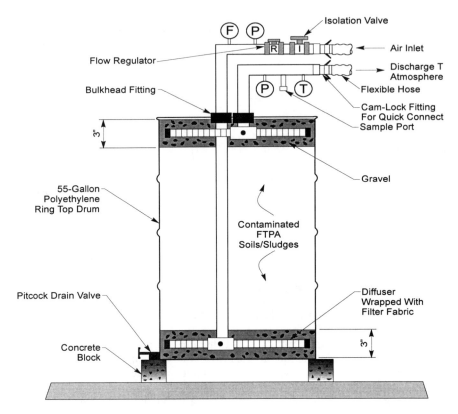

FIGURE 1. Bench-Scale Apparatus.

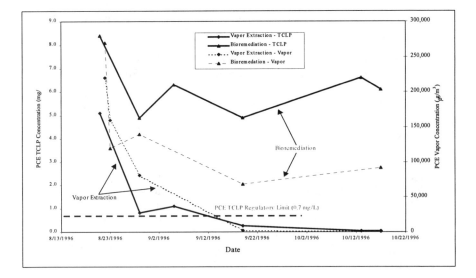

FIGURE 2. Bench-Scale Testing Results.

Results of the bench-scale testing were used to estimate mass removal rates and required treatment duration for the full-scale treatment cell design. Transient extracted air quality from the treatment cell was estimated as a function of operation time of the treatment system.

It was assumed that the initial influent concentration to the treatment system would be equivalent to that measured during the bench-scale testing. PCE and TCE were the only two VOCs that exceeded TCLP toxicity criteria prior to initiating the test. PCE concentrations for physical drying/controlled aeration were reduced below TCLP regulatory limits after approximately 1,550 to 1,600 pore volumes of air had been passed through the test cell (approximately 25 days) as determined by graphical analysis of the data. A three-order of magnitude reduction in PCE influent vapor concentration was realized prior to meeting TCLP regulatory limits (560,000 micrograms per cubic meter [$\mu g/m^3$] to 480 $\mu g/m^3$). TCE concentrations for physical drying/controlled aeration were reduced below TCLP regulatory limits after approximately 190 to 200 pore volumes of air (3 days) had been passed though the test cell. For all VOCs tested, the influent vapor concentrations were reduced appreciably after 450 pore volume exchanges. These results indicated that the reduction of PCE concentrations in the soil would dictate anticipated treatment duration for the full-scale system.

One significant difference between the test cell and the full-scale treatment cell, is the likelihood of non-uniform air flow at full-scale. The results would likely be a longer time frame to reach TCLP goals and a flattening of the mass removal curve (i.e., the initial removal rates will not be as high as predicted but will be more consistent over the duration of treatment).

The total mass of each compound removed per day to achieve non-hazardous levels was calculated on a unit volume of soil basis estimated from data collected during the bench-scale test. This was used to aid in the selection of the appropriate vapor treatment technology and to estimate the duration of system operation. As discussed previously, an estimated 1,550 to 1,600 pore volumes of air were required to reduce PCE concentrations in the soil/sludges prior to below TCLP regulatory limits. Assuming the same mass removal rates and the actual calculated pore volume exchange rate at full-scale (approximately 6 pore volumes per day), the anticipated treatment duration would be between 265 and 300 days.

Pilot-Scale Testing. Because of uncertainties associated with the range of geotechnical properties (densities, moisture content, clay content, air permeability) of the bulked waste material that would be anticipated during full-scale construction of the treatment cell, and because of uncertainties associated with the range and type of VOCs which may volatilize from the waste material within the cell, a pilot-scale constructability test was conducted to test the validity of the design parameters established during bench-scale testing. This test was performed during the initial phase of full-scale cell construction. The primary objectives of the constructability test were to confirm:
- Pipe network spacing;
- Vapor extraction flow rate and vacuum; and
- Treatment cell air quality.

Details of the pilot-testing parameters are presented in Table 1. All construction details of the test cell were identical to those of the full-scale treatment cell. The bulking agent used in full-scale was shredded tires available from the site as opposed to wood chips used during the bench-scale tests. Vapor monitoring of the extracted soil vapor and treatment cell soil gas provided the empirical information required to verify design parameters.

Several conclusions were drawn from the data collected during the pilot-scale testing:

- Based on soil gas response within the pilot-scale treatment cell, the in place treatment cell solids are sufficiently permeable to treat the solids with vapor extraction. This permeability is facilitated by the incorporation of tire chips into the waste pit solids, and by the large volume of woody debris in the pilot-scale treatment cell.
- Based on the increase in soil gas response during testing, active air injection is more effective in facilitating vapor movement within the treatment cell than passive air injection.
- The horizontal and vertical spacing of the extraction/injection laterals as designed are sufficient to allow treatment of the entire cell.
- The injection and extraction flow rates and pressures/vacuums as designed are sufficient to allow treatment of the entire cell.
- A significant rise in soil gas temperature within the treatment cell was measured. This may increase mass transfer of the VOCs adsorbed on the solid phase to the gas phase, which will increase treatment efficiency. However, such temperature rise, if not properly monitored and controlled, could be problematic for ignitable waste pit material.
- A large volume of water can be expected to accumulate during operation of the full-scale treatment system.
- The influent loading to the temporary treatment system was much less than anticipated during the design phase.

Full-Scale Operation. An illustration of the full-scale treatment cell design is presented on Figure 3. Full-scale construction consisted of geomembrane/geotextile liner and cover, leachate collection system, air distribution and collection network at 20-feet spacings, and conditioned waste material. Bench-

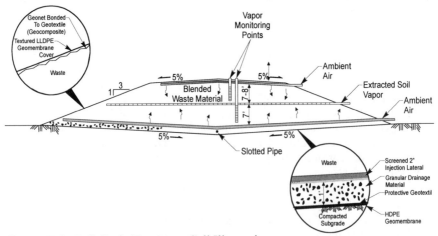

FIGURE 3. Full-Scale Treatment Cell Illustration.

scale geotechnical testing of waste pit material was performed to determine site-specific design parameters (bulk density, specific gravity, composite porosity, air permeability at optimum moisture content and blending ratios, and shear strength) for the treatment cell.

Characteristics of the waste materials excavated varied greatly within the waste pit during excavation. Waste materials included: visibly contaminated soils, wood debris, paint sludge, whole drums, drum carcasses, gas cylinders, and municipal debris. The waste pit was excavated using a conventional track mounted excavator and an all- terrain dump truck. Segregation of larger size material (greater than 12-inches diameter) was accomplished via the excavator. The excavated waste was loaded and was transported to the treatment cell. At the treatment cell, a blending area was built using pre-cast concrete traffic barriers to contain the waste. The traffic barriers were positioned in a "U" shape and elevated using road base. Tire chips were added and blended at a 10:1 ratio using a dedicated tracked excavator for mixing. The blended waste was then loaded into a feed hopper that placed the waste on a standard conveyor belt system that terminated to a radial telescoping stacking conveyor. The hopper was fixed with an 8-inch "grizzly" to remove oversized material. As the waste material was placed in the treatment cell by the conveyor, one section of the conveyor belt would be removed as necessary and the stacker would slide back and start building the next section of the cell. The waste pit solids were placed in this fashion to minimize compaction within the pile. The final shaping and grading was accomplished using a small, wide track, and low ground-bearing dozer on top of the cell before the final liner was installed.

Several construction challenges associated with the excavation, blending, and placement of the waste materials included:

- High VOC's of contaminated soils;
- Perched ground water;
- Daily temporary cover of excavated materials to control VOC emissions;
- Blending of excavated materials to dry (25-perent moisture)
- Allowance for consolidation of stacked waste materials, 3:1 slopes and 15 feet height

Subsequent excavation and full-scale characterization of the treatment cell indicated that approximately 33-percent of the solids failed TCLP for VOCs with PCE at levels of 1.0 to 3.0 mg/L after placement. Full-scale treatment commenced in January 2000 and is ongoing. Extracted soil vapor concentrations are mirroring those measured during bench-scale testing.

Routine temperature monitoring of the treatment cell was initiated subsequent completion of the pilot-scale testing phase. This monitoring was a direct result of the temperature increased witnessed during that phase of testing. In-situ treatment cell temperatures are detailed in Figure 4. Average soil concentrations ranged from 70.4 to 93.3 °F during the monitoring period. Maximum soil concentrations ranged from 100.2 to 127.4 °F during the monitoring period. These maximum values are consistent with maximum values usually witnessed during composting efforts (USEPA, 1998). This temperature rise indicates some abiotic or biotic process was occurring in situ. With the large amount of carbon sources (VOCs and woody debris) within the treatment cell, high moisture content, and the introduction of oxygen through injected air, it is speculated that the increase is biotic in nature. Such rise in temperature is beneficial because it increases the rate of VOC transfer from soils to the vapor

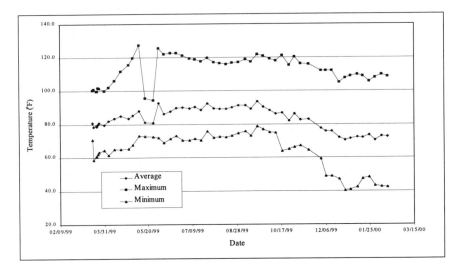

FIGURE 4. Treatment Cell Temperatures.

phase. If the temperature is maintained, the increased diffusion rates of the contaminants into the vapor phase could reduce the duration required for cleanup as estimated during bench-scale testing. However, if the rise in temperature becomes excessive, it could be detrimental to the treatment cell because if the temperature exceeds the flash-point of the waste material being treated, spontaneous combustion could result.

CONCLUSIONS

The use of multiple testing phases during the design and construction of an on-site treatment cell has been invaluable in implementing the remedy. The sequential steps used in evaluating the treatability of the waste pit solids and the constructability of the on site treatment cell as designed allowed for real time decision making by all interested parties. In addition, the use of a ex situ treatment method (as opposed to in situ) allowed for a more engineered approach to the application of soil vapor extraction. By homogenizing the waste material to the extent practical it is hoped that the level of uncertainty associated with in situ applications as it pertains to mass removal and treatment duration are decreased. By decreasing the uncertainly, financial burdens and level of efforts to complete the remedy are more accurately estimated to the benefit of all.

REFERENCES

U.S. Environmental Protection Agency (USEPA). 1998. *An Analysis of Composting As an Environmental Remediation Technology.*

LONG-TERM MONITORING FOR SVE EXIT STRATEGIES[1]

Gustavious P. Williams, Ph.D., David Tomasko, Ph.D. and Zhenhua Jiang, Ph.D.,
(Argonne National Laboratory, Argonne, Illinois)
Ira May (Army Environmental Center, Aberdeen, Maryland

ABSTRACT: The concentrations of volatile organic compound (VOC) vapors removed by soil vapor extraction (SVE) cleanup systems often approach an asymptotic value after a long period of operation. Even after long time periods, measurable amounts of contaminants are still removed by these systems. Since the total amount of contaminant in the subsurface is often not known completely, it is difficult to compose a meaningful exit strategy. Researchers at Argonne National Laboratory, using computer models validated with field samples, demonstrated that tailing behavior observed in SVE wells at a former U.S. Army munitions manufacturing facility is caused by the extraction of VOCs from contaminated groundwater. This process has not been previously discussed in the literature as significant and has been omitted from previous design studies. This raises questions about when these systems should be terminated, or more realistically, when should additional studies be performed to support the decision to terminate these systems. This modeling study concluded that SVE system operations should be monitored and reviewed when extraction rates drop to 5% of the initial extraction rate.

INTRODUCTION

The concentrations of volatile organic compound (VOC) vapors removed by *in-situ* soil vapor extraction (SVE) systems often approach an asymptotic value after a long period of pumping. This phenomenon is often referred to as "tailing." Researchers at Argonne National Laboratory recently completed a study that demonstrates that the tailing behavior seen in VOC concentrations from SVE wells at a former U.S. Army munitions manufacturing facility (Twin Cities Army Ammunition Plant [TCAAP] located near New Brighton, Minnesota) is caused by by the extraction of volatile contaminant vapors from contaminated groundwater in the saturated (phreatic) zone. Although this process apparently can be a significant source of contaminant vapors collected in an SVE system, it has not been previously discussed in the literature and has been omitted from previous modeling studies.

---

[1]The submitted manuscript has been created by the University of Chicago as Operator of Argonne National Laboratory ("Argonne") under Contract No. W-31-109-ENG-38 with the U.S. Department of Energy. The U.S. Government retains for itself, and others acting on its behalf, a paid-up, nonexclusive, irrevocable worldwide license in said article to reproduce, prepare derivative works, distribute copies to the public, and perform publicly and display publicly, by or on behalf of the Government.

BACKGROUND

VOC contamination in groundwater and overlying unsaturated soils is widespread throughout the United States. More than 15% of community drinking water supplies in this country are contaminated with chlorinated hydrocarbons (Gerdes and Steele, 1999). As of June 1994, state and local environmental agencies across the country reported more than 260,000 releases of organic materials from underground storage tanks. The number of confirmed releases continues to grow, with an estimated 1,000 new releases reported each week (U.S. Environmental Protection Agency,1995). As many as 18% of the current Superfund sites employ SVE, also known as soil venting or vacuum extraction, remediation technologies (U.S. National Report to IUGG, 1991-1994. 1995).

SVE is an *in-situ* remedial technology that reduces the VOC concentrations in soils in the unsaturated (vadose) zone. SVE has proven to be effective in reducing concentrations of VOCs and of certain semi-volatile organic compounds contaminating the vadose zone. The initial rates of extraction are, in general, very high provided that the permeability of the soil is high and its clay content is low. With time, the rate of contaminant removal rapidly decreases, but rarely goes to zero. Instead, the extraction rate approaches a small value asymptotically. Additional pumping under these conditions does not produce any additional reduction in the rate of mass extraction.

This tailing effect observed in field SVE systems has been extensively studied by the scientific community. Some researchers believe that the effect is caused by complex chemical and physical processes (e.g., nonequilibrium partitioning of the gas phase) (Conant et al., 1996; Fisher et al., 1996; Poulsen et al., 1996). Others feel that the tailing is produced by the potentially complex geology of the field sites. These complexities include layered systems composed of alternating layers of clay and more permeable material (Kaleris and Croise, 1999), and small pockets of nonaqueous phase liquids (NAPL) trapped at residual saturations (Garg and Rixey, 1999). In either case, numerical modeling of the entire cleanup process with minimal simplifying assumptions, from the start of operations until the system is shut down, is difficult to perform. In fact, in many cases, this modeling can not be performed at all because of limitations of today's computers. Because of these limitations, SVE systems are generally designed using empirical correlations or engineering judgment and thus the designs are subject to inefficiency and uncertainty. SVE operations frequently are continued during the asymptotic period in the belief that residual VOCs are still being removed from the soil matrix. Most current working hypotheses support this belief.

STUDY DATA

This study uses 15 years of data taken from the operation of two large SVE systems installed at the Twin Cities Army Ammunition Plant (TCAAP). The systems contain 39 (Site D) and 143 (Site G) vents and began operations in 1985. Data to support this study were obtained from the SVE system at Site D. The Site D system was designed to remove trichloroethylene (TCE) and other VOCs of concern (Wenck Associates, Inc., 1995). Initial extraction rates were on the order of 800 to 1,000 lbs/day. Current rates are a fraction of that between 1-2 lb/day. This paper discusses the causes of this tailing behavior, and examines a number of different monitoring techniques that can be used to make termination decisions.

Site D at TCAAP was used from approximately 1952 to 1968, and possibly as late as 1973, for disposing of and/or burning wastes, including oil, solvents, rags, maizo, neutralized cyanide, mercurous nitrate, powder water, and scrap propellant powder in five disposal pits (Argonne National Laboratory, 1988). Examination of soil borings revealed the presence of VOC contamination (TCE, 1,1,1-trichloroethane [TCA], and trans-1,2-dichloroethene [T12DCE) in the Arsenal Sand Formation beneath the site. In the vicinity of the disposal pits, the soil was stained and residues extended from 4.5 ft to a maximum of 39 ft (1.4 to 11.9 m) below the ground surface. Additional VOC contamination was found in an extensive groundwater plume under the site. The depth to the water table is about 150 ft (46 m). As part of the remediation, a clay cap was installed over the area to limit infiltration of precipitation and prevent additional contaminants from reaching the underlying groundwater. An SVE system with 39 extraction wells was installed at the site and began operation in 1986. This system was empirically designed and did not account for improvements in the extraction efficiency created by the presence of the clay cap. Approximately 100,000 pounds (45,350 kg) of VOCs have been removed from the site to date. The peak mass extraction rate was 1,300 pounds (590 kg) of VOC per day, which occurred on the third day of operations. By 1994, the mass removal rate approached a near asymptotic value of about 2 pounds (0.9 kg) per day, nearly three orders of magnitude less than the initial value; however, operation of the system continues.

MODELING

SVE operational data, along with field geologic and contaminant characterization data were used to develop a series of simulations using the computer code T2VOC. The simulations were used to examine long-term SVE behavior and to determine how commonly available monitoring data could be used to develop and impose meaningful exit strategies.

Argonne researchers used a slightly modified version of the computer model T2VOC (Falta et. at., 1992) on an IBM SP computer (Argonne National Laboratory, 1999). T2VOC is a numerical simulator that can compute three-phase, three-component, non-isothermal, heterogeneous flow and transport in three dimensions. The code, developed at Lawrence Berkeley Laboratory, was designed to model near-surface remediation processes, such as steam-sweeping and SVE.

The Argonne study investigated the extraction of TCE from the Arsenal Sand Formation by using several geometries designed to simulate the effects of a homogeneous clean sand similar to the Arsenal Sand Formation, alternating layers of clay and sand, and randomly generated blocky heterogeneities with a volume of approximately 27 ft^3 (0.8 m^3). Initial conditions for the T2VOC runs were developed to simulate field conditions in the zone of contamination. A number of calculations were also performed for an initially clean unsaturated zone and contamination in the underlying groundwater. Another set of runs was made with clean groundwater and initial vadose zone contamination. In total, more than 50 individual geometries and contaminant distributions were simulated with T2VOC. The results from a typical multidimensional remediation process predicted by T2VOC are shown in Figure 1.

Many of the geometries and contaminant distributions simulated with T2VOC produced mass removal curves that could approximate the magnitude and shape of the observed extraction curve for limited periods of time (either early or late time). However, the simulations that exhibited long tails did not duplicate the observed initial rapid drop in extraction rates, and those that matched the initial rapid decline in extraction rates did not match the long-term tailing. In order to match the observed results (very high initial extraction rates, followed by a rapid drop off in removal, followed by a long asymptotic tail of low, but measurable mass removal), contaminated groundwater was incorporated into the model. When contaminated groundwater was included in the model, the entire extraction curve derived from field data could be replicated.

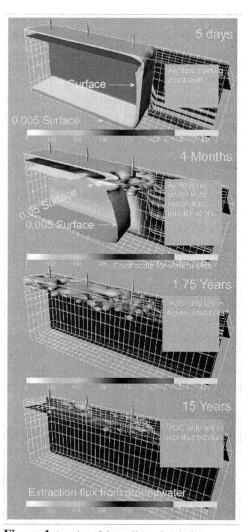

Figure 1 A series of three-dimensional plots over time showing the amount of VOC in the subsurface from operation of an SVE system. VOCs in the vadose zone are quickly removed and only small areas near the flow field stagnation points remain; not enough mass is available to generate observed long-term tails in extraction rates.

Field Data and Model Validation. In the spring of 1997, additional field work was performed at TCAAP to obtain data that could be used to resolve the tailing issue. This field work included VOC sampling and installation of a deep-well extraction system. The results of this additional field work indicated the following: only very low concentrations of TCE in the soil occurred near the water table (5 to 50 µg/kg); the permeability of the porous medium was measured at 1.7×10^{-12} m^2; and the highest concentration of VOCs in the effluent occurred in wells that were screened immediately above the water table. These results provide independent validation of the Argonne model and numerical results. That is, residual TCE contamination in deep soils is at very low concentrations and most of the initial VOC contamination in the vadose zone has been removed by the SVE system. Permeabilities used to simulate the Arsenal Sands with T2VOC are consistent with measured field values, and contaminated groundwater in the phreatic zone is contributing VOC vapors to the extraction system, creating the asymptotic tail currently observed.

Simulations showed that a number of different factors could contribute to the long-term tailing behavior observed at TCAAP and other SVE installations. The geologic setting, well spacings, and fraction of organic carbon present were all important modeling parameters. It was also found that contaminated groundwater beneath an SVE system can contribute measurable amounts of VOC's to the system, on the order of 2 lbs/day for the 39 vent system at TCAAP. This finding and has not previously been reported in the literature.

RESULTS AND CONCLUSIONS

Argonne has demonstrated that contaminated groundwater can produce the long tailing asymptotic behavior observed in SVE extraction systems. This mechanism has previously been discounted as negligible or omitted in long-term operational designs. It was not necessary to introduce fine-scale residual pockets of contamination, complex multilayered systems, or nonequilibrium effects in order to replicate the observed field performance.

The modeling study demonstrated that the percent decrease in extraction rates (a common monitoring parameter) could be reliably used as an indicator of the percentage of contaminant mass remaining in the subsurface (a value not usually known). It was found that when extraction rates dropped below 5% of the initial rates, approximately 90% of the initial contaminant had been removed. When 99% of the initial contaminant had been removed, extraction rates were approximately 0.3% of the initial rates. It was also found that it took longer to remove the last 9% than the first 90% of the contaminant mass in the subsurface.

Conclusion. This modeling study concluded that SVE system operations should be monitored and reviewed when extraction rates drop to 5% of the initial extraction rate. More details can be found in Williams and others (1999).

REFERENCES

Argonne National Laboratory 1988. Installation Restoration Program: Preliminary Assessment of the Twin Cities Army Ammunition Plant, prepared for Commander, U.S. Army Toxic and Hazardous Materials Agency, Aberdeen Proving Ground, Maryland.

Argonne National Laboratory. 1999. Mathematics and Computer Science Division. http://www.mcs.anl.gov/computing/machines/quad/.

Bryngelson, B. 1999. "The EPA Superfund Remediation of Oil Spill Sites."
http://www.hort.agri.umn.edu/h5015/98papers/bryngelson.html.

Conant, B. H., R. W. Gillham, and C.A. Mendoza. 1996. "Vapor Transport of
Trichloroethylene in the Unsaturated Zone: Field and Numerical Modeling
Investigations." *Water Resources Research.* 32(1): 9.

Fischer, U., R. Schulin, M. Keller, and F. Stauffer. 1996. *Water Resources Research.*
32(12): 3413.

Falta, R. W., K. Pruess, I. Javandel, and P. A. Witherspoon. 1992. "Numerical Modeling
of Steam Injection for the Removal of Nonaqueous Phase Liquids From the Subsurface 1:
Code Validation and Application." *Water Resources Research.* 28(2): 451.

Garg, S. and W. G. Rixey. 1999. "The Dissolution of Benzene, Toluene, —Xylene, and
Naphthalene from a Residually Trapped Non-aqueous Phase Liquid under Mass Transfer
Limited Conditions." *Journal of Contaminant Hydrology.* 36(1999): 313-331.

Gerdes, K. and J. Steele. 1999. "Volatile Organic Compounds in Non-Arid Soils
Integrated Demonstration Overview."
http://iridium.nttc.edu.env/VOCNA/VOCNA_overview.html.

Kaleris, V. and J. Croise. 1999. "Estimation of Cleanup Time in Layered Soils by Vapor
Extraction." *Journal of Contaminant Hydrology.* 36(1999) 105-129.

Poulsen, G. T., J. W. Massmann, and P. Modrup.1996. "Effects of Vapor Extraction on
Contaminant Flux to Atmosphere and Ground Water." *Journal of Environmental
Engineering.* 700-706.

U.S. Environmental Protection Agency.1995. *How To Evaluate Alternative Cleanup
Technologies for Underground Storage Tank Sites.* EPA510B95007.

U.S. National Report to IUGG, 1991-1994. 1995. American Geophysical Union, Rev.
Geophys. Vol. 33 Suppl.

Wenck Associates, Inc.1995. *Interim Remedial Action Performance Evaluation:. Soil
Vapor Extraction Systems at Sites D and G Final Report*, prepared for Commander, Twin
Cities Army Ammunition Plant, New Brighton, Minnesota by the U.S. Army Corps of
Engineers, CEMRO-ED-E, Omaha, Nebraska

Williams, G.P., Z. Jiang, and D. Tomasko. 1999 *Twin Cities Amy Ammunition Plant Soil
Vapor Extraction System: A Post-Audit Modeling Study*, Prepared for the Army
Environmental Center, Aberdeen, Maryland, Argonne National Laboratory Technical
Memorandum, Draft, August.

ACKNOWLEDGMENTS
Work supported under a military interdepartmental purchase request from the U.S.
Department of Defense Army Environmental Center, through U.S. Department of
Energy contract W-31-109-Eng-38

RECOVERING FREE PRODUCT FROM CLAYEY FORMATIONS USING HYDRAULIC FRACTURES

William W. Slack FRx, Inc., Cincinnati, Ohio
Lawrence C. Murdoch, Clemson University, Clemson, South Carolina
Dave Butler, Applied Engineering & Science, Inc., Atlanta, Georgia

Abstract: Removal of all free-phase hydrocarbon liquids is a critical first step in the remediation of contaminated soil and groundwater. Should the hydrocarbons accumulate in low permeability media, such as clay, the limited flow capacity of the soil can frustrate recovery efforts. Sand-filled hydraulic fractures have long been used to enhance flow in low permeability media. Two concepts have been advanced for the application of hydraulic fractures as enhancements for free-phase liquid recovery. The simplest relies on skimmer pumps, which separate the hydrocarbons from water in the recovery well, pumping only hydrocarbon to the surface. The approach offers the attractive benefit of avoiding the cost of treating contaminated water that might otherwise be produced. However, the physics of water coning limit the rate of hydrocarbon recovery. A more sophisticated approach uses additional fractures to control water in situ and maximize hydrocarbon recovery. We compared the approaches by examination of two projects that utilize sand-filled hydraulic fractures for recovery of fuels from clayey soils.

INTRODUCTION

Light non-aqueous phase liquids (LNAPL), such as motor fuels or lubricating oil, can be a major threat to both underlying groundwater or nearby receptors. The presence of free product at a site often precludes closure because of this potential threat. Several well designs, including special skimming pumps or in-well separators, have been developed to maximize the recovery of free product. This is typically accomplished by targeting a floating LNAPL separately from the underlying water. The designs are intended to recover free product, while minimizing the simultaneous recovery of contaminated water to avoid the expense of treating the contaminated water until after the NAPL problem has been addressed. The performance of free product recovery systems in clean sands can be quite good, however, those systems perform poorly in clayey sediments largely because the discharge of liquids is minimal.

Hydraulic fractures can improve the performance of wells completed in fine-grained sediments or rock. Sand-filled hydraulic fractures form sheet-like, highly permeable layers in the subsurface that will increase the flow rate to wells in fine-grained formations typically by one to two orders of magnitude. We have used hydraulic fractures with free product recovery systems at several sites, and consistent improvements have been observed.

Two different designs of free product recovery systems that use hydraulic fractures have been utilized. The simplest design uses one hydraulic fracture placed at the bottom of the LNAPL layer. A skimmer pump is installed in a recovery well that intersects the fracture and the pump is operated as normal. As a

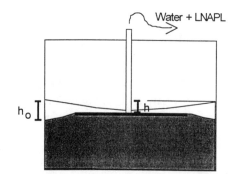

variation of this concept, additional fractures may be created at overlying or underlying elevations so that seasonal variations in water table can be accommodated. This approach (Fig. 1) is limited in drawdown to approximately 20% of the thickness of the layer. The limitation arises from the need to preserve uniform head throughout the water zone because the water is static. As a result, water will cone-up wherever the product layer is thinned. The maximum amount of coning can be determined by comparing the head at the extreme limit of the cone to the head at the extraction point. At the extreme limit, the head in the water is the

FIGURE 1. Using a fracture to skim free-phase hydrocarbon. Product with an initial thickness of h_o, can be skimmed with a maximum drawdown of h.

height of product, h_o, multiplied by product density, ρ_o. At the extraction point, no product exists, so the head must be composed of the height of the cone multiplied by the density of water, ρ_w. Since the two heads are equal, the maximum drawdown, h, of product is

$$h = h_o \left(\frac{\rho_w - \rho_o}{\rho_w} \right) \qquad (1)$$

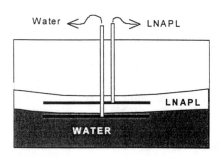

FIGURE 2. **Product recovery from one fracture while controlling water with a second fracture.**

For typical densities of hydrocarbon fuels, the relation evaluates to approximately 20% of the free-product thickness.

Another approach is to use two flat-lying hydraulic fractures stacked one on top of the other. The upper fracture is created in the zone containing LNAPL, whereas other one is in the underlying formation saturated with water. Pumping from both fractures simultaneously will limit the upward migration of the interface between water and NAPL. The concept is illustrated in Figure 2.

The approach eliminates, or limits, up-coning of water. Preliminary theoretical analyses of the well design confirm field observations and show that the two-fracture design has merit.

We have previously reported on the use of hydraulic fractures to enhance the recovery of free-phase LNAPL while controlling water coning with an additional fracture (Murdoch et al., 1994). The work was done at one of the many refineries near Beaumont, Texas. In a five-day-long test using the two-fracture design, the fracture created in the LNAPL zone produced LNAPL with negligible amounts of water at a rate more than 20 times faster than a conventional well. The well intersecting the underlying fracture produced water with negligible amounts of NAPL, so that both increased the rate of recovery and separated NAPL and water phases as they flowed to the well. We have since completed a project that implemented the more simple approach of separating the product in the recovery well with a phase-separating pump. The results provide an interesting comparison to the earlier work.

BIRMINGHAM FREE-PHASE RECOVERY PROJECT

Seventy-one sand-filled hydraulic fractures were created in a railroad yard in Birmingham, Alabama, to facilitate recovery of free-phase petroleum hydrocarbons by skimmer pumps. The contaminants consisted mostly of fuel oil and diesel, which was released from multiple points during the refueling of locomotives. Three of the fractures were created in the fall of 1997 as a pilot test. The remainder were created during the summer of 1998 in order to deploy the recovery technology across the 14-acre site.

The site is underlain by the Conasauga Formation that forms the axis of the Birmingham Anticline. The Conasauga Formation is a highly fractured dolomite/limestone approximately 4 to 8 m below ground surface (bgs). The overlying soil consists of weathered residuum containing silty clay and fragments of limestone and chert. A layer of fill materials covers the residuum. The fill, which contains silty clay, coal, cinders, building scraps and other materials consistent with 100+ years of railroad operations, varies in thickness from ½ to 3 m. Native soil samples collected by Shelby tubes in the interval 3.9 to 4.6 m bgs have greater porosity and lower hydraulic conductivity than samples collected from 1.2 to 1.8 m bgs. Hydraulic conductivity for all samples were consistent with the clay-rich, compacted soils, ranging from 1.9×10^{-6} cm/sec to 1.4×10^{-8} cm/sec.

The installation of numerous monitoring wells over the years has provided insight as to surficial hydrogeology and contaminant transport. Groundwater apparently follows preferential pathways, such as steeply dipping natural fractures and zones of higher permeability in the fill. Surface water infiltrates the fill material and percolates downward, in some places forming perched zones over the underlying native soil. Presumably the contaminants follow similar pathways. Depth to groundwater varies spatially across the site, ranging from less than 2 m to more than 6 m depth. Seasonal variations in water table have also been observed. As a result of the heterogeneous structure of the fill and native soil units and the spatial and temporal variation in water table, wells drilled within a few meters of each other can discharge at extremely different rates.

Hydrological factors frustrated recovery of free-phase hydrocarbons from conventional wells. In a pre-fracture test, recovery of water and hydrocarbon from MW-1S yielded hydrocarbon at a discharge rate of 3 liters per day. Smaller discharge rates were predicted when skimmer pumps were used in the well. Three liters per day was deemed too meager for remediation, and lesser discharge certainly would have proven impractical.

The approach for enhanced free product recovery involved creating sand-filled hydraulic fractures at multiple depths around conventional wells. Presumably, the fractures would provide pathways for migration of hydrocarbon into the well. The depths for the fractures were chosen to optimize the coverage of the then current product and groundwater interfaces while allowing for seasonal fluctuations.

Pilot Test.   The recovery approach was first implemented on a pilot scale to evaluate uncertainties about fracture form and the concerns about a fracture intersecting an existing well. Three fractures were created from locations around MW-1S. Fractures were placed at depths spanning the range of water table fluctuations to ensure propagation into current product plume.   The wells for fracture creation were placed within an expected fracture radius from MW-1S. The materials planned for each fracture were limited conservatively to quantities that would be contained in a fully subterranean, sub-horizontal fracture, i.e. a more-or-less horizontal fracture that does not vent to the surface.

The desire to have the fractures intersect well MW-1S also carried uncertainties about fracture and well interaction.   On one hand, MW-1S could have acted as a pin, inhibiting fracture propagation toward or around it. In such case, the fracture would grow substantially in other directions and could not be expected to be adequately connected to the well to enhance recovery.   On the other hand, the fracture could fully enter MW-1S during creation and exit to the surface or adversely interact with the first fractures.   If so, the propagating fracture could lose its transport liquid through the screen and consequently be of limited size and utility.   Thus the first two fractures were located 3 m from MW-1S. After uneventful creation of two fractures, the third was located 2 m away. Table 1 lists the characteristics of the fractures.

TABLE 1. Specifications for hydraulic fractures created around MW-1S for pilot testing of enhanced skimmer pumping.

FracID	Depth (m)	Location Relative to MW-1S	Volume of Sand (m^3)	Max/min Diameter (m)
IP1	3.73	3 m N	0.25	7.9 / 7.8
IP2	4.64	3 m NE	0.37	8 / 6
IP3	3.1	2 m NW	0.23	8.8 / 7.6

Data collected during creation of the fractures permitted assessment of the form of the fracture and provided a preliminary indication of their function. Pressure logs that follow a typical form, symmetric uplift domes, and absence of vents to the ground surface indicated roughly horizontal hydraulic fractures. The two upper fractures definitely propagated past MW-1S, while the lowest one probably did. Accordingly, existing wells at this site do not act as pins that suppress fracture aperture, especially if the fracture has been nucleated a few meters from the well. In addition, fracture wells presumably could be placed closer to existing wells without suffering limited propagation. Closer placement should also increase thickness of the fracture at intersection with the well, which should improve recovery of contaminants.

In any case, horizontal fractures were created and geological conditions were indicated that favor creation of substantially larger fractures, possible double or triple the diameter and two to three times thicker. Improved performance of fractures as remedial enhancements was expected from larger fractures.

Following creation of the fractures, a product recovery pump was installed in MW-1S. The pump was designed to separate hydrocarbons internally and pump only hydrocarbon to the surface. The pump was set up to achieve a maximum rate of 180 liters per day. As expected, the rate decreased to substantially smaller rates once the initial volume of hydrocarbon in the well was depleted.

The in-well separator pump was operated in MW-1S over five days for a total of 74 hours of operation. The volume of free product recovered was gauged daily. Based on the final measured pumping rate and the total volume recovered, the well discharged 14.7 liters per day. This represents an increase of over five times the discharge of hydrocarbons from the same well utilizing a total fluids pump in the pre-fracture test. In addition, the pump appeared to be suitably robust for long-term operations. In conclusion, the three fracture pilot test justified implementation of the recovery scheme site-wide.

Full-Field Deployment. The approach for field-wide deployment of fractures was revised according to the lessons learned from the pilot test. Fractures were planned to contain as much as five times the sand used in the pilot test. Pairs of overlying fractures were planned, instead of the trio of overlying fractures created at the pilot test location. Typically these pairs were created at depths separated by 2 m, e.g. 3.5 m and 5.5 m bgs. With these constraints in mind, recovery wells and sets of fractures were located within the periphery of the two plumes. In general, location of a recovery well was selected, and one pair fractures was specified to be created within 1.5 m of it. Additional pairs of fractures were then specified at locations adjacent to the limits of the central fractures. Figure 3 shows an example of how fractures were planned for RW-5. The assumption that subsurface flow among spatially contiguous fractures was made in the interest of operational economy, but did not have any justification from the pilot test.

One minor operational change was introduced to improve the efficacy of fracturing during field-wide deployment. The nominal 2-inch (5 cm) steel pipe through which the fractures were created was recovered by drilling crews after fractures were created and re-used. As a result, less pipe was used, which translated into less life-cycle waste in addition to lower cost, fewer permanent pathways were created, which should limit unwanted penetration of fluids into the subsurface, and no stickups remained to frustrate future land use.

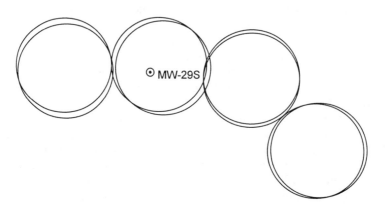

FIGURE 3. Placement of fractures around MW-29S/RW5. Two fractures, indicated by slightly misaligned circles, were created at different depths at each of four locations around MW-29S, which was renamed RW5 when the product-separating pump was installed.

The sixty-eight fractures created as part of the field-wide deployment of the fracture assisted separated product recovery system generally had characteristics predicted by the pilot test. However, in 17 instances, cross-linked gel (usually loaded with sand) reached the ground surface after substantial quantity of slurry had been injected. These vents occurred from 3 to 12 m away from the injection well and at all cardinal directions from the well. Injection was stopped when the venting rate obviously approached the injection rate. In another instance, slurry vented from a nearby recovery well, necessitating premature termination. In all of these cases, sufficient volume of material was injected to create a useable fracture, and no steps were taken to inject additional material to meet design criteria.

Two fractures vented in an adverse form. Both vented within a meter of the injection well and after injection of a modest quantity of slurry. The vent patterns suggested the hydraulic fractures may have intersected undocumented excavations near the wells. These fractures will provide limited enhancement to remedial operations at the particular locations.

One fracture could not be created to desired size. Excessive injection pressures were encountered and injection was terminated. Excessive pressure indicates that loss of gel and packing of injected sand blocked the fracture propagation pathway. Such would occur if the fracture were propagating through extremely permeable media that acted like a filter. Consequently, the fracture probably will serve to connect the recovery well to an existing preferential flow path.

Each fracture contained an average of 1.15 m^3 of sand. The shallowest was created at a depth of 3.35 m while the deepest was created at 5.8 m. THe pattern of uplift typically formed an slightly elongate dome, which is consistent with flay-lying to gently dipping hydraulic fractures at shallow depths. (Murdoch et al., 1994). The center of uplift, location of maximum uplift, and foci of the ellipse rarely coincide with each other or with the injection well. However, the volume of uplift usually correlates well with the volume of injected slurry. In contrast, the uplift volumes of several fractures were substantially less than would be expected from the quantities of injected material. Such a discrepancy has been encountered at other sites. Either of two factors can cause the effect. First, certain geological conditions can distribute the displacement of soil by the fracture over a large area, rendering measurement difficult. Depth and elasticity of the soil favor this mechanism. Although the fractures at the Birmingham site were not particularly deep, the substantial thickness of overlying fill material represents an elastic and compressible unit that can adsorb the uplift signal. Second, fracture slurry that propagates into a highly permeable zone or a zone with substantial course porosity, such as a sand and gravel bar, will not create a fracture but rather will displace fluid within the media and fill the pore space. Such features can be expected to exist at the Birmingham site either as native material or as the occasional extension of an uncharted excavation or deposit of fill.

Results of fracturing suggested the recovery system would operate successfully. Fractures of suitable size were created at chosen locations throughout the hydrocarbon plume. Fracture sand and decomposed slurry were found upon installation of several recovery wells, confirming sufficient lateral extent of the fractures. Even the variance of fracture form, as discussed in the previous paragraph, suggests a beneficial characteristic: fractures that connect to existing sand lenses will aid in recovery of product from an even greater area of the site. Finally, no deleterious processes were observed during the creation of fractures.

Recovery operations commenced in the summer of 1999. As expected, some of the recovery wells performed better than the pilot test, but average performance was remarkable similar to the pilot test. Long term assessment is not yet possible. During the late summer and fall of 1999 the area endured a drought, and seasonal variation of the water table was extreme. The water table fell below the bottom of several of the recovery wells, and representative data have not been collected in several months.

CONCLUSIONS

The recovery of free-phase hydrocarbon from low permeability soils can be enhanced and accelerated by sand-filled hydraulic fractures. The fractures effect the desired enhancement by providing low-resistance flow paths within the target soils.

The greatest contrast between recovery rates of fracture-enhanced systems and conventional fluid recovery systems apparently can be realized if water and LNAPL are recovered from separate fractures. At one site in Beaumont, Texas, the recovery rate from a pair of fractures was at least 20 time more than conventional wells. At another site where LNAPL was recovered from one well, and no water was recovered, the fractures appear to increase the product recovery rate by about a factor of 5. Other differences, such as the product thickness and the hydraulic conductivity of the native soil may also have contributed to these differences. In practice, the expense of treating the contaminated water recovered using the two-fracture system may out-weigh the efficiency of controlling water coning. Nevertheless, at both sites underlain by fine-grained materials the use of hydraulic fractures caused LNAPL recovery to increase from a meager trickle to rates where remediation was viable.

REFERENCES

Murdoch, L. C., D. Wilson, K. V. Savage, W. W. Slack, and J. E. Uber. 1994. *Handbook of Alternative Methods for Delivery and Recovery*, US EPA EPA/625/R-94/003.

MULTIPHASE EXTRACTION PROVIDES EFFICIENT REMEDIATION OF CHLORINATED COMPOUNDS

Keith M. Metzger, Richard J. Rush, and Robert M. Booth
(XCG Consultants Ltd., Kitchener, Ontario, Canada)

ABSTRACT: A Multi-Phase Extraction (MPE) system is being applied to remediate an industrial site contaminated with chlorinated solvents, especially trichloroethylene (TCE), cis-1,2-dichloroethylene (cDCE), and vinyl chloride (VC). The MPE system has been operational for approximately 1 month of a three-year project. This paper summarizes the preliminary findings. Components of the MPE system include a liquid ring vacuum pump, a photocatalytic oxidation system, and a liquid-phase GAC vessel. The average VOC mass removal rate over the first month of operation was approximately 8 kg/day. Both the treated vapour and groundwater satisfy treatment objectives. The diameter of the hydraulic capture zone is approximately 100 m.

INTRODUCTION

Chlorinated organic solvents have been widely used in North American industry since the late 1940s. High production rates combined with poor disposal practices have resulted in wide-spread subsurface contamination. If spilled into the subsurface in sufficient quantities, these dense non-aqueous phase liquids (DNAPLs) penetrate the water table where they tend to provide a long term source of contamination. The persistent nature of DNAPLs are largely due to low aqueous solubility, low degradability and ability to exist as a stable separate phase ("pure phase"). Additional DNAPL properties include high vapour pressures and low liquid viscosities. DNAPL migration in the subsurface is largely dependent on soil heterogeneities and relatively independent of groundwater flow. These properties make source delineation and remediation extremely difficult.

Typically the pump and treat remedial approach has been applied at sites contaminated with chlorinated solvents. The pump and treat approach has been relatively unsuccessful at remediating chlorinated solvents, largely because the chlorinated solvents residuals adhered to soil particles do not readily desorb into the groundwater where they can be captured by pump and treat systems.

Multi-Phase Extraction (MPE) systems provide effective remediation of sites contaminated with chlorinated solvents because they promote the natural volatility of the chlorinated solvents. The high vacuums generated by MPE systems tend to "strip" the chlorinated solvents from the residual and dissolved phases into the vapour phase where they can be extracted with the soil vapour.

A Site Specific Risk Assessment has been conducted to evaluate the public health and ecological risks associated with the contamination. The target soil vapour and groundwater clean-up criteria developed during the SSRA are summarized in Table 1.

Table 1
Risk Based Site Specific Clean-up Criteria

Compounds	Soil Vapour (mg/m^3)	Groundwater (mg/L)
Trichloroethylene	7	1.0
cis-1,2-dichloroethylene	5	1.0
Vinyl chloride	0.3	0.1

Operation of the MPE system commenced in January 2000. It has been operational for approximately 1 month. The target duration required to achieve the remediation objectives is three years. This paper summarizes the preliminary findings of this remediation project.

Objective. The objective is to remediate the subject site to satisfy the SSRA clean-up criteria.

Site Description. A 0.75 ha industrial site in Kitchener Ontario (client confidential) became contaminated with chlorinated solvents, especially trichloroethylene (TCE), cis-1,2-dichloroethylene (cDCE), and vinyl chloride (VC). The site was historically used for the manufacturing of electronic components. The contamination is the result of improper disposal of spent degreasing compounds. The former plant building has been demolished and the site is now vacant.

The key subsurface features include the following:

- The following sequence of geological units (from surface to depth) has been identified at the subject site: a surficial aquifer, an upper silt/clay layer, an intermediate aquifer, and a second clay layer. To date, the MPE system has been applied to the surficial aquifer only. The surficial aquifer consists of sand and silty sand. The thickness of the surficial aquifer ranges from 4 m to 5 m across the subject site. At the extraction well the surficial aquifer is approximately 4.4 m thick.
- The groundwater in the surficial aquifer flows predominantly to the north. The hydraulic conductivity of the surficial aquifer ranges from 7×10^{-5} cm/s to 3×10^{-4} cm/s.
- At the start of remediation the saturated thickness of the surficial aquifer was approximately 3 m.
- A conventional pumping test (not using MPE) at the extraction well revealed that the groundwater extraction rate was approximately 0.3 L/min (0.07 gpm).
- The compounds present at the highest concentrations are TCE, cDCE, and VC. The concentrations of TCE, cDCE, and VC in the surficial aquifer at the extraction well (hot spot on the site) were 22 mg/L, 57 mg/L, and 18 mg/L, respectively.

- The groundwater plume in the surficial aquifer covers the entire site and extends approximately 150 m downgradient. The TCE concentration contours are shown in Figure 1.
- Free product chlorinated solvents were extensively investigated during the subsurface investigations but were never found.
- The total mass of chlorinated solvents on the subject site is unknown; however, it is estimated to be approximately 3,000 kg.

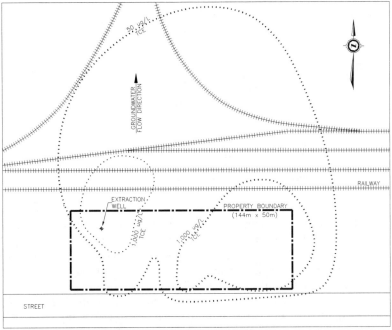

Figure 1: Site Plan Showing Extent of Contamination

MATERIALS AND METHODS

The fundamental principle of operation of a MPE system involves the application of a high vacuum (negative pressure) at an extraction well. The high vacuum causes both groundwater and soil vapour to be simultaneously removed from the extraction well. The vacuum decreases with radial distance from the extraction well. This results in an induced pressure gradient toward the extraction well. The rates at which groundwater and soil vapour migrate toward the extraction well are dependent on the physical properties of the groundwater and soil vapour (e.g. viscosity), and the characteristics of the medium through which they are moving (e.g. porosity).

The MPE equipment includes a collection system, a vacuum system, a vapour treatment system and a liquid treatment system, as shown schematically in Figure 2.

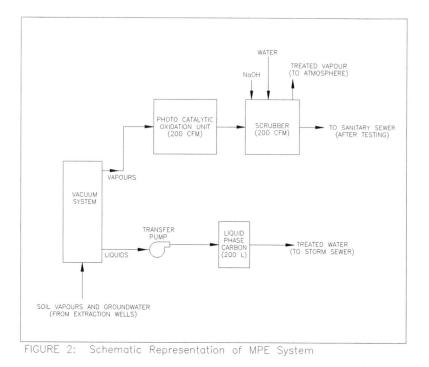

FIGURE 2:   Schematic Representation of MPE System

The collection system includes a 50 mm diameter extraction well with a 3 m long screen. The vacuum system includes a liquid ring vacuum pump (LRVP) rated at 15 kW (20 HP). The vapour treatment system consists of a photocatalytic oxidation system. The liquid treatment system includes a bag filter followed by a 200 L liquid-phase Granular Activated Carbon (GAC) vessel.

The photocatalytic oxidation system uses ultraviolet (UV) light (wavelength of 254 nm) and titanium dioxide (TiO_2) to destroy the VOCs in the extracted vapour stream. When TiO_2 is illuminated with UV light, it generates oxidative and reductive species which readily attack and destroy the target VOCs.

The chlorine atoms in the destroyed VOCs form hydrochloric acid (HCl) gas. Therefore, the off-gases from the photocatalytic oxidation unit requires treatment with a scrubber to remove the HCl. During the first month of operation, the mass of HCl generated by the photocatalytic oxidation system did not exceed the allowable discharge criteria; therefore, the scrubber has not been added to the MPE system at this time.

RESULTS AND DISCUSSION

Extraction Rates.   With respect to the soil vapours, the average vacuum generated at the extraction well was approximately 67% of an atmosphere (20 inches of mercury). The soil vapour removal rate was approximately 2 m^3/min (70 scfm). The concentrations of TCE, cDCE and VC in the extracted soil vapour were approximately 1,600 mg/m^3, 360 mg/m^3, and 7 mg/m^3, respectively. The average VOC mass removal rate in the soil vapours was approximately 8 kg/day.

With respect to the groundwater, the initial recovery rate was approximately 12 L/min (2.6 gpm); however, within the first day it had decreased to approximately 3 L/min (0.67 gpm). This is approximately 40 times greater than the groundwater extraction rate that could have been produced by conventional pumping.

After operating the MPE system for approximately one week, free product chlorinated solvents were observed in observation wells near the extraction well. Free product was removed at approximately 1 L/day. After the free product was observed, the concentrations of TCE in the groundwater increased to approximately 540 mg/L. The concentrations of the other VOCs in the groundwater remained fairly consistent. The concentrations of TCE, cDCE, and VC in the extracted water downstream of the inlet separator (a component of the MPE system) were approximately 1.6 mg/L, 1.3 mg/L, and 0.02 mg/L, respectively. Approximately 99% of the VOCs in the groundwater were transferred to the vapour phase within the inlet separator. This represents a mass transfer rate of approximately 2.7 kg/day from the groundwater to the vapour phase. This mass is approximately 33% of the mass entering the vapour treatment system.

System Performance. The soil vapour and groundwater treatment objectives for the subject site are summarized in Table 2. As shown in Table 2, the photocatalytic oxidation system removes over 99% of the VOCs in the soil vapour. The soil vapour treatment provided by the photocatalytic oxidation system satisfies the treatment objectives.

Table 2
Summary of Soil Vapour and Groundwater Treatment

Compounds	Vapours (mg/m^3)				Groundwater (mg/L)			
	Influent Conc.	Stack Discharge Concentration		Treatment Provided (%)	Influent Conc.	Effluent Concentrations		Treatment Provided (%)
		Actual	Design			Actual	Design	
Trichloroethylene	580	0.8	23,000	99.9	540	0.015	0.02	99.99
Dichloroethylene	360	ND	2,100	100	39	0.160	0.200	99.6
Vinyl Chloride	7	ND	20	100	1.3	0.007	0.4	99.5
Hydrochloric Acid	0	338	440	-	-	-	-	-

The concentration of HCl generated by the photocatalytic oxidation system does not exceed the allowable HCl discharge criteria. Therefore, the scrubber has not been added to the MPE system at this time.

As shown in Table 2, the liquid-phase carbon removes over 99% of the VOCs in the groundwater. All treated groundwater satisfies the treatment objectives.

Capture Zone. The extraction of groundwater produced a hydraulic capture zone that is roughly circular in shape (see Figure 3).

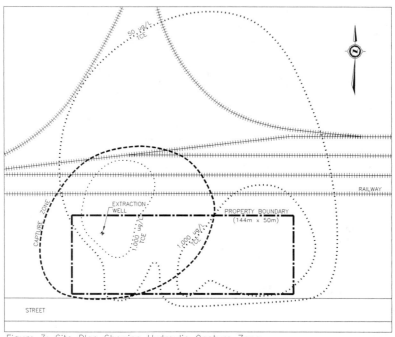

Figure 3: Site Plan Showing Hydraulic Capture Zone

The drawdown at a radius of 50 m from the extraction well was approximately 0.3 m. The drawdown, which was approximately 2 m at the extraction well, decreased relatively uniformly to the fringe of the capture zone.

All of the surficial aquifer groundwater on the western half of the site that contains VOCs at concentrations exceeding the SSRA clean-up guidelines is contained within the capture zone.

It is expected that the size of the capture zone may decrease once the frost cover disappears in the spring.

REFERENCES

United States Environmental Protection Agency (USEPA), "Multi-Phase Extraction: State of the Practice," EPA 542-R-99-004, June 1999.

CROSSHOLE FLOW/THERMAL TECHNOLOGY
USED IN REMEDIATION OF PCE DNAPL

BY *RICHARD C. DORRLER, PG*, ARCADIS GERAGHTY & MILLER,
MAHWAH, NJ; SUTHAN S. SUTHERSAN, PHD, PE, ARCADIS GERAGHTY
& MILLER, LANGHORNE, PA; AND RICHARD E.
KOMOSKI, PE, FORMERLY WITH JOHNSON & JOHNSON, NEW BRUNSWICK, NJ

ABSTRACT: Development of flow pathways between wells facilitates the application of various insitu treatment technologies. These technologies then can be directed at the specific zones targeted for remediation. In this case study, the use of forced hot air injection in combination with standard soil vapor extraction (SVE) in adjoining wells, results in greater contaminant mass removal through increased volatilization. Due to their chemical and physical properties, volatile organic contaminants are more effectively and completely removed after being converted to a vapor phase, with recovery by SVE, as opposed to being left in a liquid phase with recovery by long term pumping and treatment.

SVE is well documented as being a highly effective technology for remediation of volatile organic contaminants (such as BTEX), within permeable soils above the water table. This study was undertaken, in part, to evaluate the feasibility of extending the application of standard SVE technology for remediation of volatile organics to more complex subsurface conditions. Pilot tests recently conducted (1998-99) at a site in the Piedmont of South Carolina, have shown promise with the development of a more advanced technology. The pilot tests used SVE, in combination with hot air injection, to extend the application of standard SVE for the remediation of high concentration PCE/DNAPL in low permeability saprolite soil and fractured rock, at depths of up to 65 feet (19.8m) below the water table.

After testing various types of heating devices, an in-well heater was designed and fabricated based on heat output (>1000°F) and durability (able to withstand the rush of air up to 100 cfm at 60 psi). It was found that heating the zone targeted for remediation to temperatures at or above the boiling point of water and PCE (between 212° and 250°F) dramatically increased mass fluxes in the SVE recovery wells. In less than 30 days PCE concentrations in a primary well decreased 97% from greater than 4,000 ug/L to less than 100 ug/L. The thermal techniques used in these tests created insitu steaming, which increased volatility and decreased subsurface absorption of residual and dissolved PCE. Temperatures of purge water from the primary well remained at or near 240°F for several days and 160°F for several weeks following the test.

The use of hot air injection, in combination with standard SVE techniques, applied simultaneously in multiple wells has been demonstrated to be very effective in rapidly removing high concentration PCE/DNAPL from source areas beneath the water table. The use of multiple wells, as opposed to single well remediation technologies, induces strong pressure gradients between wells and throughout the subsurface zones targeted for remediation. The pressure gradients increase the mobility and recovery of DNAPL in low-permeability zones.

INTRODUCTION

Cleaning up contaminants that have seeped into the ground is very difficult to do. Some of the most pervasive and difficult contaminants to cleanup are the chlorinated organic solvents. These solvents often occur as dense nonaqueous phase liquids (DNAPL). According to an EPA study, as much as 60 percent of the Superfund sites across the country may have DNAPL contamination from chlorinated solvents such as tetrachloroethylene (PCE) and trichloroethylene (TCE). Since DNAPL often occurs in isolated pockets and is immiscible with water, it cannot be removed by simply pumping groundwater and therefore, the contamination can last for an indefinitely long period of time.

Most new and emerging technologies take a hands-off approach by implementing extensive containment measures or installing long-term treatment systems for remediating contaminants that emanate from DNAPL source areas. Since these approaches do not remove the central mass of contamination, they cannot be considered as a final remedy and can only prolong the cleanup efforts.

In this paper we present an alternative remediation approach, using an innovative combination of technologies designed to remove the high concentration DNAPL contamination, directly at its source, where it resides in the ground. This remediation approach utilizes multiple wells, with applied high pressure and vacuum to open pathways between wells, and establishes a crosshole flow condition throughout the subsurface zone targeted for remediation. Once the crosshole flow conditions are established, the formation is dewatered. Next an advanced, forced-hot-air injection technology is utilized to volatilize the remaining DNAPL contamination. The heat is generated with an innovative, in-well heating system known as the FAIR-WELL HEATER (patent pending). After being changed from a liquid phase to a vapor phase, the DNAPL can be more readily recovered by adjacent wells utilizing standard soil vapor extraction (SVE) techniques.

Thus by focusing intensive remediation efforts on the high concentration DNAPL area, the source of the contaminant plume is quickly removed. With the major portion of the mass of contamination removed from the ground, the remaining low level contamination, which is dispersed within the groundwater plume, can be left to attenuate naturally thereby eliminating the need for long-term containment or groundwater pumping and treatment.

DNAPL Dilemma. DNAPLs are especially difficult to cleanup since they are immiscible in, and denser than water. When a spill of DNAPL occurs, the liquid will migrate downward to the water table and continue sinking through the groundwater to the base of the aquifer where it will often accumulate in pools of separate phase liquid. The contaminants will slowly dissolve into the groundwater as it passes through the DNAPL source area, producing a spreading plume of aqueous phase contamination.

Since the primary tools for investigating and recovering contaminants from groundwater are drilled wells, the greater depths and discrete locations of DNAPL make source areas very difficult to find. Also, pumping of groundwater from wells has little to no effect on DNAPL recovery. In fact, there are many who say that DNAPL remediation can not be done. After all, even if you did know where to find it, how do you go about getting it out of the ground? Yet restoration of contaminated groundwater is required under both state and federal regulations. Herein lies the dilemma with DNAPL remediation projects. Since most groundwater is, or potentially could be, a source of drinking water, and as required by regulation, contaminated groundwater must be remediated regardless of the level of difficulty that is involved. The challenge is to figure out the best approach to remediation at your site.

In order to comply with regulatory requirements, many sites have implemented various remedial technologies intended to hydraulically control the plume and/or to treat the contaminants within the plume in such a way as to prevent the contaminants from leaving the site. However, these remedial approaches do more to treat the symptom rather than provide a cure for the cause or source of the contamination problem. Consequently, many sites where pump and treat was seen as a low cost, stop-gap alternative, are now realizing, after many years of operation, that before the pumps can be turned off something still has to be done to cleanup the source of the contamination (i.e. the DNAPL).

DIFFICULTIES WITH DNAPL REMEDIATION

What makes DNAPL so difficult to investigate and remediate is that it takes so many shapes, forms and phases as it seeps vertically downward throughout the entire column of subsurface materials. DNAPL is a sinker. Due to its density being greater than that of water, once it spreads through the vadose zone to the water table, it continues to flow downward through the groundwater to the base of the aquifer. As it is flowing under the influence of gravity, through the subsurface, it is considered to be in a free phase. While still in a free phase it will collect on the top of low permeability materials where it will accumulate in pools and pockets of varying shapes depending on the nature of the subsurface materials and on the overall volume of DNAPL.

As the free phase DNAPL continues along its path, it leaves behind immobile globules and ganglia known as residual phase DNAPL. Since DNAPL is essentially immiscible in water, both the free and residual phases are actually product and these are also considered to be a separate phase, as opposed to that which is dissolved in the groundwater.

Since the bulk of the DNAPL contamination usually exists as a product in either the free or residual phase, it is these phases that should be the main target for remediation. However, since contamination in the DNAPL source area exists in so many different forms and phases, the best remediation approach should include a combination of several innovative technologies. Each innovative technology should have the capability to remediate one or more phases of DNAPL contamination and if designed to enhance each other, a more thorough cleanup of the DNAPL source area can be achieved.

MULTIPLE WELL/CROSSHOLE FLOW TECHNOLOGY

This technology involves the use of a number of insitu cleaning processes directed at recovering the various forms and phases of DNAPL (e.g. free phase, residual, dissolved and vapor phase). These insitu cleaning processes include the following:

1. free phase product pumping
2. dissolved phase/groundwater pumping
3. hi-vacuum liquid/residual extraction
4. hot water flushing, surging, pulsing and back washing
5. hot air flushing
6. steam stripping
7. thermal desorption and volatilization
8. soil vapor extraction

All of these insitu cleaning processes are combined in a unique way by applying them simultaneously in multiple wells, thereby inducing strong push-pull forces between wells. By using multiple wells as opposed to single well remediation technologies, the aggressive cleaning action can be directed crosshole and extended throughout the zone targeted for treatment. This advanced multiple well/crosshole technology is specifically designed to take advantage of the flow characteristics of the formation (i.e. open fractures and pores). The locations and depths of the wells, the patterns of injection and extraction points and the air and water flow rates and pressures are all designed for the specific purpose of mobilizing and controlling the movement of the various phases of DNAPL into the primary recovery streams leading to the extraction wells.

The propagation of the air flow streams is dependent on the prevailing subsurface conditions and can be controlled by the design or layout of the network of remediation wells and by the operation of the air injection process. These design and operational factors will vary from site to site depending on the specific subsurface soil conditions and/or geologic structures and the extent to which they can be developed and utilized as an integral part of the overall contaminant recovery system. For example, in layered soil deposits or horizontally bedded rock formations, air flow pathways can be readily propagated for hundreds of feet. Conversely, it is our experience within saprolite soils that due to the discontinuous and sharply bending relic foliation fractures, air injection rates of up to 100 cfm will be needed to achieve minimal horizontal propagation of air flow pathways to distances of up to 30 feet (9m). Finally, the prevention of over propagation of the air flow pathways can be controlled by intermittent or pulsed air injection; by regulating the air injection pressures; by the placement of surrounding vent and/or vacuum wells to monitor and intercept or short-circuit the air flow streams; and by implementing the air injection in phases or sections (from 1 to 4 wells at a time) while operating the SVE continuously over the entire area targeted for remediation.

Besides taking advantage of subsurface fractures, etc. to develop crosshole flow pathways, the system also uses the chemical properties of the specific contaminant to assist with its recovery. For example, since PCE is highly volatile and has a low solubility, air is used as the primary carrier rather than water. Also, the diffusion of PCE is four orders of magnitude greater in air than water. Thus air is a more effective and thorough cleaning medium than is water for this type of contaminant.

FORCED AIR IN-WELL HEATING SYSTEM

The higher the air flow rates and temperatures that can be delivered to the ground where the contaminants reside, the greater will be the rate of volatilization. And so, the use of forced hot air injection results in greater mass removal through increased volatilization. Every degree of increase in heat greatly increases the vapor pressure of the volatile organic contaminant thereby exponentially increasing their rate of volatilization. Also, the boiling point of many chlorinated solvents is slightly above or below that of water. For example the boiling point for TCE is 188°F and PCE is 250°F. At these temperatures these solvents cannot exist as a liquid but must volatilize.

In order to provide the needed high air flow rates and temperatures to the subsurface zones targeted for remediation, an advanced forced air in-well heating system known as the FAIR-WELL HEATER (patent pending) was recently developed for a site in the Piedmont of South Carolina. Since the low heat carrying capacity of air can be offset by high air flow rates and temperatures, the design of this system was based primarily on convected as opposed to radiant heat. To that end, the FAIR-WELL HEATER was designed and tested, based on heat output of >1000°F at an air flow rate of 100 cfm and a pressure of 60 psi.

Due to the greater volume of air, forced air injection is more effective in opening up many more air channels along existing fractures thereby exposing more contaminant surface area to the vigorous stripping action of the flowing hot air. Therefore, once changed from a liquid to a vapor phase, these contaminants can be more readily carried upward by the rapidly flowing air stream and be recovered by SVE.

MULTI-PURPOSE WELL DESIGN

In order to provide the most flexibility for facilitating the creation of crosshole flow conditions and for accessing contamination within the zones targeted for remediation, a multi-purpose well design was needed. Also, once crosshole flow conditions are established, the multi-purpose well essentially becomes a diversified delivery system for the various insitu (DNAPL) cleaning processes.

There are two primary steps involved in establishing the crosshole flow conditions for insitu DNAPL remediation. These two steps are as follows:

Step 1: Free phase pumping and hi-vacuum dewatering

Step 2: Forced hot air injection and vapor extraction

Depending on the site specific subsurface conditions, there are often many other substeps involved (e.g. hydraulic/pneumatic fracturing), however, these two steps are the primary modes of operation. Also, it is intended in design of the crosshole flow technology to allow the flow between wells to be reversed in order to enhance the cleaning action and contaminant removal. Therefore, the design of the multi-purpose well also must accommodate both modes of operation so that any individual well can be used for either Step 1 or 2 (i.e. injection or extraction).

CASE STUDY

A site located in the Piedmont of South Carolina has been undergoing remediation of PCE contamination since 1985. Initial remediation efforts included groundwater pumping for plume containment and soil vapor extraction for remediation of residual PCE contamination within the shallow saprolite soils. Starting in 1995, remediation efforts focused on the DNAPL source areas at the upgradient end of the site.

First, the DNAPL source areas were carefully delineated, both horizontally and vertically. Residual contamination was found along relic foliation fractures within the saprolite soil, however the bulk of the free phase PCE/DNAPL was found within a transition zone (weathered zone between the soil and rock) and within the fractures in the shallow bedrock. The primary zone targeted for remediation was determined to be at a depth of between 45 to 75 feet (13.7 to 22.9m) below ground surface. Also, an intensive investigation was undertaken to determine the hydraulic properties and to map the geologic structures and subsurface characteristics of the soil and rock. The investigative techniques included oriented split-spoon sampling; rock coring; downhole video logging; isolated straddle packer sampling and pressure response testing.

Following a successful 6 month pilot test, and after obtaining the necessary State approvals, a full scale source area remediation system was implemented to address the DNAPL contamination. Pneumatic fracturing was very effective in opening up dead-end fractures, providing interconnection between wells and greatly increasing the yield of the bedrock aquifer. At some locations, up to 400 psi were needed to create crosshole flow between wells. Up to 16 existing wells were utilized within the 50 by 50 feet (15.3 by 15.3m) DNAPL source area. In many of these wells yields were increased from less than 1 gpm to greater than 10 gpm. To prevent potential drag-down of contaminants from shallow zones, the DNAPL source area was dewatered in stages (see Figure 1). After several months of free phase pumping and hi-vacuum dewatering (Step 1), the main portion of the DNAPL mass was removed and localized drawdowns of up to 80 feet (24.4m) were achieved.

Forced hot air injection and vapor extraction (Step 2) were very effective in removing high levels of contamination since these intensive cleaning steps were focused on the mass of residual PCE contamination within the localized high concentration DNAPL source area. Crosshole hot air/water flushing through fractures was very effective in recovering immobile DNAPL, especially when using air preheated to >1000°F above ambient temperatures. By increasing the temperature of the target zone from 60°F to >250°F, increased PCE vapor concentrations (from <20 ppmv to >400 ppmv), and thus recovery rates by 20 times. In less than 30 days PCE concentrations in one of the primary wells decreased 97% from greater than 4,000 ug/L to less than 100 ug/L. The thermal techniques used in these tests created insitu steaming, which increased volatility and decreased subsurface absorption of residual and dissolved PCE. Temperatures of purge water from the primary well remained at or near 240°F for several days and 160°F for several weeks following the test.

PERFORMANCE EVALUATION

As explained in the Semi-Annual and Annual Effectiveness Reports provided to the state, the overall performance of the DNAPL source area remediation system is based primarily on the amount of PCE contamination removed during the 3 year period of operation (1995-98); and on the subsequent changes in PCE concentrations as measured quarterly and semi-annually in a network of up to 87 monitoring wells installed at the site. PCE concentrations within the remaining hot spots have dropped below the 1 mg/L target level and are now at levels that are equivalent to those within the surrounding areas of the dissolved plume (i.e. hundreds of ug/L).

Also, biweekly monitoring of recovered vapor and water flow rates and PCE concentrations, allowed continuous adjustments and tuning of the injection and extraction points to optimize contaminant recovery. PCE concentration changes over time as measured in the source area extraction wells, have shown a marked decrease from 1995 to 1998. The average concentrations in 1995 ranged from 1 to 7 mg/L, whereas in 1996 the average concentrations ranged from 0.5 to 0.9 mg/L and in 1998 average concentrations are continuing to hold between 0.1 to 0.2 mg/L. Since these concentrations are from extraction wells that are continually pumping groundwater from the PCE/DNAPL source areas, they provide the best indication of the highest concentrations remaining within these areas.

By comparing the total pounds of PCE removed versus the estimate of total pounds released, an accounting of pounds remaining can be made. In this case, the estimated release was 500 gallons or approximately 7,000 pounds of PCE of which approximately 3,500 pounds were recovered by vapor extraction and 1,500 pounds were recovered by groundwater pumping from the shallow soils prior to the start-up of source remediation system. Thus, the remaining 2,000 pounds recovered from the transition zone and shallow bedrock by the source remediation system accounts for the bulk of the remaining PCE that was initially released.

Finally, since some portion of the PCE could be absorbed into the matrix of the soil or rock, there is always some concern that when remediation pumping stops and water levels recover, that concentrations will rebound due to dissolution of the PCE retained in the matrix. To evaluate this potential for rebound, a three month shutdown test was conducted in the recovery wells positioned within the center portion of the DNAPL source area. Despite a full recovery of water levels and complete flooding of the target remediation zone, PCE concentrations have remained below 200 ug/L and are similar to those in the dissolved plume as measured in the surrounding monitoring wells.

CONCLUSIONS

The use of hot air injection in combination with standard SVE techniques applied simultaneously in multiple wells, has been demonstrated to be very effective in rapidly removing high concentration PCE/DNAPL from a source area beneath the water table. The use of multiple wells, as opposed to single well remediation technologies, induces strong pressure gradients between wells and throughout the subsurface zones targeted for remediation. The pressure gradients increase the mobility and recovery of DNAPL in low-permeability zones.

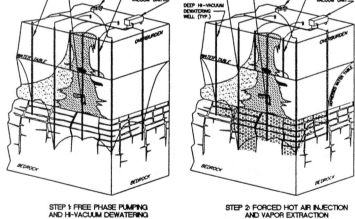

STEP 1: FREE PHASE PUMPING
AND HI-VACUUM DEWATERING

STEP 2: FORCED HOT AIR INJECTION
AND VAPOR EXTRACTION

STAGE 1: SHALLOW REMEDIATION

STEP 1: FREE PHASE PUMPING
AND HI-VACUUM DEWATERING

STEP 2: FORCED HOT AIR INJECTION
AND VAPOR EXTRACTION

STAGE 2: DEEP REMEDIATION

FIGURE 1. TWO STAGE PCE DNAPL SOURCE AREA REMEDIATION
WITH CROSSHOLE FLOW/THERMAL TECHNOLOGY

STEAM STRIPPING/HYDROUS PYROLYSIS OXIDATION FOR IN-SITU REMEDIATION OF A TCE DNAPL SPILL

Gorm Heron, Douglas LaBrecque, Deidra Beadle, Hank Sowers
(SteamTech Environmental Services, Bakersfield, CA)

ABSTRACT: A field demonstration of Steam Stripping and Hydrous Pyrolysis Oxidation (SS/HPO) was conducted at the Portsmouth DOE facility in Ohio. A trichloroethene (TCE) release site had DNAPL located in a semi-confined water bearing zone, and in the upper part of a shale layer. It was demonstrated that steam injection can heat the water-bearing zone directly, and that the TCE can be removed effectively from the layers, including the top of the shale layer located below the steam zone. Both direct vaporization and removal of TCE as a vapor, and chemical oxidation contributed to reduction of the TCE source term. No spreading of the TCE occurred. The study answered the pertinent site-specific questions and the applicability of SS/HPO to this site, and a full-scale clean-up was recommended.

INTRODUCTION

In-situ thermal remediation is a very promising technology for DNAPL removal from soils and groundwater. The thermodynamics and laboratory scale research indicates that heating the subsurface to steam temperature can lead to complete DNAPL vaporization and removal from the subsurface (Udell 1996; Heron et al. 1998a). The traditional mass-transfer limitations and problems with hydraulic contact between injected fluids and the DNAPL are prevented, since energy flow is much more uniform and predictable than fluid flow.

Steam Stripping/Hydrous Pyrolysis Oxidation is a technology where steam and air are injected into the subsurface through wells, and contaminated fluids are extracted for on-site treatment. The fundamentals of the steam injection and extraction technology were borrowed from the enhanced oil recovery industry and further developed under the name of Steam Enhanced Extraction at University of California, Berkeley (Udell and Stewart, 1989), and later patented (Udell et al. 1991). Steam Enhanced Extraction was later combined with electrical heating in the process called Dynamic Underground Stripping (DUS) by Lawrence Livermore National Laboratory and University of California (Newmark, ed. 1994; Newmark and Aines, 1997). This process has also been patented (Daily et al. 1995).

The steam delivers energy to the target area, heating it to steam temperature. Heating the subsurface to steam temperature has been shown to dramatically shift the thermodynamic equilibrium controlling how TCE and other contaminants are stored in soil and water (Heron et al. 1998a; Davis 1997). The vapor pressures of the contaminants are 10 to 100 times higher at the elevated temperatures. Thus contaminants may boil within the soil and will be readily removed by vapor extraction. The Henry's law constant, which describes the tendency of contaminants to vaporize from the dissolved state into soil gases, increases 10-fold for TCE when heating from ambient to steam temperature (Heron et al. 1998b). These effects cause contaminants to become much more mobile at elevated temperatures. Bench-scale studies have shown greater than 99% mass removal of TCE from soils during heating, both using direct steam injection and electrical heating to produce steam within the soil itself (Udell, 1996; Heron et al. 1998c).

In addition to removing contaminants, studies have shown that some contaminants are destroyed in place through a process called Hydrous Pyrolysis/Oxidation (HPO). During steam stripping, HPO destroys dissolved contaminants in place by utilizing hydrothermal oxidation (Knauss et al., 1997). In a heated, oxygenated zone, contaminants are oxidized and degraded to benign products (TCE is converted to carbon dioxide, water, and chloride ions). This process can be stimulated by injection of atmospheric air with the steam, adding oxygen to fuel the reactions.

FIELD DEMONSTRATION: SITE DESCRIPTION AND OBJECTIVES

A field demonstration of Steam Stripping/Hydrous Pyrolysis Oxidation (SS/HPO) was conducted at a TCE DNAPL site at the Portsmouth Gaseous Diffusion Plant (PORTS). PORTS is located approximately 80 miles south of Columbus, in south central Ohio, USA. The site was constructed between 1952 and 1956 and has operated since January 1955 enriching uranium for electrical power generation.

The DNAPL TCE was spilled from a holding pond (see the location on Figure 1) into a relatively permeable sandy aquifer (the lower Gallia, Figure 2) sandwiched in between two low-permeable strata (Minford clays and silts above, Sunbury Shale below).

The groundwater contamination plume extends eastward about 1,900 ft (580 m) from the holding pond. TCE concentrations in downgradient wells are as high as 970 mg/L indicating the presence of DNAPL. TCE moved through permeable sediments (Minford and Gallia Members, Figure 2) and along the top of a nearly impermeable shale layer, the Sunbury shale. Soil sampling during well installation showed that a substantial amount of DNAPL had penetrated the upper one to two feet (0.3-0.6 m) of the Sunbury Shale. This DNAPL, as well as that located within the water-bearing zone in the Gallia, was the target of the remedial effort.

The pilot project treatment area is oval in shape, 120 ft wide and 180 ft long (37 by 55 m), oriented northeast/southwest and is located at the west end of the TCE plume and the X-701B site (Figure 1). The DNAPL extent in the west-east direction is not known. Based on the elevated concentrations in groundwater as far as 1,900 ft (580 m) away from the source, it may be assumed that the DNAPL zone extends beyond the demonstration area. This creates a unique challenge for the field demonstration; groundwater drawn in from the east will likely be highly contaminated during the entire test. Thus, total removal of TCE from the source zone to trace levels was not a practical target, given the chosen target zone and the objectives of the study. Total cleanup of the area was not an objective.

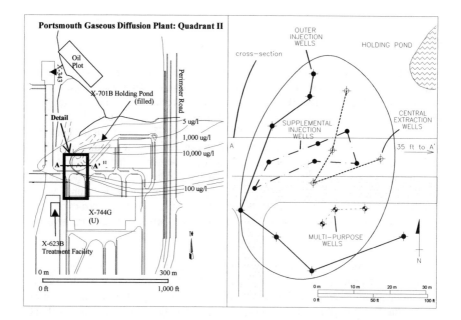

Figure 1. Site map of the TCE plume at PORTS X-701 and a well-field layout for the SS/HPO demonstration at PORTS. The holding pond to the northeast is the source of the contamination.

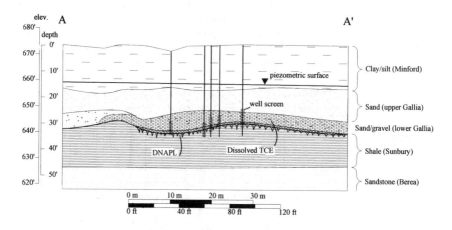

Figure 2. West-east cross-section through the demonstration area with the interpreted DNAPL distribution and location of example wells screened in the lower Gallia.

RESULTS AND DISCUSSION

Implementation overview. The project was conducted in three phases. During Phase I, existing data were reviewed and a conceptual design was made for the demonstration. During Phase II, additional site characterization was conducted, and wells were drilled and completed into the Gallia water-bearing unit, through the Sunbury Shale, and into the deeper Berea Sandstone. The soil samples collected showed that the bulk of the contamination was located within the top of the Sunbury shale. Aquifer tests were conducted in three wells screened in the lower Gallia, resulting in intrinsic permeability estimates in the 5 darcy (5 x 10^{-3} cm/s) range. Phase III involved the final engineering design, installation of additional wells, mobilization of steam and effluent treatment systems, on-site construction, and operation for approximately 4 months. Details of the implementation are given in SES (1999).

Well-field design. Based on the geology and the aquifer test results, it was concluded that heating the site from below was not possible. This would require substantial steam flow into the Berea Sandstone, and upward steam migration through the Sunbury Shale. The layers were much too tight for this to be practical. Thus, it was decided that steam injection into the lower Gallia formation was the best approach. A ring of 7 outer injection wells and four central extraction wells were installed (Figure 1). In addition, due to the relatively low permeability of the target area, three multi-purpose wells were installed. These were used both for extraction and steam injection during different phases of the implementation. Two of the main extraction wells were also used for steam injection in the later stages. Finally, after a cold channel was identified by the detailed monitoring, six additional steam injection wells were added to the well-field, for improved heat-up during the final month of operation (LaBrecque et al. 2000).

Steam injection and extraction rates. Steam was injected into a total of 19 wells at typical rates of 50 to 600 lbs/hr (23-280 kg/hr) per well, with a total injection rate of between 1,500 and 5,500 lbs/hr, averaging 2,500 lbs/hr (1,200 kg/hr). A total of approximately 4 million liters of water was injected as steam. In the same period, 6.5 million liters of water was extracted from the center of the site, ensuring hydraulic control and net extraction. The aggressive vacuum extraction system recovered 1.6 million kg of vapors, at the applied vacuum of between 6 and 18 in Hg (200-600 mbar). Overall, vapors and liquids were extracted very aggressively in order to prevent contaminant migration outward.

The initial steam injection rates were somewhat lower than the values predicted by simple analyses of the aquifer test data. The injection rates were improved by supplemental well development, increasing the steam injection pressures from 12 psig to 16 psig, and by addition of more wells.

Site heat-up and steam movement. The heat-up was monitored in detail using 314 dedicated thermocouples and a network of buried electrodes for Electrical Resistivity Tomography (ERT). The details are presented in SES (1999) and LaBrecque et al. (2000). As predicted, the steam migrated in the lower Gallia, without significant steam penetration directly to the lower Shale or the upper silts and clays. After approximately 2 months of operation, most of the lower Gallia unit had reached temperatures close to that of steam, and TCE was being recovered at the maximum rate.

A cold channel was found by the ERT monitoring, in the central part of the site. This cold zone was caused by a channel of high-permeable sands that allowed rapid water flow from the west of the treatment zone into the area, effectively quenching the steam zones in that part of the site. This situation was remediated by installation of two additional steam injection wells at the western end of the channel, and four supplemental injectors around the cold central extraction well (Figure 1).

After approximately 3 months of operation, the target lower Gallia and the upper two feet of the Sunbury Shale was heated to the desired temperature. After this, one month of cyclic steam injection and air co-injection was conducted in order to accelerate the TCE removal and destruction.

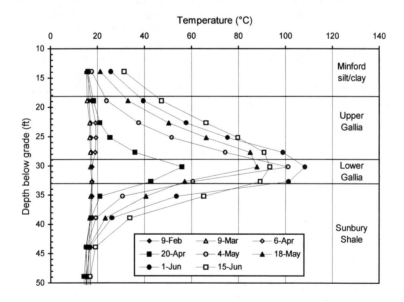

Figure 3. Example of steam breakthrough in the lower Gallia, followed by thermal conductive heating of the surrounding layers.

DNAPL removal from the underlying Sunbury Shale. The detailed monitoring confirmed that the heat-up followed the designed pattern: First the lower Gallia was heated by the steam migration convectively through the most permeable section of the formation, then the underlying shale was heated from the top down by thermal conduction. This is illustrated by the observed temperature profiles in a monitoring location (Figure 3).

The main removal mechanism is TCE DNAPL boiling and vaporization in places heated to temperatures above the in-situ boiling point. Where DNAPL is present, this temperature is approximately 72 °C, named the eutectic temperature (Heron et al. 1998a). At this temperature, the vapor pressure of the water plus the vapor pressure of the TCE add up to one atmosphere, and boiling occurs. For the TCE trapped as DNAPL in the low-permeable shale layer, this is fortunate since the energy travels faster downward by conduction than the liquids do. In other words, the shale layer is heated to above the eutectic temperature, allowing for boiling and upward migration of the evolved TCE and water vapors. As the vapors migrate up and into the steam zone, they are flushed along with the steam towards the extraction wells.

TCE mass removal and HPO reactions. A total of about 400 kg (830 lbs) of TCE was removed from the subsurface during operations. During the last month of operation, air was injected with the steam to stimulate HPO reactions. The carbon dioxide concentrations in the extracted vapors rose from around 330 ppmv (background) to between 1,400 and 2,000 ppmv. This is a strong indication of TCE mineralization, as the produced inorganic carbon leads to equilibrium shifts in the carbonate system, producing carbon dioxide and calcite. The amount of TCE destroyed by HPO was not quantified, since this requires detailed analyses of carbon isotopes and a geochemical balance for the site (e.g. Aines et al. 2000). However, the high carbon dioxide concentrations indicate that the mass of TCE destroyed by oxidation was substantial.

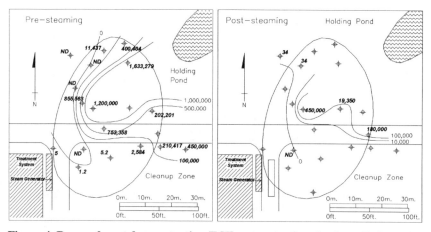

Figure 4. Pre- and post demonstration TCE concentrations in the soils from the most contaminated layer, the upper Sunbury Shale.

Pre-and post TCE distribution. The overall soil concentration of TCE in the most heavily contaminated layer is shown in Figure 4, representing pre-and post SS/HPO sampling rounds. It appears that the target layer has been largely remediated, leaving relatively low TCE soil concentrations behind. The reported levels are well below the pre-demonstration values, supporting the fact that TCE was removed from the layer below the steam zone. Three-dimensional kriging of all the collected soil concentrations indicated that total masses of 550 kg (1,200 lbs) and 115 kg (247 lbs) were present in the demonstration volume pre- and post the SS/HPO demonstration, respectively. These numbers agree well with the recovered TCE mass.

Answers to the key questions. The key questions and answers are as follows:

- *Can SS/HPO be effective in heating the site and removing the TCE?* Yes, both the steaming of the Gallia formation, conductive heating of the upper Sunbury Shale, and stimulated HPO reactions contribute to effective TCE removal from the target layers.
- *Can steam be injected without spreading of the TCE, and how can it be monitored?* Yes, pneumatic and hydraulic control was achieved and applied safely to this site. Downward TCE mobilization was not observed, and ERT and temperature monitoring proved very efficient in measuring steam propagation and identifying problem areas such as the cold channel found by ERT.
- *What are the limitations of the SS/HPO technology at this site?* The main limitations identified was the relatively low steam injection rates achieved at the safe injection pressures. The intrinsic permeability of the lower Gallia water-bearing zone was found to be around 5 darcy (5×10^{-3} cm/sec), which means that the injection wells need to be placed at about 12 m (40 ft) distance, which leads to a predicted total of 150 injection wells for the full-scale clean-up. This is a significant drilling and installation cost.
- *How should full-scale remediation using SS/HPO be conducted, and what is the cost?* The recommended approach is described in SES (1999), involving the proposed well-field, sizing of all major equipment, a schedule and budget.

Overall, it was concluded that SS/HPO is a viable technology for removal of TCE DNAPL from the semi-confined water bearing zone as well as the upper, relatively impermeable Sunbury Shale.

REFERENCES

Aines, R.D. et al. 2000. "Tracking inorganic carbon compounds to quantify in-situ oxidation of polycyclic aromatic hydrocarbons during the Visalia Pole Yard Hydrous Pyrolysis/Oxidation field test". Manuscript submitted to Environmental Science and Technology, in review.

Daily, W.D., A.L. Ramirez, R.L. Newmark, K.S. Udell, H.M. Buettner, and R.D. Aines. 1995. *Dynamic Underground Stripping: Steam and electric heating for in situ decontamination of soils and groundwater.* US Patent # 5,449,251.

Davis, E.L. 1997. *How heat can accelerate in-situ soil and aquifer remediation: important chemical properties and guidance on choosing the appropriate technique*. US EPA Issue paper EPA/540/S-97/502.

Heron, G., T.H. Christensen, T. Heron, and T. Larsen. 1998a. "Thermally enhanced remediation at DNAPL sites: The competition between downward mobilization and upward volatilization". In proceedings of The First International Conference on Remediation of Chlorinated and Recalcitrant Compounds, May 18-21, Monterey, CA. Battelle Press, 1(2), pp. 193-198.

Heron, G., T.H. Christensen, and C.G. Enfield. 1998b. "Henry's Law Constant for Trichloroethylene between 10 and 95 °C". Environmental Science and Technology, 32 (10), 1433-1437.

Heron, G., M. van Zutphen, M.; T.H. Christensen, and C.G. Enfield. 1998c. "Soil heating for enhanced remediation of chlorinated solvents: A laboratory study on resistive heating and vapor extraction in a silty, low-permeable soil contaminated with trichloroethylene". Environmental Science and Technology, 32 (10), 1474-1481.

Knauss, K., R.D. Aines, M.J. Dibley, R.N. Leif, and D.A. Mew. 1997. *Hydrous pyrolysis/oxidation: In-ground thermal destruction of organic contaminants*. UCRL-JC-126636, Lawrence Livermore National Laboratory, CA.

LaBrecque, D.J., G. Heron, X. Yang, D.D. Beadle, and H.J. Sowers. 2000. "The role of advanced monitoring in steam stripping in-situ remediation of DNAPL". This volume.

Newmark, R.L. (ed.) 1994. *Demonstration of Dynamic Underground Stripping at the LLNL Gasoline Spill Site*. Final Report UCRL-ID-116964, Vol. 1-4. Lawrence Livermore National Laboratory, Livermore, California.

Newmark, R.L., and R.D. Aines 1997. *Dumping pump and treat: rapid cleanups using thermal technology*. Lawrence Livermore National Laboratory report UCRL-JC-126637, Livermore, California.

SteamTech Environmental Services. 1999. *Steam Stripping and Hydrous Pyrolysis Pilot Project for the Portsmouth Gaseous Diffusion Plant, Portsmouth, Ohio*. Final report for DOE # DOE/OR/11-3032&D1.

Udell, K.S., and L.D. Stewart. 1989. *Field study of in situ steam injection and vacuum extraction for recovery of volatile organic solvents*. UCB-SEEHRL Report No. 89-2, University of California, Berkeley, CA.

Udell, K.S., N. Sitar, J.R. Hunt, and L.D. Stewart. 1991. *Process for In Situ Decontamination of Subsurface Soil and Groundwater*. United States Patent # 5,018,576.

Udell, K.S. 1996. "Heat and mass transfer in clean-up of underground toxic wastes". In Annual Reviews of Heat Transfer, Vol. 7, Chang-Lin Tien, Ed.; Begell House, Inc.: New York, Wallingford, UK, pp. 333-405.

INDIRECT THERMAL DESORPTION AT THE
LOCKHEED MARTIN SITE

John Dupras, P.E. (Maxymillian Technologies, Pittsfield, MA, USA)

Abstract: The Indirect thermal Desorption System (IDS) is a non-contact, condense and collect technology, with a compact footprint of 70 x 80 feet. A mobile, Series IIIA Water Treatment System accompanies the IDS to treat condensate resulting from the soil treatment operations. Treated water is used to cool and remoisturize the treated soil, thereby providing a closed loop for the process water.

Maxymillian Technologies, Inc. (MT) was contracted by Foster Wheeler Environmental Corporation to remediate 21,342 tons of VOC and PCB contaminated soil at the Lockheed Martin Astronautics Site in Littleton, Colorado, using the transportable Indirect Thermal Desorption System.

Maxymillian Technologies received the Notice to Proceed on August 28, 1998 and mobilized the IDS on-site for soil treatment operations. MT assembled the IDS, performed clean and contaminated shakedown, and completed the Proof of Performance Test by September 16, 1998. The feed rate demonstrated was 18.5 tons per hour (TPH) at a TCE concentration of 2091 ppm. Treated soil complied with site treatment standards and all emissions measurements complied with requirements of the Colorado Department of Public Health and Environment Air Permit for the IDS. MT completed site operations January 15, 1999 after treating 21,342 tons of contaminated soil in four month's time.

INTRODUCTION

Based upon established scientific data and market trends, MT designed and constructed a thermal desorption system with the following design goals:

- Indirectly fired - The desorption chamber is heated by a distinct, indirect-fired source.
- Compact footprint – This mobile system has a footprint of 70'x100'.
- Cost effective - The system will remediate PCB contaminated soils to <2 ppm at a feed rate of 10-20 tons per hour.

The IDS indirectly heats soil in an enclosed rotary drum desorber where contaminants are volatilized from the soil. Remediated soils are passed through a soil remoisturizing and discharge system and are then stockpiled and tested to verify decontamination. Contaminants are both filtered and condensed from the carrier gas and are treated and removed from the liquid stream. Vapors are carried through a HEPA filter for particulate removal and then through a vapor phase carbon system before a second HEPA filter/polymer tray system. An induced draft (ID) fan maintains negative pressure throughout the system during operations. Remediated soils are passed through a soil remoisturizing and discharge system and then stockpiled and tested to verify decontamination.

Waste Applicability. MT designed the IDS to process a variety of soil types and consistencies at a high feed rate. The IDS has successfully treated thousands of tons of soil contaminated with PCBs, volatile organic compounds (VOCs) and semi-volatile organic compounds (SVOCs). Additionally, bench-scale tests have proven the process to be applicable to a wide range of contaminated soil types. The IDS is not limited by soil Btu values or contaminant concentrations and can accommodate feed material with a moisture content of 20%. It has been MT's experience, however, that waste streams with higher moisture contents can be more effectively desorbed with appropriate materials handling prior to thermal treatment.

PROCESS AND EQUIPMENT DESCRIPTION

System Components. The IDS is comprised of multiple components to treat contaminated soils (feed materials) and the associated vapor stream that results from the process. The system consists of:

- materials feed system, including feed screen, conveyor, weigh scale and feed pugmill
- indirectly heated desorber to volatilize contaminants from feed materials
- high temperature baghouse
- soil discharge system
- vapor treatment system, to condense and collect liquid and vapor phase contaminants
- liquid treatment system to collect, and treat contaminants from the liquid stream.

Refer to Figure 1, the Indirect System Process Flow Diagram. The Indirect System components are described in the following subsections.

Material Feed System. Soil is loaded using a front-end loader into a self-contained hydraulic screen/feed hopper. The soil is passed through a 2" bar grizzly-screen, then into a hopper and up a feed conveyor that includes a Ramsey weigh scale (calibrated daily). The conveyor drops soil into another hopper mounted with a feed pugmill that empties soils into the calciner. A soil seal is maintained in the feed pugmill to ensure that ambient air does not enter the indirectly heated desorber. The feed rate into the desorber is monitored from the control room and controlled by variable speed drives. The operator can view the material as it moves up the main conveyor and discharges to the feed pugmill hopper. If necessary, the soil feed into the desorber can be slowed or stopped from the control room.

Indirectly Heated Desorber. From the feed pugmill, contaminated materials are fed through a sealed end plate into the thermal desorber (or calciner). Once in the thermal desorber, the soil is heated to the temperatures required for sufficient desorption of the soil contaminants. No flame comes in contact with the soil during the heating process. Rather, the heat is transferred indirectly from an outer desorber shell to the inner heating chamber containing the contaminated soil.

The desorber consists of an inner rotating shell surrounded by an outer refractory insulated shell. The soil is moved through the 5-foot diameter by 48-foot long rotating shell, which is indirectly heated to the required temperature. Six burners are situated along the outside of the shell fire into the annular space between the inner and outer shells. Heat is transferred through the inner shell providing heat indirectly to the soil. The steel outer shell has been insulated to minimize potential heat loss. Shell temperatures throughout the length of the thermal desorber are carefully monitored and controlled. The burner management system operates the burners in three distinct zones to allow precise heating of soil to required temperatures. Both ends of the inner shell are sealed to prevent ambient airflow or cross-contamination into the system.

Vapor Treatment System. The gas stream containing volatilized contaminants, vaporized water, and entrained particulate, exits the IDS breach and enters the baghouse. The baghouse is capable of filtering high temperature gases to remove particulate from the gas stream. Maintaining a high temperature in the baghouse prevents condensation and allows clean particulate to be removed from the volatilized waste stream. The clean particulate falls into the soil discharge breach where it is combined with clean material exiting the desorber.

The gases, which are now filtered and essentially free of particulate, exit the high temperature baghouse and enter the quench. Gases are first overspray saturated and then further cooled by recirculated water. Water is drawn from the bottom of the quench sump and recirculated through nozzles at the top of the quench to sub-cool the gases below vaporization temperatures. By sub-cooling the gas stream, steam contained in the gas stream is changed to the liquid phase. This liquid accumulates in the quench sump, and is drawn off by gravity, as required to maintain an adequate quench water level. The water that is drawn off from the quench is sent through a filter press then to the liquid treatment system. The IDS is designed to condense and collect the majority of contaminants in the liquid phase at the quench.

Gases leaving the quench are now free of bulk contaminants and contain only trace amounts of contaminants. The remaining air pollution control devices are designed to separate the trace contaminants from the gas stream. Gases leave the quench and enter the condenser. The condenser is a single pass shell and tube design. Non-contact coolant fluid consisting of a water and glycol mixture is supplied by the chiller and routed through the system's condenser. The purpose of the condenser is to reduce the moisture content of the gas stream by further sub-cooling the process gas stream. Any moisture collected from the condenser is routed back to the liquid treatment system. The condenser is designed with an identical standby unit, which can be placed into service without interruption of soil feed, should it be necessary to take the lead condenser off-line.

Upon exiting the condenser, the gases are routed through a pre-filter box into a coalescing filter. The pre-filter box helps to protect and extend coalescing filter life by trapping any trace, very fine particulate that may be suspended in the gas stream. The coalescing filter is designed to trap very fine oil droplets which might be suspended in the gas stream following the quench and condenser. Any mist or droplets collected by the pre-filter or coalescing filter are routed to the liquid treatment system. Filters are removed and replaced from the pre-filter box, as required, and disposed of with other site residuals.

Process gases exit the coalescing filter cabinet and pass through a High Efficiency Particulate Air (HEPA) filter. This HEPA filter is designed to further polish the gas stream and remove any entrained particulate.

Gases now pass through the vapor phase carbon pre-heater. The pre-heater is a non-contact fin and tube heat exchanger that uses a glycol solution to raise the gas stream temperature. The pre-heater is used under cold weather conditions. The pre-heater reduces the relative humidity of the gas stream entering the vapor phase carbon bed, eliminating any moisture carryover into the vapor phase carbon bed, thereby optimizing carbon efficiency.

The vapor phase carbon beds consist of three vessels; each containing 1000 pounds of activated carbon. The vessels are arranged to provide continuous filtering through lead and lag vessels. The third vessel provides a further polish for the backstream.

Upon exiting the final vapor phase carbon vessel, the gases are routed through a HEPA filter to capture any particulate escaping the final carbon vessel. Included in the final HEPA filter cabinet is a polymer bed designed to provide a final gas stream polish in the event that there are trace contaminants not absorbed by the GAC system. The gas stream is moved through the system by the ID fan.

The air pollution control equipment and all interconnecting pipes are grounded against static electricity. The ID fan is constructed of non-sparking materials and has an explosion proof drive system. These safety measures prevent the possible ignition of any vapors in the gas stream.

Process gases exit the Air Pollution Control (APC) devices and are rerouted through the desorber furnace to further control carbon monoxide (CO). Using this system configuration for the Lockheed Martin Site Remediation operations allowed the IDS to comply with the strict CO emission limits contained within the Lockheed Martin Synthetic Minor Air Permit. Compliance with all air emission standards was demonstrated at the exit of the vapor phase carbon vessels prior to the desorber furnace.

Treated Soil Handling System. The decontaminated soil exits the thermal desorber inner shell and enters a soil exit breach hopper. This treated soil, combined with clean particulate removed from the gas stream by the baghouse, travels down the soil discharge chute to the base of the bucket elevator. The soil discharge chute is fully enclosed and sealed to prevent fugitive emissions. The bucket elevator vertically transports the soil to the soil-cooling silo. The elevator is designed to handle a maximum of 50 tons of soil per hour. From the elevator discharge chute, the soil enters the top of the soil-cooling silo. Soil travels vertically down through the silo and is discharged via a double dump valve at the bottom.

The soil-cooling silo has an inside diameter shell of 8 feet and an overall height of 38 feet. The interior of the silo contains several 8-inch diameter carbon steel pipes. Non-contact ambient air is forced upward through the pipes. The heat of the soil is transferred through the pipes to the non-contact air.

The treated materials exit the cooling silo and enter soil cooling and remoisturization components. Water nozzles located in the pugmill provide soil cooling, and use recycled water that has been treated in the water treatment system. The IDS is designed to return original water to the soil: water and contaminants are driven from the soil, the water is cleaned, and then returned to the soil.

Treated materials exit the discharge pugmill via a hooded radial-stacking conveyor to a processed soil stockpile area to await decontamination verification. Water sprays are located in the hooded portion of the conveyor to provide additional dust control, when needed.

Process Control Systems. The IDS is operated from the control room. The system's instrumentation controls and Data Acquisition System (DAS) are automatically controlled and overseen by the operator.

A strip chart recorder, in conjunction with the computer DAS, comprises the system for performing feedback functions to maintain operational control of the IDS.

The strip chart recorder acts as an interface between operational field points and the computer. In addition, the strip chart acknowledges soil feed cutoff conditions for applicable operating parameters. Operational control is maintained through relay outputs. Strip chart functions include:

- Scanning the input data for alarm conditions, annunciating alarms and initiating actions
- Processing data
- Displaying the results on a screen
- Allowing access to the data from the computer.

The computerized control and data acquisition software serves as a data acquisition and management system. Communications with the strip chart recorder occur through an input/output driver software system and a serial communications port. The computer monitors operational field point data and duplicates alarms from the strip chart recorder on to the computer screen. This facilitates visual interpretation of the data. Data is archived on the hard drive. The software package is used to print reports of data, averages, and display trends.

The IDS is equipped with a number of controls to maintain safe reliable and consistent system operations. Monitoring points have been selected to provide applicable regulatory information and process operation data for use by the system operator. Alarms and delays have been incorporated to minimize disruptions in soil feed operations.

Dust and steam from the soil discharge system are the primary potential sources of IDS fugitive emissions. Although both dust and steam from this point are clean, MT implements additional measures to ensure proper operations. Treated materials are discharged from the soil-cooling silo in a controlled manner. Water sprays and de-misters are located within the discharge pugmill and radial stacking conveyor mitigates dust generation and can be used when necessary.

The secondary potential source of fugitive emissions occurs during the desorption process. Negative pressure generated from the ID fan, prevents fugitive gas and particulate emissions. In addition, highly efficient seals are used at the feed and discharge ends of the thermal desorber.

Liquid Treatment System. As previously stated, liquid generated from soil treatment consists of a water and contaminant mixture and is treated using a water treatment system. Prior to treatment, the liquid is stored in an on-site tank. Contaminated liquid is then pumped to the system for treatment in a flocculator. A pH adjustment and flocculant is added to the mixture, prior to the liquid entering the clarifier.

In the clarifier, material settles and is pumped to a filter press onto which a filter cake is formed. The filtrate is returned back to the flocculator. The water then passes from the clarifier through dual particulate filters. MT has positioned the filter so that flow can be directed to one filter, while the other filter is replaced. During the final treatment step, the liquid enters a lead/lag liquid phase granular activated carbon system that consists of two carbon vessels. Each vessel contains 1,500 pounds of granular activated carbon. The clean effluent is recycled to the discharge pugmill for use in cooling and remoisturization of clean discharged materials.

Waste Streams. The IDS maximizes the use of non-contact and closed loop systems for heating and cooling process materials. These closed loop systems isolate the contaminated material, thereby minimizing waste streams requiring treatment or disposal. IDS operations generate four primary waste streams, which are collected, stored and recycled or disposed of properly. Wastes produced from the IDS treatment process include:

1. Filter cake from the filter press

2. HEPA filters

3. Vapor and liquid phase carbon that has been expended

4. Condensed oils.

LOCKHEED MARTIN SITE REMEDIATION - CASE STUDY

MT was contracted by Foster Wheeler Environmental to remediate 21,342 tons of VOC and PCB contaminated soil. MT completed soil remediation operations at the Lockheed Martin Astronautics Site in Littleton, Colorado in less than 5 months.

After receiving the Notice to Proceed on August 28, 1998, it took 15 calendar days to assemble the IDS, perform clean and contaminated shakedown, and complete the Proof of Performance Test (POP). Treated soil and IDS emission measurements surpassed all site treatment standards and requirements of the Colorado Department of Public Health and Environment Air Permit for the IDS. The emissions limits were especially stringent because they were added to existing emissions of the Lockheed Martin facility that required the entire facility to stay within their permit as a synthetic minor source. The IDS on-line factor was 90% and the average feed-rate was 14 tons per hour during the Lockheed project.

Performance Results. During full-scale operational testing, the IDS met or exceeded the established objectives and criteria for thermal desorption and the site· specifications. Following are two tables that identify key parameters measured during the IDS testing phase at the Proof of Performance test for the Lockheed Martin Site Remediation. The first table includes the results of the pre and post-treatment PCB concentration in the soil. The second table characterizes the emissions results during treatment operations.

CONCLUSION

The Indirect Thermal Desorption System is an indirect-fired rotary desorber, with collection of organics in the off-gas by condensation and adsorption. The full-scale treatment system includes:

- Rapid mobilization period due to small footprint and structural mobility
- High feed rate of 10 to 20 tons per hour
- Solid record of successful contaminant removal from soil, gases and water
- Condense and collect technology; non-destructive
- Limited waste products; closed-loop system
- Commercially available, proven technology.

The IDS successfully remediated 21,342 tons of VOC and PCB contaminated soil at the Lockheed Martin Littleton, Colorado site in less than 5 months time. An on-line factor of 90% and a soil feed rate of 14 tons per hour (TPH) resulted in fast and reliable treatment operations. The IDS demonstrated highly effective emissions results, exemplified by a 99.99989% VOC Control Removal Efficiency (CRE).

TABLE 1. Lockheed Martin Site Data

Parameter	Units	Results
Soil Feed Rate	tph	18.5
Treated Soil Exit Temperature	°F	540
Waste Feed Soil TCE Concentration	ppm	2091
Post Treatment Soil TCE Concentration	ppm	ND

TABLE 2. Lockheed Martin Site Emissions Data

Emissions	Test Objective	Flowrate	Test Results
Particulate Matter[a]	1.73 lb/hr	N/A	0.0016 gr/dscf
HCl Emissions	N/A	138 dscfm,177 acfm	<0.00025 lb/hr
SVOCs	N/A	N/A	<0.00028 lb/hr
VOCs	N/A	N/A	0.0283 lb/hr
CO[a]	N/A	N/A	3.4 ppmv
PCB Emission Rate	N/A	141 dscfm [b]	<7.77 E-08 lb/hr
VOC Control Removal Efficiency	93%	N/A	99.99989
THC[a]	N/A	N/A	1.7 ppmv
PCDD/PCDF Emission Rate	N/A	140 dscfm, 179 acfm	9.73×10^{-12} g/sec
PCDD/PCDF Concentration[a]	TEQ <1	N/A	0.034 ng/dscm
PCDD/PCDF TEQ Emission Rate	N/A	N/A	7.27×10^{-15} g/sec

[a]Corrected to 7% oxygen, dry basis.
[b]Stack gas volumetric flow rate.

FIGURE 1.
IDS Process Flow Diagram

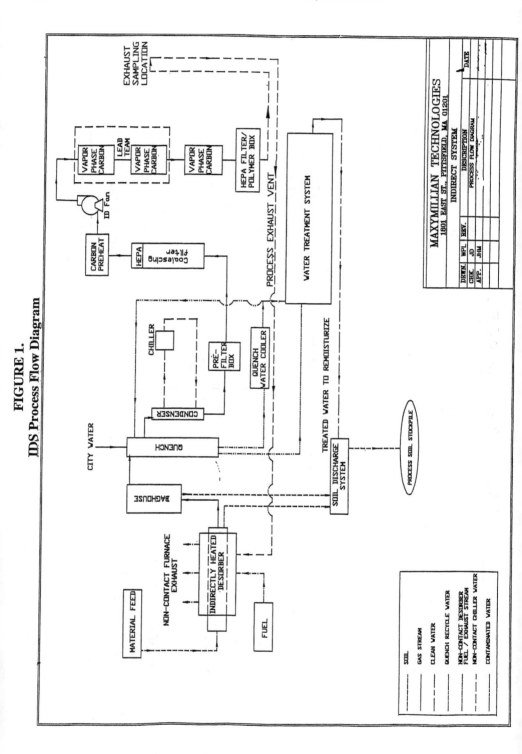

CLOSURE OF TRICHLOROETHENE AND 1,1,1-TRICHLOROETHANE DNAPL REMEDIATION USING THERMAL TECHNOLOGIES

Gregory Smith, (URS Corporation, Rolling Meadows, Illinois)
David Fleming, (Current Environmental Solutions, Bellevue, Washington)
Valdis Jurka, (Lucent Technologies, Inc., Morristown, New Jersey)
Tim Adams, (ENSR Consulting & Engineering, Warrenville, Illinois)

ABSTRACT: A soil and groundwater remediation, cleaning up trichloroethene (TCE) and 1,1,1-trichlorethene (TCA) has been completed, receiving a no further remediation letter from Illinois EPA. The chlorinated aliphatic compounds were present as a dense non-aqueous phase liquid (DNAPL) measuring up to 2.4 m thick in one of the monitoring wells.

A number of techniques were used to accomplish this. Initially, steam with enhanced biodegradation[1] was utilized from 1991 to 1998. This process also incorporated a "huff and puff" mode of operation to optimize removal of the TCE and TCA by vapor extraction. Redox manipulation was also performed to provide optimum conditions for the biodegradation of the residual compounds in the groundwater. From June 1998 to the end of April 1999, six phase electrical resistive heating was used to treat the areas where the steam was being limited by subsurface man-made features.

The remediation was terminated after concentrations of chlorinated aliphatic compounds were reduced below site-specific risk-based criteria, concentrations of chlorinated compounds in the groundwater declined to below detectable levels and vapor concentrations were in the low part per million levels. Post closure monitoring shows no rebound for TCE and TCA, and biodegradation of residual compounds.

The remediation removed or biodegraded approximately 32,000 kg of chlorinated aliphatic compounds. Of this, it is estimated that in the order of 18,000 kg were removed through vapor or groundwater extraction, while 12,000 kg were biodegraded. Active remediation is complete, treatment equipment has been removed and redevelopment is underway as part of a movie theater complex.

BACKGROUND

The work described herein was conducted at a former electronics manufacturing facility in the Chicago, Illinois suburbs. Chlorinated aliphatic compounds in the form of TCE and TCA were discovered in soils and groundwater beneath the plant. Pools of these compounds were discovered in 8 areas, measuring up to 2.4 m thick in monitoring wells. The impacted area represented approximately 1.3 hectares.

The manufacturing facility is located on a lacustrine sequence of sediments, consisting of fine sands, and silts, with some clay lenses to a depth of 5.5 to 6.1 m (18 to 20 feet deep). The lacustrine sediments have hydraulic conductivities measured from 1.05×10^{-5} to 1.23×10^{-4} cm/sec. The lacustrine sequence is underlain by a dense glacial clay till. The

---

[1] U.S. Patent No. 5,279,740

clay till is a silty clay formation known as the Tinley Groundmoraine, which is continuous throughout a significant portion of the Chicago area. The groundmoraine extends to depths of 16.8 m (55 feet). The water table is encountered at 2.1 m below the plant floor.

STEAM AND ENHANCED BIOTRANSFORMATION PROCESS

Remediation operations began in 1991. The initial approach was to utilize the intrinsic biodegradation taking place to the extent practical. Complicating this was the discovery of pools of chlorinated solvents below the water table measuring up to 2.4 m thick. These pools represented non-aqueous phase liquids that are denser than water, commonly known as DNAPL. The dissolved or aqueous phase liquid (APL) potion of the plume was undergoing a natural dehalogenation producing di- and mono-chlorinated ethenes and ethanes. Testing conducted in 1989 to optimize the intrinsic biodegradation resulted in significant production of dichlorinated ethanes and ethenes, from 1,1,1-trichloroethane and trichloroethene, but in the timeframe of the field testing, little mono-chlorinated compounds were observed. It was recognized that the production of the di-chlorinated compounds resulted in generation of chemicals with different chemical characteristics than the parent compounds. These di-chlorinated compounds move more readily with the groundwater, are more volatile than the parent compounds, and as such are more easily removed from the subsurface than the parent compounds. This relationship is illustrated in Figure 1. Figure 1 (modified from Jackson and Patterson, 1989) presents an empirical relationship between the number of pore volumes to attain 90% reduction in concentration through pumping versus the octanol water partition coefficient (presented on a log scale). This biotransformation was used as a means of foreshortening the timeframe for removal of the chlorinated solvents using vapor and groundwater extraction

Figure 1: Estimation of Number of Pore Volumes to Attain 90% Aquifer Decontamination

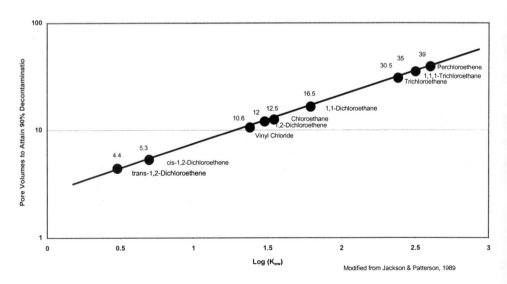

Modified from Jackson & Patterson, 1989

To address the DNAPL, steam injection was utilized based on the work of Hunt, et al., (1988). The steam injection was performed simultaneously with the enhancement of the biotransformation to accelerate the rate of removal. Figure 2 illustrates the hydrocarbon removal over time for the remediation. It can be seen that the hydrocarbon removal rate was initially very high and then rapidly declined, consistent with conventional vapor and groundwater extraction techniques. The effect of the steam injection in increasing the rate of mass removal is observed beginning in July 1991 and peaking in June 1992. Modifications to system operations occurred in April 1992, April 1993, and January 1994 involving modification of well arrangements and additions of wells to improve system operations.

Figure 2: Hydrocarbon Removal Data

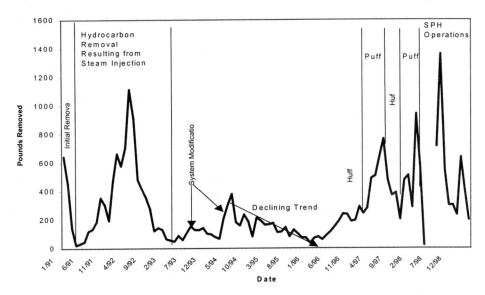

Beginning in December 1994 and continuing through January 1996, it was observed that there was a generally declining trend in hydrocarbon removal. However, the mean concentration of trichloroethene ranged between 7 to 37 mg/l (Figure 3), mainly based on high concentrations in 7 wells. At this time, modifications to existing operations were evaluated and a risk-based closure procedure under 35 IL Administrative Code Part 742 was initiated. This procedure resulted in closure of a portion of the site remediation operations, receiving a "no further remediation" letter from Illinois EPA in April 1998 (Smith, et. al., 1998a).

"HUFF AND PUFF" MODE OF OPERATION

Udell and Itamura, (1998) describe the theory behind the increases in mass removal through vapor extraction which can achieved through reducing the water content in porous media under steam injection. Operationally, this involves dewatering the porous media to facilitate introducing steam. After pressure buildup is achieved (huff cycle), the treatment zone is dewatered to facilitate mass removal through vapor extraction (puff cycle).

Dewatering for the first puff began in early 1996 and continued to September 1996. Steam injection took place from September 1996 to January 1992, whereupon a second puff cycle was initiated. Figure 2 shows the increases in hydrocarbon removal achieved for the puff cycles. An additional huff took place from July 1997 to November 1997, with the subsequent puff taking place from December 1997 through March 1998. The system was shut down for the installation of the six phase treatment system (briefly described below).

Figure 2 shows the increases in hydrocarbon removal that can be achieved using this mode of operation. Further, this mode of operation reduced the quantity of water requiring treatment. The treated groundwater was disposed to the local POTW, which involved a user charge. This mode of operations reduced user charges.

Prior to initiation of the huff and puff mode of operation, hydrocarbon removal rates ranged between 60.4 and 176.8 kg per month in 1995. After the huff and puff mode of operations took place, the removal rates averaged 229.3 kg per month. Based on average rates of removal, the huff and puff mode increased the rate of removal by 60%. It is estimated that this would a resulted in a time savings of 22 months (if the huff and puff mode was initiated in the fall of 1992, when removal rates had been observed to have significantly declined) to achieve the same total mass removal to March 1998. Therefore efficiencies can be realized in the operation of steam-based remediation. This mode of operation was applied to a second steam based remediation (Adams, et. al., 1998) to achieve greater efficiency.

SIX PHASE ELECTRICAL RESISTIVE HEATING

As mentioned above, 7 wells were responsible for high average concentrations at the end of 1998. Investigations in the areas where the wells were located determined that the steam injection was ineffective for DNAPL removal as a result of subsurface man-made features. These features included a subsurface void associated with an abandoned closed-end catch basin; preferential steam flow along backfill material around building foundations; and, compressed native clay, limiting the removal of vapors. The compressed clay was believed to be a result of the proof-rolling of native soils prior to backfilling for the construction of the building. These conditions required a change in the mode of operations to complete the site remediation closure.

Based on the demonstrated effectiveness of thermal methods in DNAPL removal for this project, alternate thermal methods were evaluated to treat these remaining areas. There was a deadline to facilitate site redevelopment, such that implementability, time to achieve cleanup, compatibility with existing treatment equipment and the treatment systems impact on existing structures (which were to remain) were evaluated. Six Phase Heating (SPH) was selected based on these criteria. More detail on SPH treatment is provided in Beyke, et. al. (2000a) and Beyke, et, al. (2000b).

SPH treatment began on June 4, 1998 and continued to December 1998. In December 1998 it was decided, (based on the successes achieved with SPH) to expand the treatment area to provide against the degradation of residual concentrations of chlorinated aliphatic compounds resulting in the production of vinyl chloride above the cleanup criteria. This was done to ensure that after the active treatment was complete, the treatment equipment could be removed and site re-development could take place.

SPH treatment resulted in the removal of 3,940 kg of solvents, and resulted in the reduction of the mean concentration of TCE from 4.94 mg/l in March 1998 when the system was installed to 0.037 mg/l in May 1999 after active treatment was discontinued. Mean TCA concentrations reduced from 4.37 mg/l to 0.002 mg/l over the same period.

REDOX MANIPULATION

In 1996, experimentation was begun into the use of oxidants to determine if this represented a viable treatment technology to foreshorten the remedial timeframe. Bench scale testing using chlorine dioxide determined that no significant concentration reductions were observed in the groundwater in test cells. However, there were significant reductions observed in the soil concentrations. It was hypothesized that oxidants react with soil mineral surfaces before reacting with organic compounds (Smith, et. al., 1998b). The benefit to this was that it increased partitioning into the groundwater, which would allow for removal through conventional groundwater pumping. It was believed that because of the partitioning of chlorinated solvents to soils and groundwater, modification of soil mineral surfaces to promote partitioning to groundwater would be advantageous to achieve groundwater cleanup.

Chlorine dioxide was used at 26 wells to facilitate partitioning of the TCE and TCA into the groundwater promoting the removal of these compounds through groundwater pumping. This treatment was conducted in a series of tests beginning in April 1996, continuing through March 1997. During this period, hydrocarbon removal through both vapor and groundwater extraction increased. A more thorough discussion is found in Smith et. al., (1998b).

REMEDIATION CLOSURE DATA

Hydrocarbon removal was tracked using a total hydrocarbon analyzer for the air stripper and vapor extraction system throughout the remedial operations. The quantity of hydrocarbon remediated through biodegradation had not been tracked during the remedial operations, but can be estimated based on the chloride concentrations in the water over time, and the volume of extracted and treated. Further, the residual chloride concentrations in groundwater at the completion of active remediation can be used to estimate the degraded TCE and TCA in groundwater. Chlorinated aliphatic hydrocarbon has also been removed as a DNAPL present in the sludge removed from the treatment system. This quantity is more difficult to estimate, since disposal acceptance samples are taken infrequently. Estimations of the quantities of hydrocarbon removed through the various mechanisms are presented below.

Table 1: Estimated Mass of Hydrocarbon Removed (in kilograms)

	Steam Process	SPH	Totals
Groundwater Pumping	5,420	500	5,920
Vapor Extraction	10,400	2,140	12,540
Biodegradation	9,110	1,310	10,420
Totals	24,930	3,940	28,880

An estimated 581 tons of soil cuttings and sludges were also removed and disposed offsite over the eight years of operation. The sludges contained concentrations of TCE and TCA requiring incineration. The soils were landfilled, or thermally desorbed and landfilled. It is expected that this soil and sludge represented an estimated 5% of the total mass of chlorinated solvents removed. Therefore, the total quantity of TCE and TCA remediated is approximately 32,000 kg. This translates into 116 drums of solvent over the thirty-year operating period of the plant, representing a loss of less than one-third of a drum per month. Given the quantity of solvents being used by the plant, this rate of loss would not likely have been recognized through inventory tracking.

Figure 3 shows the average concentrations versus time for trichloroethene and 1,1,1-trichloroethane. It can be seen that the mean concentrations declined initially, but increased in April 1995. The peak mean concentration observed on April 1996 was due to one well having extremely high concentrations, which later declined. The period from December 1994 to January 1996, where declining removal rates were observed, showed fairly consistent mean concentrations. Further, when huff and puff operations were taking place (early 1996 through March 1998), the mean concentrations remained fairly consistent. The greatest reductions is mean concentration occurred when the DNAPL removal operations focussed on identified source areas. This observation shows a dichotomy in groundwater remediation: the measure of success is based on concentration in groundwater, yet the means to achieve that success is through removal of contaminant mass. It appears that for sites where DNAPL is present, there is no relationship between concentration reduction and mass removal until a mass removal threshold is achieved.

Figure 3: Mean TCE and TCA Concentrations

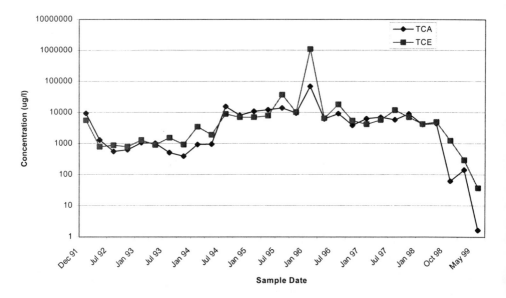

According to Wiedemeier, et. al., 1996, each milligram of dissolved organic carbon that is oxidized via reductive dehalogenation requires the consumption of 5.65 milligrams of organic chloride. As such this establishes the organic carbon demand to be satisfied for reductive dehalogenation to occur. Using this relationship, the authors determined the organic carbon demand for the residual concentrations of chlorinated aliphatic hydrocarbon at each well. Organic carbon demand ranged from 0.001 mg/l to 1.4 mg/l (where there were detectable concentrations of chlorinated aliphatic compounds), while organic carbon concentrations ranged from 13 to 1,600 mg/l. This indicates ample concentrations of organic carbon available to promote the breakdown of the residual concentrations of chlorinated aliphatic compounds.

According to Bouwer, (1994), optimal conditions for the reductive dehalogenation of chlorinated aliphatic compounds takes place under reducing conditions, with Eh ranging from –220 to –240 millivolts. The average Eh measured for the May 1999 groundwater was –252.6 millivolts, with a range of –346 to –131 millivolts. Throughout most of the remediation area, redox conditions are optimal for the reductive dehalogenation of the residual chlorinated aliphatic compounds.

POST CLOSURE MONITORING

At the time of preparing this paper, the post closure baseline sampling (May 1999), and a sampling in December 1999 was conducted. This is presented in Table 2. Two of these wells (Ba3 and D9) showed significant increases in vinyl chloride, while the third well (F9) showed a minor increases in cis-1,2-dichloroethene and 1,1-dichloroethane. No rebound is observed for TCE and TCA. Five wells are below Illinois Class II criteria. Four of these five wells flowed separate phase DNAPL for a period of up to 2 months during a period of the remediation.

Table 2: Summary of Post Closure Monitoring

Well	Redox (millivolts)		Concentration (mg/l)							
			TCE		TCA		Cis-1,2-DCE		VC	
	5/99	12/99	5/99	12/99	5/99	12/99	5/99	12/99	5/99	12/99
Aa6	-257	-297	1	ND	ND	0.007	1.1	0.005	0.064	ND
Ba3	-253	-226	ND	ND	ND	ND	0.39	0.15	0.002	2.9
D3	-269	-185	0.032	ND	ND	ND	0.003	0.019	ND	ND
D7	-242	-185	0.25	ND	ND	ND	1.3	0.16	0.012	ND
D9	-282	-147	0.001	0.001	ND	ND	0.3	0.14	0.01	1.5
E5	-266	-204	0.004	ND	ND	ND	0.74	0.43	0.005	ND
F9	-238	-214	0.016	0.006	ND	ND	0.76	0.9	0.2	0.08
F13	-238	-184	0.001	0.003	ND	ND	0.012	0.026	0.006	0.018
G3	-261	-242	0.012	ND	ND	ND	0.46	0.012	ND	ND
Ga8	-281	-192	ND	ND	ND	ND	0.15	0.027	0.033	0.027
Ga13	-213	-211	0.019	0.003	ND	ND	0.09	0.01	0.095	0.72
Ja4	-216	-208	ND	0.002	ND	ND	0.18	0.094	0.072	0.33
Ja9	-237	48	ND	ND	ND	ND	0.1	0.053	0.019	0.007

ND = not detected

SUMMARY AND CONCLUSIONS

A remediation of TCE and TCA DNAPL (with pools measuring up to 2.4 m thick) has been completed, receiving 2 no further remediation letters from Illinois EPA. The closure of remedial operations took place using a risk-based closure procedure per 35 Illinois Administrative Code Part 742.

Thermal methods (steam injection and Six Phase Heating) have been proven effective for DNAPL removal. No rebound of TCE or TCA has been observed.

Optimal conditions for bio-attenuation of residual concentrations of chlorinated aliphatic compounds exist. This is corroborated by the post closure monitoring data, showing significant reductions, and rebound only in the di and mono-chlorinated compounds.

REFERENCES

Beyke, G. B., G. J. Smith, and V. Jurka (2000a). "Application of Six Phase Heating to Achieve Closure at a Trichloroethene and 1,1,1-Trichloroethane DNAPL Site." *Proceedings of the Second International Conference on Remediation of Chlorinated and Recalcitrant Compounds, Monterey, California, May 17 – 20, 2000.*

Bouwer, E. J. (1994) "Bioremediation of chlorinated solvents using alternate electron acceptors". In *Handbook of Bioremediation:* (Norris, R. D., Hinchee, R. E., Brown, R., McCarty, P. L., Semprini, W., Wilson, J. T., Kampbell, D. H., Reinhard, M., Bouwer, E. J., Borden, R. C., Vogel, T. M., Thomas, J. M., and Ward, C. H., Eds.), Lewis Publishers, Boca Raton, FL, p. 149-175.

Hunt, J. R. N. Sitar, and K. S. Udell (1988). "Nonaqueous Phase Liquid Transport and Cleanup 1. Analysis of Mechanisms." Water Resources Research, Vol. 24, No. 8.

Jackson, R. A. and R. J. Patterson (1989). "A Remedial Investigation of an Organically Polluted Aquifer." Groundwater Monitoring Review 3(ix).

Smith, G. J., T. V. Adams, and V. Jurka (1998a) "Closing a DNAPL Site Through Source Removal and Natural Attenuation". ." *Proceedings of the First International Conference on Remediation of Chlorinated and Recalcitrant Compounds, Monterey, California, May 18 – 21, 1998.*

Smith, G. J., B. Dumdei, and V. Jurka (1998b). "Modification of Sorption Characteristics in Aquifers to Improve Groundwater Remediation." *Proceedings of the First International Conference on Remediation of Chlorinated and Recalcitrant Compounds, Monterey, California, May 18 – 21, 1998.*

Beyke, G., G. J. Smith, and V. Jurka (2000b) "DNAPL Remediation Closure with Six Phase Heating". *Proceedings of the Second International Conference on Remediation of Chlorinated and Recalcitrant Compounds, Monterey, California, May 17 – 20, 2000.*

Udell, K. S. and M. T. Itamura (1998). "Removal of Dissolved Solvents from Heated Heterogeneous Soils During Depressurization." *Proceedings of the First International Conference on Remediation of Chlorinated and Recalcitrant Compounds, Monterey, California, May 18 – 21, 1998.*

Wiedemeier, T.H., M.A. Swanson, D.E. Moutoux, E.K. Gordon, J.T. Wilson, B.H. Wilson, D.H. Kampbell, J.E. Hansen, P. Haas, and F.H. Chapelle (1996). *Technical Protocol for Evaluating Natural Attenuation of Chlorinated Solvents in Groundwater (Draft, Revision 1).* Air Force Center for Environmental Excellence, Brooks Air Force Base, San Antonio, TX.

FIELD SCALE IMPLEMENTATION OF IN SITU THERMAL DESORPTION THERMAL WELL TECHNOLOGY

Denis M. Conley, Senior Scientist, Haley & Aldrich Inc., Rochester, NY
Christopher M. Lonie, P.E. Naval Facilites Engineering Command,
Port Hueneme, CA

Abstract: In Situ Thermal Desorption (ISTD) technology using Thermal Wells was implemented at the former Naval Facility Centerville Beach (NFCB), Ferndale, CA. The project was conducted from September 1998 through February 1999 by TerraTherm Environmental Services as part of the facility restoration program managed by the U.S. Navy Engineering Field Activity West (EFA-West), San Bruno, CA. The ISTD Thermal Well technology deployed at the site used electrical heating elements suspended within stainless steel well casings to raise the temperature of soils in situ and remove targeted contaminants by volatilization and/or in situ destruction reactions. Process gases and contaminant vapors created during heating were drawn through the hot soil to an array of soil vapor extraction wells. The extracted contaminant vapors received secondary and tertiary treatment within a flameless thermal oxidizer (FTO) and/or adsorbed by vapor phase granulated activated carbon (GAC). Subsurface soils at the NFCB were impacted with polychlorinated biphenyls (PCBs) identified as Aroclor 1254 by USEPA Method 8080. A network of 57 Thermal Wells was installed to a depth of 17 feet (5.2 m) within and around the area of known contamination. After achieving an in situ soil temperature of >600°F, the treatment system was shutdown and soils were allowed to cool prior to confirmation sampling activities. The results of post treatment soil sampling indicated that the target treatment zone contaminant concentrations were reduced to <1.0 mg/kg total PCBs and exhibited a 2,3,7,8 TCDD toxicity equivalent quotient (TEQ) of <0.1 µg/kg.

INTRODUCTION

The Naval Facility Centerville Beach (NFCB) is located on County Road 100, in Ferndale, Humboldt County, California. The NFCB was operated as a research facility by the Department of the Navy, West Division, from 1956 to 1993. Naval operations at the facility included oceanographic research of the northern Pacific coastline. As a contingency to local shore power, the facility was equipped with a back-up diesel powered generator and 1,000 KVA electrical transformer station.

Contaminants of Concern: Environment contaminants of concern detected within overburden soils at NFCB included polychlorinated biphenyls (PCBs) from an apparent release of electrical transformer fluids, and petroleum hydrocarbons from underground storage tank releases. The former transformer area had undergone significant investigative and remedial activities. The sampling and analysis performed during the investigative activities indicated that elevated levels (>500 mg/kg) of PCBs (Aroclor 1254) were present in the soil beneath the building foundation (Radian 1997). Data also indicated that PCB concentrations >2.0 mg/kg extended to depths of approximately 15 feet (5 m) beneath the former transformer pad location (Leedshill-Herkenhoff, 1991).

Site Geology: Overburden soils at the NFCB consist of terraced silty and clayey collovial materials that vary in thickness across the site. The terraced deposit resides above the Pleistocene Hookton formation which is comprised of indurated sand, gravel with minor silt and clay layers. The Hookton formation is orange and yellow in color and is greater than 100 feet (33 m) in thickness (Dames & Moore, 1985).

Site Hydrogeology: Depth to groundwater in the vicinity of former transformer pad location had been measured from 99 to 145 feet (33 to 48 m) below ground surface (bgs) while perched water had been detected as shallow as 6 feet (2 m) bgs (Radian, 1997).

MATERIALS AND METHODS

Thermal Well System: The ISTD Thermal well system was designed based on the geophysical properties of the site soils, and the physical properties of the contaminants present. The thermal vapor extraction well casings and thermal well heaters were constructed of corrosion resistant metallurgy. Stainless steel wire wrapped extraction well screens were installed from 2 feet to 15 feet (0.6 to 5 m) bgs to promote efficient recovery of process gases during heating. Thermal Wells were installed using standard drilling techniques on a grid of equilateral triangles spaced 6 feet (2 m) apart forming a hexagonal pattern. The thermal vapor extraction wells were located at the center of each hexagonal pattern of heater wells. The spacing between each extraction well was approximately 10 feet (3 m). Surface completions of the thermal wells used an assembly constructed of carbon steel that was mechanically attached to a vapor barrier consisting of a high temperature resistant coated fabric. Insulated manifold piping was installed and connected to the thermal vapor extraction wells using brass sealed flanges. Inconel sheathed Type K thermocouples were placed within the heaters to monitor the temperature of the heating elements. Thermocouple arrays were also installed at numerous locations between wells within the treatment zone to record in situ soil temperatures. The vapor extraction manifolds were interfaced with a common header leading to the process and control system for secondary and tertiary treatment prior to atmospheric discharge. The manifold piping and common header were equipped with magnehelic gauges to monitor applied vacuum. The integrated process and control system, constructed on a 40 foot (13 m) flatbed trailer, was delivered to the site by a commercial tractor trailer rig.

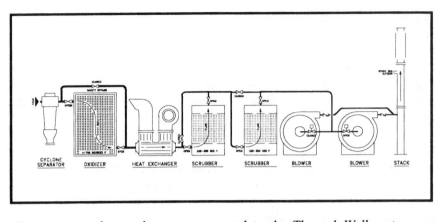

The process and control system connected to the Thermal Well system was designed to control electrical energy applied to the heating elements; monitor contaminant vapor and process gas concentrations; collect and treat vapors extracted from the treatment area; and to provide emergency power in case of loss of shore power.

Control System: - The thermal well and Air Quality Control (AQC) systems were monitored by a programmable logic controller (PLC) located within the control room of the process trailer. The PLC displayed the operating status of each thermal well heating element and the AQC system components through a personal computer. The personal computer provided data logging and graphing capabilities for trend analysis and assisted in scheduling of preventative maintenance on system components.

Process Vapor Extraction & Treatment: Two (2) sealed vacuum blowers, configured in parallel, were used to maintain negative pressure on the soils and the AQC system components. The vacuum blower system pulled the process gas stream through the secondary and tertiary treatment processes to prevent uncontrolled fugitive emissions. Secondary treatment was provided by a flameless thermal oxidizer (FTO) with a demonstrated treatment efficiency of <99.99%. Two (2) granulated activated carbon (GAC) beds, configured in series, provided tertiary treatment.

Air Emissions Monitoring: The process effluent was monitored using a calibrated continuous emission monitoring (CEM) system manufactured by Rosemount Analytical. The CEM utilized an extractive sample probe and conditioning system. The sample stream was introduced to a non-destructive infrared analyzer for the quantification of carbon monoxide (CO) and carbon dioxide (CO_2) prior to the analysis of oxygen (O_2) using a zirconium oxide detector, and total hydrocarbons (THC) using a flame ionization detector. The data was stored using computer software designed for the project and could be retrieved remotely.

Stack emissions were sampled following EPA Methods and analyzed to demonstrate compliance with local air permit criteria. Emission samples were collected from sample ports located in the stack and were representative of the final atmospheric discharge.

Electrical Power Generator: In the event of an interruption of shore power, an emergency generator was available to operate the vacuum blowers and FTO to assure that vapors were processed through the secondary and tertiary treatment.

RESULTS AND DISCUSSION

Final approval of the project work plan by the Regional Water Quality Control Board, California EPA, and Humboldt County Dept. of Health was received in late July and early August 1998. Prior to the installation of the Thermal Well system, pre-treatment soil sampling was conducted in early September, 1998 by TetraTech EMI. In mid September, TerraTherm installed the ISTD thermal well system utilizing local skilled labor for site construction. By early October, TerraTherm installed the process and control system for remedial phase activities. The integrated system shakedown was conducted from 10-16 October 1998 by performing Thermal Well system control checks, CEM calibration, and establishing daily system operations procedures. The soil heating was conducted from November 1998 through January 1999. Interim soil sampling was conducted in early February with the system shutdown on February 26, 1999.

Project Performance Data: Performance data collected and analyzed during the project consisted of soil sampling and analysis data; process monitoring information including in situ and process vapor temperatures; emissions monitoring information including stack testing results, and CEM system data; and ambient air monitoring data.

Soil Data: Soil samples were analyzed for total PCBs using EPA Method 8082, and PCDD/Fs by EPA Method 8290 (USEPA, 1986). Pre-treatment analyses indicated the presence of PCB Aroclor 1254 at a concentration of >500 mg/kg. Pre-treatment concentrations of polychlorinated dibenzodioxins/furans (PCDD/Fs) indicated a 2,3,7,8–TCDD Toxicity Equivalent Quotient (TEQ) of >1.0 ug/kg for the PCB impacted soils under the building foundation (TTEMi, 1998). Post-treatment soil sampling was conducted by TetraTech EMI in early April 1999. Discreet soil samples were collected using a hydraulic drive rig equipped with a stainless steel macro-core (3.0 inch; 7.6 cm interior diameter) soil sampler. Post-treatment soil analysis data indicate that the total PCB detected was

<1.0 mg/kg, and the total PCDD/Fs detected was <0.10 ug/kg 2,3,7,8 TCDD TEQ within the target treatment zone. (TTEMi, 1999)

In Situ Temperature Monitoring Data: The boiling point of soil pore water (212°F) was reached at 15 feet (5 m) bgs in approximately 7 days of heating. The average soil temperature at the center of each triangle of thermal wells exceeded 212°F at approximately 25 days from the application of full power to the thermal wells. The soil temperature increased at a rate of approximately 20°F per day. When heating ceased, soil temperatures ranged from 675°F (7 feet; 2.1 m bgs) to 950°F (15 feet; 5 m bgs) at the thermocouple array placed in the center of the target treatment area (TC-6).

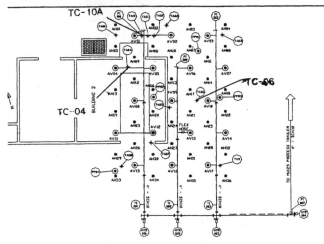

Figure 1.   Thermal Well and Thermocouple Layout

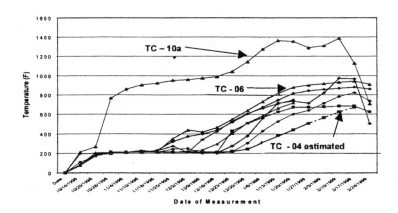

Figure 2. Thermocouple Measurements – NFCB ISTD Demonstration

Emissions Data: Emissions data collected during the execution of the project included: effluent CO, CO_2, O_2, THC concentrations, and source testing for total PCBs, PCDD/Fs, hydrogen chloride, & total particulates. Carbon monoxide (CO) emissions were monitored by the CEM system. Generally, CO emissions were below 10 ppmV with a 3 minute lag throughout the soil treatment. The mean concentration was approximately 2 ppmV. Carbon dioxide (CO_2) emissions were recorded by the CEM system and were observed generally at <2.0%. The concentration was generally higher during the initial heating of the soil with a gradual decrease as the target treatment temperature (600 °F) was achieved. Excess oxygen (O_2) as indicated by the percent of oxygen in the system emission measured on a wet basis was at or above 7%. Relative humidity (RH) of the process vapor stream peaked of approximately 45%. The RH dropped to a low of approximately 15% as the in situ soil thermocouples achieved the target treatment temperature of >600 °F. Total hydrocarbon (THC) emissions were monitored continuously by the CEM system using a flame ionization detector (FID) calibrated to methane (CH_4). THC readings observed during the treatment were generally below 10 ppmV or <0.005 lb/hr as CH_4. Exhaust temperature was monitored by a thermocouple placed in the CEM sample port and recorded by the PLC. The temperature was maintained at approximately 200 °F through adjustment of the heat exchanger flow. The GAC beds were maintained at >220 °F to minimize the formation of condensation within the vessels. Source testing was performed by an approved California Air Resources Board independent contractor following USEPA sampling procedures.

TABLE 1.   Independent Stack Testing Results – NFCB ISTD Demonstration

Parameter	Run#1	Run#2	Run#3	Average
PCBs	**(ng/dscm)**	**(ng/dscm)**	**(ng/dscm)**	**(ng/dscm)**
Dichlorobiphenyl	8.74E-04	3.52E-03	4.90E-03	3.10E-03
Trichlorobiphenyl	6.01E-04	<5.42E-04	<5.44E-04	5.62E-04
Pentachlorobiphenyl	2.13E-03	<5.42E-04	<1.14E-03	1.27E-03
				Average
PCDD/Fs (TEQ)	**(ng/dscm)**	**(ng/dscm)**	**(ng/dscm)**	**(ng/dscm)**
Total 2,3,7,8 Dioxin	2.96E-03	2.88E-03	2.88E-03	2.89E-03
Total 2,3,7,8 Furans	2.50E-03	2.58E-03	2.61E-03	2.56E-03
Total 2,3,7,8-TCDD Equivalents (TEQ)	5.47E-03	5.46E-03	5.44E-03	5.46E-03
				Average
Particulates	**(gr/dscf)**	**(gr/dscf)**	**(gr/dscf)**	**(gr/dscf)**
Total PM 10	1.72E-03	2.15E-03	2.16E-03	2.01E-03
				Average
Hydrogen Chloride	**(lb/hr)**	**(lb/hr)**	**(lb/hr)**	**(lh/hr)**
HCl Vapor	5.8E-03	2.18E-03	6.31E-03	4.76E-03

Sampling methods included Methods 1, 2C, 3 & 4 for the determination of stack gas velocity, flow, temperature, molecular weight and moisture content (USEPA, 1988) respectively. Isokinetic sampling using an EPA Method 005 sampling train was conducted for the measurement of particulate matter, hydrogen chloride, total PCBs, and PCDD/Fs isomers. Samples were shipped under chain of custody for off-site analyses conducted by a CalEPA certified air testing laboratory. The PCB and PCDD/Fs isomers were collected upon an XAD-2 sorbent trap in accordance with EPA Method 23. Aqueous condensate was collected within glass impingers prior to the sorbent trap. The sorbent trap and impinger condensate samples were submitted for analysis by High Resolution combined Gas Chromatograph /Mass Spectrometer (HR GC/MS) in accordance with EPA Method 8290.

Ambient Air Monitoring: Two ambient air monitoring procedures were conducted throughout the project to ensure protection of on-site worker health and safety, and the environment. Initial survey monitoring was conducted using a handheld Model 503B Organic Vapor Monitor (OVM) manufactured by Thermo Instruments. Surveys were conducted approximately every four (4) hours by site workers in and around the treatment system. No measurement of volatile organic compounds (VOCs) above background (>1.0 ppmV) were noted. Ambient air samples were collected periodically using calibrated personal air sampling pumps manufactured by MSA to confirm these observations. The pumps were fitted with polyurethane foam tube samplers in accordance with EPA Method 0010 and attached to the treatment area exclusion zone fencing in up wind and downwind locations. Samples were collected for 24 hours at a sampling rate sufficient to filter approximately 5.0 cubic meters of ambient air. Samples were submitted under chain of custody to an off-site CalEPA certified air testing laboratory for the analysis of Total PCBs by EPA Method 8082. All samples were below the NIOSH Recommended Exposure Limit (REL) of 1.0 microgram per cubic meter (ug/M^3).

CONCLUSIONS

Confirmatory soil sampling data indicate that the Total PCB concentration within the target treatment area was reduced from greater than 500 to below 1.0 mg/kg dry weight of soil. Total 2,3,7,8 - TCDD toxicity equivalents were reduced from greater than 1.0 to less than 0.100 ug/kg dry weight of soil. The ISTD system delivered sufficient thermal energy to raise the in-situ soil temperature to >900°F. Sampling and analysis data from source testing, emissions monitoring, and ambient air testing indicate that the soil remediation was completed without exposure of on-site personnel or the environment to the contaminants of concern and/or oxidation by products. These measurements indicate that the combined Thermal Well and AQC system operated at greater than 99.9999%

destruction/removal efficiency (DRE) and treated the PCB contaminants to acceptable discharge levels for Total PCBs, hydrogen chloride, and total particulates. The calculated discharge of PCDD/Fs isomers as 2,3,7,8,-TCDD TEQ was approximately 100 times less than the Maximum Acceptable Control Technology standard of 0.2 ng/dscm established by the USEPA for the treatment of dioxin-like substances.

ACKNOWLEDGEMENTS

The author would like to thank the dedicated personnel at TerraTherm Environmental Services for their commitment to the completion of this project, and Drs. George Stegemeier and Harold J. Vinegar for their encouragement, guidance, and for their courage in the development of this valuable technology.

REFERENCES

1. Dames & Moore, 1985, *Geotechnical Investigation, Physical Security Improvements*, US Naval Facility Centerville Beach, Ferndale, CA.

2. Genium Publishing Corporation, September, 1992, *Material Safety Data Sheet #683, Polychlorinated Biphenyls.*

3. Harza Kaldeer Consultants, 1993, *Preliminary Soil Investigation for Special Project C-191*, Naval Facility Centerville Beach, Humboldt County, CA.

4. Leedshill – Herkenhoff, Inc., 1991, *Sampling & Analysis Activities Report for PCB Contamination Investigation*, Naval Facility Centerville Beach.

5. Radian Corporation, 1997, *Site Investigation, Field Sampling, and Laboratory Analysis Report, Sites 4, 6, 7*, Naval Facility Centerville Beach.

6. Regulations for Source Testing, 40 Code of Federal Regulations Part 60, Appendix B.

7. TerraTherm Environmental Services, Inc., August, 1998, *Final Work Plan, In Situ Thermal Desorption Demonstration*, Naval Facility Centerville Beach (NFCB), Centerville CA.

8. TetraTech EM, Inc., January 12, 1998, *Site Summary, Site 06 Former Transformer/PCB Spill Area*, NFCB, Ferndale, CA.

9. TetraTech EM, Inc., November, 1999, *Site 06 Summary of Validated Soil Sample Results, PCBs and PCDD/Fs,* NFCB, Ferndale, CA.

DNAPL REMEDIATION CLOSURE WITH SIX-PHASE HEATING™

Gregory Beyke, Current Environmental Solutions, Marietta, GA
Gregory Smith, Radian International, Rolling Meadows, IL
Valdis Jurka, Lucent Technologies Construction Services, Inc., Morristown, NJ

ABSTRACT: Six-Phase Heating (SPH) was used to close out a DNAPL remediation site in the metropolitan Chicago area. The site had been undergoing remediation using steam and enhanced biodegradation[1] for approximately 7 years. Manmade subsurface features were determined to be limiting the effectiveness of the steam injection. SPH was used to heat the soils and groundwater to remove DNAPL where the steam injection process was being limited. SPH was selected for use because it relies on conduction of electricity through soil moisture, and as a result is not influenced by physical obstructions or variations in hydraulic conductivity in the subsurface.

The application of SPH was performed in two areas: the first began in June 1998, and the second application in December 1998. The treatment was discontinued on April 30, 1999, and all active remediation operations at the site now complete. The SPH treatment was terminated after: (1) concentrations of chlorinated aliphatic compounds in groundwater were reduced below site-specific risk-based criteria, (2) concentrations of chlorinated compounds in the extracted groundwater were reduced to below detectable levels; and (3) vapor extraction discharge concentrations were reduced to low parts per million concentration. Illinois EPA has issued a No Further Remediation letter and all treatment equipment has been removed from the site. The remediated area is undergoing redevelopment as part of a movie theater complex.

BACKGROUND

This work was conducted at a former electronics manufacturing where machining, plating, soldering, semiconductor manufacture, and silicon chip production took place. In 1987, several underground storage tanks were removed from the site. One of the tanks, originally believed to be empty, was found to contain a mixture of trichloroethene (TCE) and 1,1,1-trichloroethane (TCA). The tank was corroded and the surrounding soils and groundwater were found to contain related constituents. Pools of these compounds were later discovered in eight areas, measuring up to 2.4 m thick in monitoring wells. The impacted area represented approximately 1.3 hectares.

The manufacturing facility is located on a lacustrine sequence of sediments, consisting of fine sands, and silts, with some clay lenses to a depth of 5.5 to 6.1 m. The lacustrine sediments have hydraulic conductivities measured from 1.05×10^{-5} to 1.23×10^{-4} cm/sec. The lacustrine sequence is underlain by dense glacial clay till. The clay till is a silty clay formation known as the Tinley Groundmoraine, which is continuous throughout a significant portion of the Chicago

---

[1] U.S. Patent No. 5,279,740

area. The groundmoraine extends to depths of 16.8 m. The water table is encountered at 2.1 m below the plant floor.

SELECTION OF SPH

From January 1991 to March. 1998, the site had been undergoing soil and groundwater treatment using a process combining steam injection with simultaneous enhanced biotransformation and biodegradation. This process had been successful in three of the five areas, achieving a no further remediation letter from Illinois EPA in April 1998 (Smith, et al., 1998). It was observed during the steam and enhanced biotransformation process that five source areas were having difficulty achieving cleanup. Investigations were conducted to determine if the reason for this difficulty. From this investigation, it was discovered that:

- In one area, a closed end catch basin open to the atmosphere was likely acting as a heat sink, especially during winter months.
- A subsurface void was present at the location of a former closed end catch basin, and was limiting the ability to maintain pressure (and hence temperature) from the steam injection process.
- Two areas had a very dense soil layer at the base of the structural fill beneath the building that was being utilized for vapor extraction. It was interpreted that this soil layer was limiting effective vapor extraction.
- An additional area was adjacent to a wall with a deep foundation. It is believed that the injected steam was following the preferential pathway through the foundation backfill, and not reaching the TCE and TCA impacted native soils.

In late 1997, the remedial strategy was reassessed (from the existing process of steam injection with simultaneous enhanced biotransformation) to achieve a goal of completing the active remedial operations in 1998. The steam injection process had been successful in remediating DNAPL from several areas onsite. Therefore, SPH was selected based on the successes previously achieved using thermal methods.

SIX-PHASE HEATING

SPH was developed by Battelle Memorial Institute for the U.S. Department of Energy, and involves conducting electricity through the subsurface via electrodes. The resistance to the flow of electrical current results in the generation of heat. Electrodes are installed in a pattern where a central neutral electrode is surrounded by six charged electrodes. The surrounding electrodes are sequentially charged 60-degrees out of phase from each other. This results in an even distribution of heat throughout the treatment zone, creating an in situ source of steam to strip the volatile compounds from the soil and groundwater.

SPH was performed in two phases. Implementation of the first phase of SPH system began March 30, 1998, and was installed and operational on June 4, 1998. The first phase of SPH treatment was conducted until October 1, 1998. The SPH system was shut down for maintenance for most of the month of October. The SPH treatment was restarted at the end of October 1998 and continued through December 1998. In December 1998, the SPH treatment area was expanded to reduce the residual concentrations in groundwater to levels where biodegradation daughter compounds would not rebound above the target cleanup levels developed for the site per 35 Illinois Administrative Code Part 742. This was done to ensure that the area could be redeveloped within a reasonable timeframe without being impeded by the remediation system.

The system was expanded over a period from December 1998 to January 1999, with the actual treatment brought on-line as areas of electrode installations were completed. Treatment continued until April 30, 1999, when sampling conducted on April 23, 1999 confirmed that all wells showed concentrations below the RBTCL concentrations, with many wells below Illinois Class II standards. All the extraction wells were sampled during the week of May 10, 1999 to provide a baseline for post closure monitoring.

GROUNDWATER CONCENTRATION REDUCTION OBSERVED

SPH began on June 4, 1998. Within 60 days, temperatures throughout the entire 24,000 cubic yard treatment volume had reached the boiling point of water. With another 70 days of heating, concentration indications of separate phase of DNAPL had been removed and TCE/TCA groundwater concentrations reduced to below the risk based target cleanup levels. Cleanup results are shown in Table 1, and depicted graphically in Figures 1 and 2.

TABLE 1: Groundwater Concentrations During Six-Phase Heating

Well	Compound	March 1998 (µg/l)	October 1998 (µg/l)	November 1998 (µg/l)	Reduction
B3	cis 1,2-DCE	49,000	780	140	99.7%
	1,1,1-TCA	82,000	<500	31	99.96%
	TCE	34,000	790	120	99.6%
Ba6	cis 1,2-DCE	9,800	200	1,200	87.8%
	1,1,1-TCA	66,000	<50	<100	>99.8%
	TCE	7,000	510	470	93.3%
C4	cis 1,2-DCE	43,000	1,300	450	99.0%
	1,1,1-TCA	11,000	<100	15	99.9%
	TCE	76,000	1,600	100	99.9%
Ca6	cis 1,2-DCE	15,000	4,100	250	98.3%
	1,1,1-TCA	16,000	14	<20	>99.9%
	TCE	63,000	81,000	1,600	97.5%
Da2	cis 1,2-DCE	18,000	120	51	99.7%
	1,1,1-TCA	28,000	290	<100	>99.9%
	TCE	47,000	8,800	320	99.3%
F13	cis 1,2-DCE	510	480	38	92.5%
	1,1,1-TCA	18,000	<250	<10	>99.9%
	TCE	800	260	12	98.5%
Fa2	cis 1,2-DCE	3,900	470	180	95.4%
	1,1,1-TCA	24,000	<50	24	99.9%
	TCE	22,000	1,200	70	99.7%
Average	cis 1,2-DCE	19,900	1,060	330	98.3%
	1,1,1-TCA	35,000	110	26	99.9%
	TCE	35,700	13,500	380	98.9%

Physical and Thermal Technologies

It can be seen that the monitoring wells achieved the Tier 3 cleanup criteria for TCA in late August or early September 1998, approximately 4 months after startup. For TCE, the Tier 3 cleanup criteria was achieved in mid to late October, 1993, approximately 6 months after startup. Concentrations of TCA achieved the Tier 1 regulatory criteria for all monitoring points, while only 2 monitoring points achieved the Tier 1 regulatory criteria during SPH treatment.

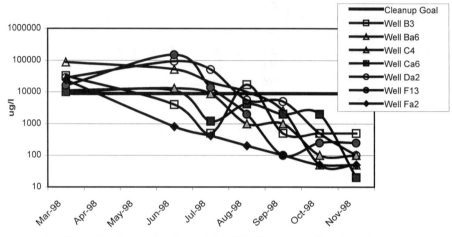

FIGURE 1: Groundwater 1,1,1-TCA Concentrations During SPH Remediation

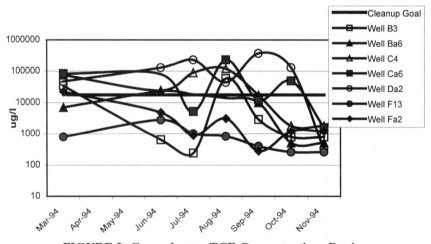

FIGURE 2: Groundwater TCE Concentrations During SPH Remediation

CHANGES IN CONCENTRATION INFLUENT TO THE TREATMENT SYSTEM

Concentrations of chlorinated solvents influent to the treatment system were monitored on a monthly basis since the remediation project began. The sample obtained on April 5, 1999 showed 5 µg/l TCE and 5 µg/l cis-1,2-DCE; all other compounds were below detectable limits. As a result, further groundwater pumping was expected to have minimal effect on groundwater cleanup although monitoring points had not achieved similarly low concentrations. As such, continued active remediation has become ineffective in removing chlorinated solvents from the subsurface, and intrinsic remedial processes will be far more cost-effective in reducing the residual concentrations.

CHANGES IN VAPOR CONCENTRATION

The air discharge monitoring showed some varying patterns over the second SPH treatment period. An air sample was collected on February 25, 1999 and analyzed using an HP 5980s gas chromatograph using an electron capture device. Methane gas was analyzed using a Foxboro® Model 128 organic vapor analyzer calibrated to methane. This analysis showed that of the total hydrocarbon reading of 70 ppm, only 17.3 ppm were chlorinated compounds. An additional air sample was obtained from the stack on March 18, 1999 and analyzed using EPA Methods. The results of this analysis are summarized with the February 25 sample and presented in Table 2.

TABLE 2: Summary of Vapor Extraction Stack Sampling Results

Compound	25 February 1999		18 March 1999	
	ppmV	% of total	ppmV	% of total
cis 1,2-dichloroethene	6.1	8.7%	2.43	7%
1,1,1-trichloroethane	1.0	1.4%	0.212	0.6%
trichloroethene	10.2	14.6%	1.16	3.3%
other chlorinated VOCs	NI		2.35	6.8%
carbon disulfide	NI		0.01	0.03%
methane	40	57.1%	28	80.8%
other hydrocarbons	12.7	18.1%	0.5	1.4%

NI = not identified

It can be seen that the concentrations of the chlorinated aliphatic compounds decreased in overall concentration and in relative amounts in the vapor stream from February 25 to March 18, 1999. Trichloroethene decreased from 10.2 ppmv to 1.16 ppmv, while the relative percent decreased from 14.6 to 3.3%. The relative percent of methane increased from 57.1 to 80.8%. Further, the absolute concentrations are decreasing and the relative percent concentrations of the chlorinated aliphatic conditions are also decreasing, while the relative concentrations of methane are increasing. The production of methane indicates methanogenic conditions in the groundwater, which are highly conducive to the bacterial mediated reductive dechlorination of the chlorinated aliphatic compounds Wiedemeier, et al., 1998). As such, it was determined that continued active treatments would provide for little significant removal of chlorinated aliphatic compounds and it was appropriate to shut down remedial operations and let the bacterial activity take over. The site has been redeveloped as a cinema.

POST CLOSURE MONITORING

At the time this paper was prepared, the post closure baseline sampling (May 1999), and a sampling in December 1999 had been conducted. Data from the nine remaining wells in the SPH remediation area are presented in Table 3. Two of the wells (Ba3 and D9) showed significant increases in vinyl chloride; this is not surprising in light of continued reductive de-chlorination without soil venting. Vinyl chloride is above Illinois Class II standards (0.01 mg/l) in four of the nine wells. No rebound has been observed for TCE and TCA and concentrations of these compounds are below Illinois Class II standards (0.025 and 1.0 mg/l respectively). Four wells are below Illinois Class II criteria for all monitored compounds. Three of these four wells flowed separate phase DNAPL for a period of up to 2 months during a period of the remediation.

Three of the wells have shown a significant increase in organic chloride concentration, five of the wells have shown a significant decrease in organic chloride concentration, and one well is essentially unchanged. Data indicates that anaerobic dechlorination is very active and further decreases seem likely.

TABLE 3: Summary of Post Closure Monitoring for Rebound

May-99 Well	Groundwater Concentration (mg/l)					Organic Chloride
	TCE	1,1,1-TCA	cis 1,2-DCE	1,1-DCA	VC	
Aa6	1.000	ND	1.100	2.700	0.064	3.588
Ba3	ND	ND	0.390	0.011	0.002	0.295
D3	0.032	ND	0.003	ND	ND	0.029
D7	0.250	ND	1.300	ND	0.012	1.162
D9	0.001	ND	0.330	ND	0.010	0.249
E5	0.004	ND	0.740	ND	0.005	0.548
F9	0.016	ND	0.760	0.001	0.200	0.684
F13	0.001	ND	0.012	0.016	0.006	0.025
G3	0.012	ND	0.460	0.069	ND	0.397
Average	0.146	0.0005	0.566	0.311	0.033	0.775

Dec-99 Well	Groundwater Concentration (mg/l)					Organic Chloride	Change From May
	TCE	1,1,1-TCA	cis1,2-DCE	1,1-DCA	VC		
Aa6	ND	0.007	0.005	0.006	ND	0.014	-99.6%
Ba3	ND	ND	0.150	0.190	2.900	1.894	541%
D3	ND	ND	0.019	ND	ND	0.015	-47%
D7	ND	ND	0.160	ND	ND	0.119	-90%
D9	0.001	ND	0.140	ND	1.500	0.956	284%
E5	ND	ND	0.430	ND	ND	0.316	-42%
F9	0.006	ND	0.900	0.005	0.080	0.713	4%
F13	0.003	ND	0.026	0.002	0.018	0.034	35%
G3	ND	ND	0.012	ND	ND	0.010	-97%
average	0.001	0.001	0.205	0.023	0.500	0.452	-42%

SUMMARY AND CONCLUSIONS

SPH provided very rapid heating of the subsurface. This promoted TCE and TCA DNAPL removal via vapor extraction. Further, the heating apparently stimulated the bacterial activity in the subsurface, increasing the methane concentration in the vapor discharge. SPH treatment resulted in a reduction in concentration below the site-specific Tier 3 cleanup criteria. However, within the time frame of the treatment, SPH was not effective throughout the treatment area in achieving the generic Tier 1 regulatory criteria. This is believed to he the result of variations in geology and a result of Raoult's Law: i.e., as the concentrations of the volatile organic compounds tend towards the regulatory criteria, the vapor pressure of the solution tends towards that of water. As a result, achieving regulatory criteria may require boiling away a significant portion of the water, This is not practical, for SPH requires conduction of electricity through the water. The work at this site shows that this is not required, for the heating stimulated bacterial activity, resulting in conditions conducive to the bio-attenuation of the residual compounds.

If present trends continue, this will demonstrate that sites having TCE and TCA DNAPL present can be cleaned to regulatory criteria using thermal methods for source removal followed by bio-attenuation processes.

REFERENCES

Smith, G. J., T. V. Adams, and V. Jurka (1998) "Closing a DNAPL Site Through Source Removal and Natural Attenuation", *Proceedings of the First International Conference on Remediation of Chlorinated and Recalcitrant Compounds*, Monterey, California, May 18-21, 1998

Wiedemeier, T.H., M.A. Swanson, D.B. Moutoux, E.K. Gordon, JET. Wilson, B.H. Wilson, D.R. Kampbell, P. Haas, R.N. Miller, J.E. Hansen, and F.H. Chapelle (1998), *Technical Protocol for Evaluating Natural Attenuation of Chlorinated Solvents in Groundwater*. EPA/600/R-98/128.

ENHANCED REMOVAL OF SEPARATE PHASE VISCOUS FUEL BY SIX-PHASE HEATING™

Gregory Beyke, P.E., Current Environmental Solutions, Marietta, GA
Trish Reifenberger, P.E., Brown and Caldwell, Atlanta, GA
Eric Maki, P.E., Current Environmental Solutions, Marietta, GA

ABSTRACT: A floating hydrocarbon plume at a manufacturing facility in Georgia was remediated using Six-Phase Heating™ (SPH) and multi-phase extraction. The hydrocarbon was a specialty fuel similar to kerosene or diesel fuel. Initially, hydrocarbon covered an area of 4900 ft^2 (500 m^2) and was up to l0 ft (3 m) thick, with most wells containing 1-3 ft (0.5-1 m) of hydrocarbon. Most of the floating hydrocarbon was beneath the manufacturing building. The soil from the floor to a depth of about 50 ft (15 m) is composed of sandy clay saprolite with moderately low permeability and high heterogeneity. The static water table is about 24 ft (7 m) below grade. Remediation began on 27 May 1999. Remediation to less than 1/8-inch (4 mm) hydrocarbon was completed on 10 December 1999. The SPH system relied on several mechanisms to remove hydrocarbon: 1) heating to reduce hydrocarbon viscosity, 2) hydrocarbon floatation/agitation by rising steam bubbles, 3) thermally enhanced vaporization (fuel boiling), and 4) vacuum-enhanced pumping.

BACKGROUND

The site was a former manufacturing facility. A large release of a specialty fuel occurred from a hydrocarbon pipeline where it passed beneath the exterior facility wall. The plume was composed of a specialty fuel with boiling point (228°C) and viscosity (2 mm^2/s St) between those of jet fuel and diesel fuel.

Most of the floating hydrocarbon was beneath the manufacturing facility. The soil from the floor to a depth of about 50 ft (15 m) is composed of sandy clay saprolite with moderately low permeability and high heterogeneity. The static water table is about 24 ft (7 m) below grade.

Initially, hydrocarbon covered an area of 4900 ft^2 (500 m^2) and was up to 10 ft (3 m) thick, with most wells containing 1-3 ft (0.5-1 m) of hydrocarbon. Due to site heterogeneity, wells separated by only a few feet varied in hydrocarbon thickness by several feet.

TECHNOLOGY SELECTION

Conventional product pumping or multiphase extraction can be used to remove such fuels; however, these processes are impeded by soil heterogeneity, hydrocarbon interfacial tension, and low fuel vapor pressure. Thus, conventional in-situ techniques would typically require in excess of one year of operation and have a greater chance of later hydrocarbon rebound. SPH is a polyphase electrical technology that uses in situ resistive heating and steam stripping to accomplish subsurface remediation.

SPH was developed by Battelle Northwest Laboratories (Battelle) for the U.S. Department of Energy to enhance the removal of volatile contaminants from low-permeability soil. The technology has now proven capable of remediating both dense and light non-aqueous phase liquids (DNAPL and LNAPL) from both the vadose and saturated zones, regardless of permeability or heterogeneity.

The SPH Power Supply uses sets of conventional 60-hertz utility transformers to convert the three-phase electricity from standard power lines into six electrical phases. These electrical phases are then delivered throughout the subsurface treatment volume by vertical, angled, or horizontal electrodes installed using standard drilling techniques.

Typically, electrodes are connected in arrays, a grouping of six electrodes that are in electrical contact, but out of phase, with each other (Figure 1). At sites

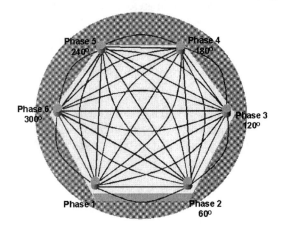

FIGURE 1 - SPH array, showing current

with multiple electrode arrays, electrodes at identical phases are all connected to the same Power Supply output. A control computer monitors the operational status of the SPH Power Supply and the conditions in the subsurface. CES operations personnel can access these computers and control the Power Supplies either directly or remotely by phone line.

SPH uses electrical resistive heating to increase subsurface temperatures to the boiling point of water. The technology is equally effective in the vadose and saturated zones. Because the SPH electrodes are electrically out of phase with each other, electrical current flows from each electrode to all of the other out of phase electrodes adjacent to it. In this manner, a volume of subsurface surrounded by SPH electrodes is saturated by the electrical current moving between the electrodes. It is the resistance of the subsurface to this current movement that causes heating.

All soils in the treatment region are heated; however, electricity prefers to take pathways of lower resistance when moving between electrodes, and these pathways are heated slightly faster. Examples of low resistance pathways in the subsurface include silt or clay lenses and areas of higher free ion content. As chlorinated compounds sink through the lithology, they tend to become trapped on these same silt and clay lenses. Over time, trapped solvents undergo biological dehalogenation producing daughter compounds and free chloride ions. Thus, at chlorinated hydrocarbon sites, the most impacted portions of the subsurface are also the low resistance electrical pathways that are preferentially treated by SPH. In a similar fashion, rock fractures and weathered rock are more electrically conductive, and heat faster, than competent rock. Thus, low permeability soils, bedrock fractures, and solvent hot spots heat, and clean up, slightly faster than other soils.

By increasing subsurface temperatures to the boiling point of water, SPH speeds the removal of contaminants by two primary mechanisms: increased volatilization and steam stripping. As subsurface temperatures begin to climb, contaminant vapor pressure, and the corresponding rate of contaminant extraction, increases by a factor of about 25. However, it is the ability to produce steam in situ that represents a significant advantage of SPH. Through preferential heating, SPH creates steam from within silt and clay stringers and lenses. The physical action of steam escaping these tight soil lenses drives contaminants out of those portions of the soil matrix that tend to lock in contamination via low permeability or capillary forces. Released steam then acts as a carrier gas, sweeping contaminants to the soil vapor extraction wells.

At the surface, a condenser separates the mixture of soil vapors, steam, and contaminants which is extracted from the subsurface into condensate and contaminant laden vapor. If these waste streams require pre-treatment before discharge, standard air abatement and water treatment technologies are used.

SPH AND DUAL PHASE EXTRACTION SYSTEM INSTALLATION

The impacted region inside the facility had low ceilings, 11 ft (3.3 m) high. The low ceilings required the use of special limited-access drilling equipment that could not turn large diameter augers. For this reason, smaller, but more numerous electrode/wells were installed than is most commonly used for SPH.

The electrode/wells were installed in 8-inch (200 mm) boreholes. A 2-inch (50 mm) steel casing and screen were inserted; this casing served as both as an electrical conductor and as a conduit for hydrocarbon extraction. A backfill of steel shot was used as the well gravel pack and electrode conductive region. The borehole above 22 ft (6.5 m) was backfilled with neat cement grout (Figure 2).

A total of 50 combination extraction/monitoring wells and SPH electrodes were installed. The electrodes directed electrical heating into the region from 20 to 30 ft (6 to 9 m) below grade. The wells extracted hydrocarbon and vapor from 22 to 27 ft (6.5 to 8.5 m) below grade.

A positive displacement vacuum blower was used to apply a vacuum to the subsurface. Vapor and steam flow from the wells passed through a steam condenser to cool the vapor and remove steam. A thermal oxidizer destroyed hydrocarbon vapors before emission to the atmosphere. An oil-water separator was used to remove separate phase hydrocarbon from the condensed steam and extracted groundwater. Most of the liquid hydrocarbon was pumped to the oxidizer for destruction. Extracted water was used to cool existing PVC monitoring wells.

FIGURE 2. Electrode/wells inside building

SPH AND DUAL PHASE EXTRACTION OPERATION

Upon start-up of the SPH system, an extensive voltage survey was performed to ensure that no hazardous voltages were present at the surface (Figure 3). CES uses a standard of less than 15 volts in areas that are accessible to personnel.

FIGURE 3. Verifying safe voltages

Usually, a well/electrode was alternately used for either heating or for extraction; however, well/electrodes occasionally performed both functions simultaneously to optimize hydrocarbon removal. Typically, 11 inches of mercury vacuum (0.35 bars absolute) was applied to the extraction wells, resulting in a vapor extraction rate of about 12 scfm (20 m³/hr) per well and a liquid (groundwater/fuel) extraction rate of about 0.25 gpm (1 l/min) per well.

Remediation began on 27 May 1999. Remediation to less than 1/8 inch (4 mm) hydrocarbon was completed on 10 December 1999. The system utilized several mechanisms to remove hydrocarbon: 1) heating to reduce hydrocarbon viscosity, 2) steam bubble floatation and hydrocarbon agitation by rising steam bubbles, 3) thermally enhanced vaporization (fuel boiling), and 4) vacuum-enhanced pumping. The relative importance of each is shown in Figure 4:

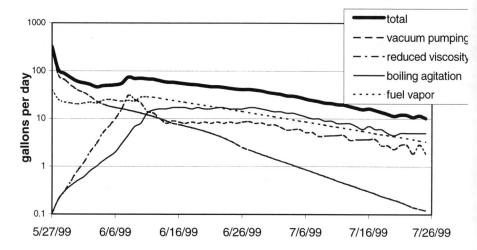

FIGURE 4: Estimated hydrocarbon removal by mechanism

Prior to beginning remediation, the importance of steam generation as a separate phase hydrocarbon removal mechanism was not properly appreciated. It is now theorized that a large fraction of NAPL can be held in the soil formation as dispersed droplets, often below the water table. These droplets are not strongly influenced by conventional pumping or dual phase removal techniques. However, during SPH, small steam bubbles are formed throughout the remediation volume. Some of these bubbles collect at the water/hydrocarbon interface due to surface tension, increasing the effective hydrocarbon buoyancy. This increased buoyancy, in conjunction with the agitation provided by the rising steam bubbles, greatly increases the upward migration rate of the small hydrocarbon droplets that were trapped below the water table. At the water surface, the hydrocarbons coalesce into a separate layer that can be removed by pumping. Once at the water surface, the agitation provided by the rising steam bubbles once again reduces the hydrocarbon effective viscosity and increases the rate of hydrocarbon flow toward the extraction wells.

CONCLUSION

A separate phase viscous hydrocarbon plume was reduced from a thickness of 10 ft (3 m) to less than 1/8 inch (4 mm) in six and a half months of SPH operation coupled with dual phase extraction. SPH enhanced product removal by the expected mechanisms: enhanced vapor recovery and reduced product viscosity. However, with steam generated in situ, the product removal was also enhanced by an unexpected mechanism, steam bubble floatation and agitation.

Occasionally, the temperature variations in the NAPL physical characteristics (viscosity, density, interfacial tension) indicate that little remedial benefit should be expected in heating the NAPL above 80°C. However, the strong NAPL recovery enhancement offered by steam bubble floatation and agitation generally makes NAPL recovery more cost effective if the NAPL remediation volume is heated to the boiling point of water - even if the NAPL physical characteristics indicate that more limited heating would be sufficient.

IN SITU THERMAL DESORPTION OF REFINED PETROLEUM HYDROCARBONS FROM SATURATED SOIL

Denis M. Conley, Haley & Aldrich of New York, Rochester, NY, USA
Kirk S. Hansen, Ground Technology, Inc., Houston, TX, USA
George L. Stegemeier, GLS Engineering Inc., Houston, TX, USA
Harold J. Vinegar, Shell E&P Technology Company, Houston, TX, USA
Frank R. Fossati, Shell Oil Company, Lake Forest, CA, USA
Fred G. Carl, Shell Technology Ventures, Houston, TX, USA
Herbert F. Clough, Hart Crowser, Inc., Lake Oswego, OR, USA

Abstract: In Situ Thermal Desorption (ISTD) is a remediation process in which heat and vacuum are applied simultaneously to subsurface soils to vaporize and destroy contaminants in situ or transport them to the surface for additional treatment. The paper presents the largest commercial application of ISTD to date (covering about 3/4 acre) and was the first project to remediate contaminated soil underneath buildings by inserting thermal wells through the building floor. This was also the first ISTD application to specifically target the removal of free product from atop groundwater in addition to soil remediation. The site was treated with 761 thermal wells containing electric heaters installed to depths of 10 to 12 feet below ground surface on a triangular grid pattern with 7-foot well spacing. Heating of the soil continued for nearly 120 days, during which time 134 tons of hydrocarbons were removed from the soil and groundwater. All free product was removed and concentration of benzene, the primary constituent of concern, was reduced to below state regulatory levels in both soil and groundwater. Quarterly groundwater monitoring has confirmed the remedial effectiveness of the technology, and site closure is pending.

INTRODUCTION:

The site was a former bulk fuel terminal located within a mixed commercial-residential neighborhood in Eugene, Oregon. The site was used for the storage and sale of bulk petroleum fuels from the 1920s through early 1970s (PEG, 1996). The facility was equipped with several above and underground storage tanks. Prior to implementation of the ISTD system, all tanks were excavated and removed. Soil and groundwater were impacted by refined petroleum hydrocarbons. Site characterization delineated areas of the site that contained free product identified as gasoline- and diesel-range organics (GRO/DRO). Soil concentrations ranged up to 3500 mg/kg GRO and 9300 mg/kg DRO (PEG, 1996). Groundwater concentrations ranged up to 25 mg/L BTEX and 15 mg/L DRO (PEG, 1995). In addition, free product had been observed as light non-aqueous phase liquids (LNAPL) in monitoring wells located on roughly two-thirds of the site. Thickness of free product ranged from trace to several feet.

Hydrogeology: The site geology consists primarily of consolidated and semi-consolidated, marine and non-marine sediments overlain by unconsolidated alluvium. A gravel layer consisting of primarily crushed and/or pea gravel covers the surface of the site to a depth of 1 to 4 feet (0.3 to 1.2 m) below ground surface (bgs). A silt layer underlies the gravel and extends to approximately 11 to 16 feet (3.4 to 4.9 m) bgs. In some areas, at typical depths of 7 to 8 feet, there is a zone of silty sand or silty gravel (PEG, 1996). Beneath the silt layer is a lower gravel layer consisting of gravels in a sand-to-clay matrix underlain by a permeable gravel layer at 18 to 19 feet (5.5 to 5.8 m) bgs. Depth to groundwater at the site exhibited seasonal fluctuations ranging from less than 2.0 feet (0.6 m) bgs in early spring (February-April) to below 10 feet (3.0 m) bgs in late summer (August-September). Hydraulic permeability was determined to range from 1 to 10 millidarcy (md) (0.1 to 1.0E-10 cm^2) within the upper silt layer. A confining layer was identified at approximately 14 to 17 feet (4.3 to 5.2 m) bgs with permeability less than 1.0 md. However, a second water-bearing unit was identified between 18 and 19 feet (5.5 and 5.8 m) bgs with a measured permeability of approximately 5000 md (5.0E-08 cm^2) (Core Petrophysics, 1997). The two aquifers were not hydraulically connected. Groundwater flow within the upper unconfined aquifer was affected by groundwater mounding in the northwest and southeastern portions of the site (PEG, 1996). The center of the site formed a groundwater trough that may have been affected by the past extensive excavation activities.

MATERIALS AND METHODS

The ISTD remedial system was designed with 761 thermal wells installed to depths of 10 to 12 feet (3.0 to 3.7 m) bgs on a triangular grid pattern of 7-foot (2.1-m) spacing (Figure 1). A total of 277 wells were configured as thermal vapor extraction wells to deliver heat while evacuating contaminant vapors and process gases. The spacing of vacuum extraction or heater/vacuum wells was approximately 7 to 12 feet (2.1 to 3.7 m). The heater/vacuum wells were connected to a central pressure blower capable of 3000 standard cubic feet per minute (SCFM) (85 m^3/min) total flow. Extracted contaminant vapors and process gases discharged from the blower received additional treatment from a regenerative thermal oxidizer (RTO) and one granulated activated carbon (GAC) bed. Remedial system monitoring included in situ thermocouples for recording soil temperatures; process piping vacuum measurements; and effluent stack monitoring of carbon monoxide (CO), carbon dioxide (CO_2), and total hydrocarbons (THC) using a CEM system.

Thermal Well Design: Thermal well design for the Eugene site introduced innovations in the selection of well patterns and spacing, in the well completion design, and in the methods of installation. The basic unit of heating was a triangular array of wells, which in previous studies proved to be the most efficient arrangement for heating soil. In previous projects, however, all heater wells were equipped to produce vaporized products by vacuum. Field experiments at Shell's Gasmer Road Test Facility in 1996 (Vinegar et al., 1997) demonstrated that some wells could be heated without being produced, provided that other vacuum wells

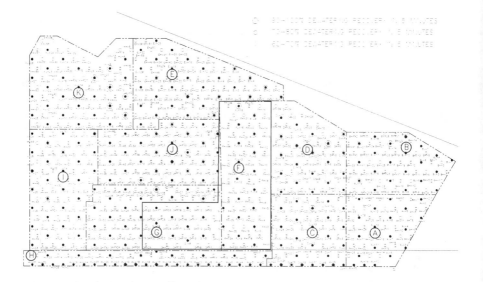

FIGURE 1. Locations of heater (open circle) and vacuum/heater (closed circle) wells on 7-foot triangular spacing. Heavy line shows L-shaped office (left) and warehouse (right). Light lines delineate property boundaries. Dashed lines indicate separate process control areas A-I for ISTD well fields.

nearby captured all the vaporized products. Subsequently, the heater-only well concept was demonstrated at a project site located in Portland, IN in late 1997 (Vinegar et al., 1999). At Eugene, triangular spaced wells were arranged into about 250 contiguous hexagonal patterns, each with a heater/vacuum well at the center surrounded by six heater-only wells. Except at the edges of the project, the heater-only wells were shared by the six surrounding hexagonal patterns, making a nominal ratio of two heater-only wells to one heater/vacuum well. Heater/vacuum wells were used in the outer locations of some of the peripheral patterns to assure complete capture of the vaporized products. A total of 102 thermal wells were drilled inside buildings through the concrete floors. Other wells were installed adjacent to the site straddling a sewer line located at a depth of 7 to 8 feet (2.1 to 2.5 m) bgs.

Energy Balance: The optimum spacing of wells in the triangular units was based on considerations of the time required for heating (operating costs) vs. the number of wells (capital costs of wells and facilities). The time needed for heating is dependent on the minimum average soil temperature required to meet the site remedial objectives, and the rate of movement of the thermal conduction heat fronts. The well spacing determines how much soil each well must heat, which includes the latent heat and the heat capacity of the fluids and the soil minerals. Since heat is supplied by electrical power, the amount of heat that can be injected depends on the power that can be injected per linear foot of heater (KW/ft; 1 KWH = 3413 BTU). Closer well spacing increases the number of wells, which nearly linearly increases the rate of heat injection. At Eugene, a 7 foot (2.1 m) spacing between heater elements was anticipated to require 60 days of heating at an injection rate of 0.3 KW/ft of heater or approximately 12,000 BTU per hour per well.

Groundwater Control: The need for site dewatering and groundwater control was recognized early in the planning and design phase. Field hydrology tests and laboratory measurements on core plugs recovered from soil borings identified two separate aquifers, as discussed previously. The initial groundwater volume in the treatment zone was estimated to be 800,000 gallons (3.03×10^6 L) using a surface area of 32,000 square feet (2,970 m^2), an average saturated thickness of 7.5 feet (2.29 m) with porosity 0.35, and an average vadose-zone thickness of 4 feet (1.22 m) with water saturation 0.50 and porosity 0.35. Based on well recovery tests, significant recharge was anticipated from rainfall directly on the site and shallow lateral flow. The vacuum/heater wells were installed with ports that permitted liquids to be pumped from the wells prior to heating. During the 4 months prior to heating (January 19 to May 20, 1998), a total of 257,000 gallons (970,000 L) of groundwater was pumped from the targeted treatment zone, representing 32% of the estimated in-place volume. Higher pump rates were achieved during and after rainstorms, confirming significant recharge of the surface aquifer. The underlying aquifer was dewatered prior to and during thermal treatment to prevent recharge from below by using 39 perimeter wells screened from 14 to 19 feet (4.3 to 5.8 m) bgs. During mid December 1997 to mid July 1998, over 1.5×10^6 gallons (5.7×10^6 L) of water was pumped from the confined aquifer. This maintained potentiometric surface of the lower aquifer at 14 to 15 feet (4.3 to 4.6 m) bgs throughout the period, about 4 feet (1.2 m) below normal.

Thermal Well Installation: Heater/vacuum well design consisted of a 5.5-inch (14-cm) diameter steel well screen. The wells were drilled to depths of 10 to 12 feet (3.0 to 4.7 m) bgs to avoid the underlying high-permeability gravel bed. Inside each well screen, a 3.5-inch (8.9-cm) heater can, welded shut at the bottom, isolated the heater elements from the flow of product in the annular space. A small-diameter tube (1/4-inch; 0.64-cm) was placed in the annular space for pumping off liquids before and during initial heating. These tubes were later used to monitor vacuum pressure in the wells. The heater-only wells consisted of a simple steel casing (3.5-inch, 8.9-cm diameter) with welded bottom into which the heaters were inserted. The cans were installed by direct push methods to depths ranging from 10 to 12 feet (3.0 to 3.7 m) bgs.

Process Control System: The process and control system was designed to control electrical energy applied to the heating elements; to monitor contaminant vapor and process gas concentrations; and to collect and treat vapors extracted from the treatment area. The thermal well field and Air Quality Control (AQC) system were monitored by a programmable logic controller (PLC) manufactured by Allen-Bradley. The PLC displayed the operating status of each thermal well heating element and the AQC system components through a personal computer (Figure 2).

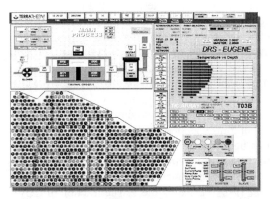

Figure 2. View of Process Control Screen after 45 days of Heating

Process Vapor Extraction & Treatment: The vacuum blower system pulled the process gas stream from the thermal well field manifold system and discharged into a regenerative thermal oxidizer (RTO) manufactured by Airex Corporation. Effluent from the RTO passed through a granulated activated carbon (GAC) bed prior to atmospheric discharge through an elevated small-diameter stack. The process effluent was monitored using a calibrated continuous emission monitoring (CEM) system manufactured by Rosemount Analytical. The quantification of carbon monoxide (CO) and carbon dioxide (CO_2) was performed with a non-destructive infrared analyzer prior to the analysis of total hydrocarbons (THC) using a flame ionization detector (FID). Total THC emissions were calculated every 30 minutes during treatment to demonstrate compliance with local air permit criteria. Emission samples collected from the stack were representative of the final atmospheric discharge.

RESULTS AND DISCUSSION

Initial Groundwater Production: Beginning in January 1998, groundwater was produced from approximately 150 heater-vacuum wells that were equipped with the small-diameter dewatering tubes. Initially, each of these wells produced over 10 gallons/day (38 L/day) for a total production of approximately 2000 gallons/day (7600 L/day) from the soil zone targeted for heating. As the target zone was dewatered, the rate fell to about 500 gallons/day (1900 L/day).

Electrical Heating: The heaters were energized on May 7, 1998. Heat was initially applied to the heater/vacuum wells, and then slowly applied to the heater-only wells. Start up was accomplished without evidence of escape of process vapors. During the next two weeks, the wells were brought up to a total energy use of approximately 2.5 megawatts of power or an average injection rate of 290 watts/foot for the 8625 linear feet (260 m) of well heaters in place. With the exception of a nine-day shut down from July 25 to August 3 for interim soil sampling, injection was continued at 2.5 megawatts for approximately 120 days. The energy input was fairly evenly distributed between the west and the east sides of the site (Figure 3). A total of 5.76 x 10^6 kilowatt hours (KWH) of electrical energy, or 19.7 x 10^9 BTU of heat, was injected into the soil during the project.

Production Response: Liquid Water: During heater testing in the first two weeks of May, heavy rains increased the liquid water production from the heater/vacuum wells to about 2000 gallons (7600 L) per day. The rains continued during the next month of heating, and precipitation continued to recharge flow into the heating zone. Vapors: The vacuum blower and vapor treatment facility were designed to handle 3000 SCFM (85 m^3/min) from the heater/vacuum wells. Figure 4 presents the total process vapor flow rate maintained during heating. A short interruption in withdrawal occurred during the nine-day heating shutdown for soil sampling. No escape of vapors or pressure build-up was observed. The vapor stream was analyzed for water content, and carbon dioxide (CO_2) to monitor the removal of water and hydrocarbons. The fraction of water in the flowing stream (f_{H2O}) was calculated from analysis of the wet oxygen content (f_{Owet}) and the dry oxygen content (f_{Odry}): (f_{H2O}) = (f_{Odry} - f_{Owet})/(f_{Odry}), see Figure 5. The fraction of CO_2 in the flowing stream is shown in Figure 6. Carbon monoxide (CO) was negligible, therefore, the amount of hydrocarbons represented by the CO_2 can be calculated approximately from the stoichiometric ratio of CH_2/CO_2, (Stegemeier and Vinegar, 2000). Masses and equivalent liquid volumes of hydrocarbons produced can be calculated from molar volume (379 cubic ft/mole), molecular weight (14 lbs/mole), and liquid density (280 lbs/bbl) of the product as diesel fuel. This calculated volume is based on petroleum barrels (1 bbl = 42 gallons) and includes any connate organic materials originally in the soil. The amount of connate CO_2 is small. The material balances can be derived from the component parts of the total vapor stream. Briefly, if all of the pore fluids are produced, the cumulative production in the vapor stream equals the amount of fluids initially in place plus the amount of convected air and water that passes through the heated region during operations. Table 1 summarizes amount of soil and fluids initially in place.

TABLE 1. Calculation of initial in-place water volume and soil masses.

Zone	Heated Area of Site =	32,000	sq ft		
	Average Soil Porosity =	0.35			
	Average Dry Density =	107.5	lbs/cu ft		
Zone	**Depth (ft bgs)**	**Water Saturation**	**Water (bbl)**	**Dry Soil (tons)**	**HC in Soil (tons)**
Vadose	0-4	0.5	3,990	6,880	4.2
Upper	4-11.5	1.0	14,970	12,900	18.0
Total Target	0-11.5		18,960	19,780	22.2
Lower	11.5-14	1.0	4,990	4,300	1.3
Total Heated	0-14		23,950	24,080	23.5

Cumulative volume of produced vapor was approximately 0.464 x10^9 SCF of vapor was produced. This nearly half-billion cubic feet of vapor represents over 14,600 tons of material removed from the soil, of which 9200 tons was convected air. Total water produced as vapor was approximately 30,400 barrels or 5300 tons. Total equivalent liquid hydrocarbons produced was approximately 957 barrels or 134 tons (Figure 7). The energy balance can be used as an independent check of the production estimates. About 19.7 x 10^9 BTU of heat

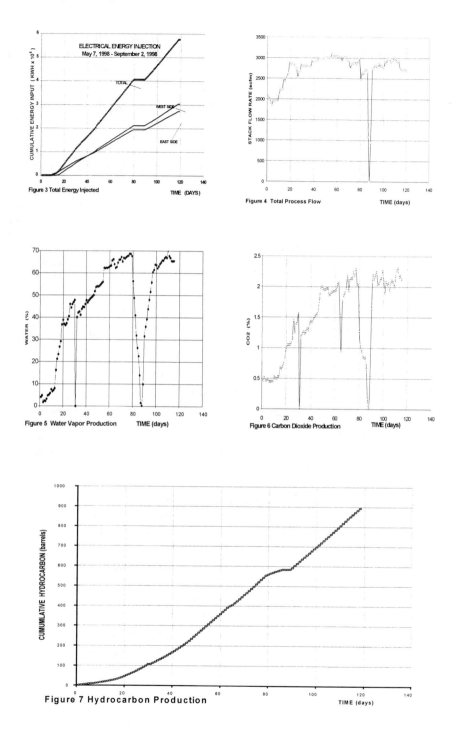

Figure 3 Total Energy Injected

Figure 4 Total Process Flow

Figure 5 Water Vapor Production

Figure 6 Carbon Dioxide Production

Figure 7 Hydrocarbon Production

was injected into the soil during the project, and an additional 5.0×10^9 BTU was released by the thermal conversion of hydrocarbons to CO_2, assuming that 90% of this conversion occurred within the soil. The amount of injected plus released heat equals that required to (1) heat the soil pore fluids and mineral grains, (2) heat the inflow water, (3) heat the inflow air, and (4) boil pore fluids plus inflow water. Table 2 gives the distribution of heat, based on the total amounts of soil and fluids removed.

TABLE 2. Heat balance calculations for Eugene, Oregon, ISTD Project

Source of Available Heat				Heat $(10^9$ BTU)
Injected heat				19.70
Heat Released by HC				5.0
Total Available Heat				24.70
Thermal Process (Initial T = 59°F, ΔT = 400 F°)	Heat Capacity or Enthalpy (BTU/lbs/F°)	Initial Weight (tons)	Produced Weight (tons)	Heat Used $(10^9$ BTU)
Heat Target Soil	0.21	19,780		3.32
Heat Liquid Water	1.00008		5,300	1.62
Latent Heat	969.0234		(As Above)	10.27
Superheat Steam	0.48		(As Above)	1.26
Heat Produced Air	0.25		9,200	1.85
Heat Produced HC	0.50		134	0.05
Heat Loss (Lower)	0.21	4,300		0.72
Heat Loss (Top/Sides)	(28% of Heat	Injected)		5.62
Total Heat Used		24,080	14,634	24.71

This balance shows that over half of the heat was used to boil and superheat the large amount of water as needed to dry the soil. About 10% of the heat was used to heat the air, and very little heat was needed to vaporize the hydrocarbons.

Soil Data: Soil samples were analyzed for total hydrocarbons using ODEQ Method NWTPH-Dx, BTEX by EPA Method 8020, and polyaromatic hydrocarbons (PAHs) by EPA Method 8270. Pre-treatment analyses indicated the presence of DRO as high as 9,300 mg/kg and concentrations of noncarcinogenic PAHs greater than 1.0 part per million (PEG, 1996). Post-treatment soil sampling was conducted by Hart Crowser (1999a) in September-December of 1998; composite soil samples were collected using a push-type soil sampler. A summary of the pre- and post-treatment soil analysis data is provided in Table 3 together with representative post-treatment groundwater analyses (Hart Crowser, 1999a). Groundwater BTEX and TPH values have remained below regulatory requirements for three subsequent quarters of monitoring (Hart Crowser, 1999b).

TABLE 3. Summary of representative soil and groundwater sample results.*

Parameter	TPH-DRO	TPH-GRO	cPAHs	Benzene	Toluene	Ethyl-benzene	Xylenes
Pre-treatment Soil Analyses (mg/kg)			1994-1995				
MW-5 (8)	9,300	3,500	NA	3.3	3.6	11	21
GP-2 (5)	2,300	380	NA	0.16	<0.025	0.65	1.7
GP-3 (3)	3,300	490	NA	<0.025	0.51	4.6	25
SGP 11(13)	2,600	150	NA	0.13	0.11	0.24	0.30
SGP-13 (3)	1,200	1,100	NA	0.53	1.9	8.4	14
MW-11 (5)	1,100	660	NA	0.11	<0.025	2.0	2.5
MW-12 (5)	1,200	120	NA	0.033	<0.025	0.42	0.65
Post-treatment Soil Analyses (mg/kg)			September-December 1998				
VGP-7A	280	NA	<0.06	<0.03	<0.03	<0.03	<0.009
VGP-11	5,600	NA	<0.06	<0.03	<0.03	<0.03	<0.009
VGP-14A	990	NA	<0.06	<0.03	<0.03	<0.03	<0.009
VGP-16A	1,100	NA	<0.06	<0.03	0.19	0.29	0.35
MW-13	510	NA	<0.06	<0.006	<0.006	<0.006	<0.006
Post-Treatment Ground-Water Analyses (mg/L)			December 1998				
MW-13	NA	<0.1	<0.0001	<0.0005	<0.0005	<0.0005	<0.0015
MW-14	NA	0.27	<0.0001	0.0025	0.0016	<0.0005	0.0027
MW-15	NA	<0.1	<0.0001	<0.0005	<0.0005	<0.0005	<0.0015
MW-16	NA	0.14	<0.0001	0.0021	0.00074	<0.0005	<0.0015

*TPH- DRO/GRO – Total petroleum hydrocarbons, diesel range/gasoline range organics, cPAHs – carcinogenic polyaromatic hydrocarbons (i.e. benzo (a) pyrene).

CONCLUSIONS

The Eugene project successfully demonstrated extension of ISTD technology to several new areas of application, including:

1. Thermal treatment of impacted groundwater and soil below the water table;
2. Use of heater/vacuum wells for partial dewatering prior to thermal treatment;
3. Remediation of groundwater and soil underneath an existing building by placing thermal treatment wells through the concrete building floor;
4. Removal of all free product and reduction of benzene in groundwater to below regulatory requirements;
5. Reduction of BTEX and carcinogenic PAHs to nondetectable levels in the soil and significant reduction of diesel-range TPH;
6. Reduction of manufacturing, installation, and operation costs by using one (1) heater/vacuum well in the center of each hexagon surrounded by six heater-only wells.

REFERENCES

Core Petrophysics, 1997, *Liquid Permeability Study, Soil Remediation Project, Eugene, Oregon,* File No. 2-970909-2SC.

Hart Crowser, 1999a, *Quarterly Activity Report, Former Jackson Street Bulk Terminal, Eugene Oregon*, January 27.

Hart Crowser, 1999b, *Proposal for Site Closure, Former Jackson Street Bulk Terminal, Eugene Oregon*, DEQ File No. 20-94-4004, December 2.

Pacific Environmental Group, 1995, *Environemntal Site Assessment Remedial Status Report, Former Shell Bulk Fuel Facility, Eugene Oregon.*

Pacific Environmental Group, 1996, *Site Characterization Report, Former Shell Bulk Fuel Facility, Eugene, Oregon.*

Stegemeier, George L. and Vinegar, Harold J., (2000), "Thermal Conduction Heating for In-situ Thermal Desorption of Soils", *Handbook of Mixed Waste Management Technology*, CRC Press, L.L.C., in review, to be published in 2000.

Vinegar, H. J, Menotti, J.L., Coles, J.M., Stegemeier, G.L., Sheldon, R.B., and Edelstein, W.A. (1997), "Remediation of Deep Soil Contamination Using Thermal Vacuum Wells ", SPE 39291, 1997 Soc. Pet. Eng. Annual Technical Conference, San Antonio, Texas, October 5-8.

Vinegar, Harold J., Stegemeier, George L., Carl, Fred G., Stevenson, Jeffery D., and Dudley, Ronald J., (1999), "In-Situ Thermal, Desorption of Soils Impacted with Chlorinated Solvents", Paper #99-450, presented at the Air and Waste Management Association Annual Meeting, St. Louis, June 22.

ELECTRICAL HEATING FOR THE REMOVAL OF RECALCITRANT ORGANIC COMPOUNDS

Bruce C. W. McGee (McMillan-McGee Corp., Calgary, Alberta, Canada)
Barry Nevokshonoff (Sequoia Environmental Remediation Inc.,
Calgary, Alberta, Canada)
Randall J. Warren (Shell Canada Products Ltd., Calgary, Alberta, Canada)

ABSTRACT: This paper presents a remediation technology that combines electrical heating of the soil with extraction to achieve removal of vapour pressure sensitive compounds, such as chlorinated solvents, volatile and semi-volatile organic compounds, and heavier hydrocarbons. This technology is commercially known as the *Electro-Thermal Dynamic Stripping Process* (**ET-DSP**). As well the results of a Shell operated field test (the **CFB Pilot**) are presented to demonstrate the feasibility and effectiveness of the technology in a commercial environment.

   Electrical heating technology has been used in the past (Buettner and Daily, 1995, U.S. DOE, 1995, McGee et. al., 1994). In a typical application of the **ET-DSP** process, electrodes are strategically placed into the contaminated zone. The pattern of electrodes is designed so that conventional three-phase power can be used to heat the soil. Also, the distance between electrodes and their location is determined from the heat transfer mechanisms associated with vapour extraction, electrical heating and fluid movement in the contaminated zone. Without consideration of all the heat transfer mechanisms, a less effective heating process will result. To determine the ideal pattern of electrode and extraction wells, a multi-phase, multi-component, three-dimensional thermal model is used to simulate the process.

   Operational data were monitored and compared to the numerical simulation of the process. Excellent agreement between field temperature, electrical operating, and energy consumption data and the numerical simulation predictions was observed. Additionally numerical modeling was used to design the power delivery system, the power requirements from the utility, and the project capital requirements.

   Several sites have been remediated using this technology. At all locations removal of contaminates was achieved, typically in less than four months of heating. This is a direct result of a substantial temperature increase in the contaminated soil and concurrent increase in the vapour pressure. The increase in vapour pressure of the contaminants makes it easier to extract them from the soil. Although the data from site to site vary, the typical cost for three phase electrical power is a minor component of the overall cleanup costs for the project. Greater cost reductions are further realized as a result of the significant decrease in time required to complete the remediation for a recalcitrant site.

INTRODUCTION

Objective. The objective of this paper is to:
1. describe a method for the in-situ recovery of recalcitrant compounds that combines electrical heating technology with soil vapour extraction (the technology is referred to as *Electro-Thermal Dynamic Stripping Process*).
2. present the results of this technology for the remediation of a soil contaminated with volatile organic compounds that leaked from an underground storage tank, referred to as the **CFB Pilot.**
3. present the measured thermal response and energy requirements and some numerical simulation calculations, and
4. summarize the economic performance of the technology.

Introduction to the Technology. Removing contaminants in-situ can be a long and costly operation. It has been demonstrated that heating the soil can greatly accelerate the removal of recalcitrant compounds (Buettner et. al., 1992, Scientific American, 1999). Thermal techniques for removing volatile organic compounds from soils include *in-situ* vapor stripping, dynamic underground stripping (Buettner and Daily, 1995), hot air injection, electromagnetic (Dev et. al., 1988) and electrical heating (DOE, 1995, McGee et. al., 1995).

Most underground storage sites that have leaked are contaminated with non-chlorinated solvents like benzene and acetone, and volatile organic compounds like gasoline. The conventional remediation technology is *soil vapour extraction* for remediation of the soil between the surface of the land and the aquifer water table and *pump-and-treat* for contaminated zones within the ground-water system. These processes are limited by retardation of the contaminants in-situ, especially in fine sediments and clays. Therefore, primary technologies must be operated for long periods of time and may not achieve cleanup standards in low-permeability soils or where the vapour pressure of the contaminant is low.

Electrical heating increases the temperature of the soil by conducting current through the resistive connate water that fills the porosity of the soil. Maximum temperatures are limited to the boiling temperature of the connate water otherwise the electrical path is boiled off. The increase in temperature raises the vapor pressure of volatile and semi-volatile contaminants, thus increasing their volatilization and removal from the soil using vapour extraction. For example, Figure 1 shows the vapour pressure relationship for benzene (C_6H_6). The curve represents the phase boundary of benzene. Above the curve, benzene naturally exits in liquid phase and in the gas phase for conditions below the curve. An increase in temperature from standard conditions of 15 °C to 80 °C changes the phase of benzene from liquid to gas at atmospheric pressure. Normal operating conditions of the **ET-DSP** process are indicated on the plot. The average pressure in the soil is reduced to one third of an atmosphere as a result of the vapour extraction wells. The average temperature in the soil is increased to 80 °C as a result of electrical heating. Under these conditions, the benzene exists in the gas phase. *The average temperature only has to exceed 50 °C for benzene to occur in the gas phase during soil vapour extraction operations*. Once the contaminant is in the gas phase, it is easily removed from the soil at the vapour extraction wells.

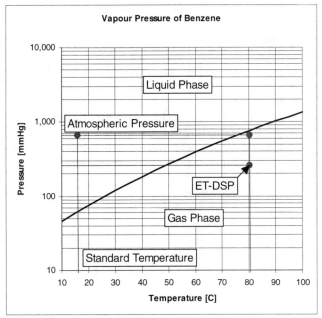

FIGURE 1. Vapour pressure plot of benzene (Hirata et. al., 1975)

The heating will tend to dry the soil by producing steam vapour, which will result in an increase in the permeability and dynamic stripping of the contaminants that may not be removed using primary soil vapor extraction. The increase in temperature will also expand the applicability of vapor extraction to less volatile contaminants, and allow cost-effective remediation of lower permeability and more heterogeneous soils.

DISCUSSION OF PILOT, METHODS, AND MATERIALS

CFB Pilot Site Description. The pilot test was conducted on the Canadian Forces Base located in Calgary, Alberta, Canada. It has been referred in this paper as the **CFB Pilot.** The pilot test was performed on a portion of a vacant property which was the location of a former service station. The subsurface soil and groundwater beneath the site was heavily impacted with gasoline and the primary compound of concern was benzene. The area of the pilot study encompassed approximately 165 m^2 and targeted 700 m^3 of soil of which approximately 250 m^3 were contaminated with gasoline. Benzene was highly adsorbed to an organic layer (>5% total organic carbon) approximately 1.5 meters thick, 3.5 meters below grade. The depth to groundwater was measured within the organic layer at approximately 3.5 meters below grade.

Site Specific Conditions. This site was selected to conduct the pilot test based on the following site characteristics:

1. The site was located within an open field, clear of any subsurface utilities or substructures.
2. There was no automobile or pedestrian traffic making the site easy to fence off from the public.
3. The subsurface lithology and the characteristics of the hydrocarbon-affected soils were understood, as numerous subsurface investigations had been completed at the site.
4. Other proven in-situ remediation techniques were attempted in this area of the plume and were only moderately successful.
5. The hydrocarbon contamination present beneath the site is primarily benzene. Benzene is a volatile compound whose vapor pressure increases substantially with increase temperatures.
6. The maximum benzene concentrations measured in the subsurface soil beneath the site are located within a shallow (three meters below grade) organic layer (formerly a slough bottom).
7. The relatively high total organic content (5% or greater) of the organic soil layer makes it very difficult to desorb the benzene from the subsurface soil using conventional in-situ techniques at standard temperatures.
8. Based on our evaluations and pilot testing, other in-situ remediation technologies were considered to be ineffective due to long time frames required to remediate.

CFB Pilot Electrical Characterization. In addition to characterization of the site for concentration levels of contaminant, electrical conductivity of the soil and its distribution also had to be measured. These involve measurements of the electrical properties of the soil as a function of temperature and water saturation. The data is important for the design of the Power Delivery System, estimate of the time required to heat the soil, determination of the power requirements, and numerical simulation of the heating process.

Based on the initial site characterization, the **ET-DSP** process was simulated for the CFB Calgary site. The simulation results are used for the design of the overall system. Based on the simulation results, the following is an optimum design for the **CFB Pilot**.

1. The distance between electrodes in a row is seven meters,
2. The distance between rows of electrodes is nine meters,
3. The minimum heating duration is 90 days,
4. Maximum operating current was estimated at 100 amps per phase, and
5. Maximum operating voltage was estimated at 280 volts phase to neutral.

ET-DSP Process. The typical pattern of electrode and vapour extraction wells and system for the **ET-DSP** is shown in Figure 2. This is the same layout that was used at the **CFP Pilot**.

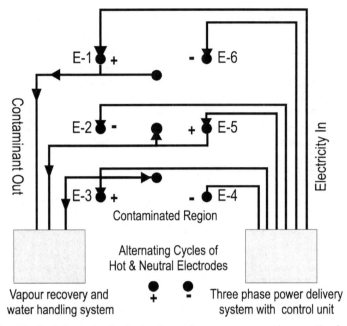

FIGURE 2. Typical layout of electrode and vapour extraction wells for the Electro-Thermal Dynamic Stripping Process.

The electrodes are arranged so that the contaminated volume of soil is contained inside the periphery of the electrodes. The vapour extraction wells are located within the contaminated soil. The position of the extraction wells relative to the electrodes is determined so that heat transfer by convection within the porous soil is maximized, thus minimizing heat losses and increasing the uniformity of the temperature distribution in the soil.

A conventional water handling and vapour recovery system are installed as part of the process. The water handling system is required to provide water injection into the electrode wells to prevent the wells from overheating. The electrode wells are designed with fluid injection capacity. Therefore some of the injected water flows from the electrode wells towards the vapour extraction wells. The heat transported by fluid movement tends to heat the soil rapidly and more uniformly. The produced fluids increase with temperature over time. These fluids are re-injected and the overall thermal efficiency is improved.

The current path is shared between the electrodes connected to the three phase power supply and is through the connate water in the porous soil. The temperature is controlled to minimize drying out of the soil until the latter stages of the heating process. As the soil changes in temperature, the resistivity of the connate water will typically decrease. Also, as the soil dries out, the resistivity will increase. A computer control system is installed to ensure that the maximum current is injected into the electrode wells at all times.

The six electrodes are connected to a three phase power delivery system. The power delivery system is equipped with computer controls so that the power from the three phases can be alternated between the six electrodes. Depending on the rate of heating, the electrodes are interchanged from hot and neutral every five to ten days. Figure 3 shows the calculated temperature contours at the end of the electrical heating phase. The contours were determined using a three dimensional, multi-phase numerical simulation program. As will be discussed, these calculations are in good agreement with measured data.

Note the higher temperatures around three of the electrodes. This is consistent with these electrodes connected to the power lines of the power conditioning unit, and the other electrodes connected to the neutral.

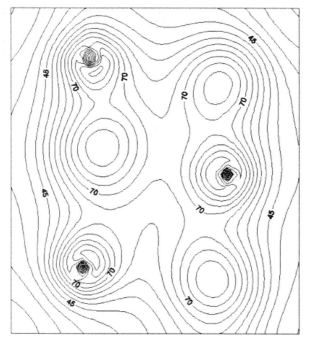

FIGURE 3.   Numerical calculation of the temperature distribution in the contaminated soil after 110 days of continuous electrical heating. The contour interval is 5 °C and the average temperature in the soil is about 60 °C.

ET-DSP Monitoring Equipment Installation. In order to monitor the change in temperature of the subsurface soil and groundwater beneath the site, 21 multi-level thermocouples were installed within the pilot testing area. Each thermocouple was installed to approximately 3.5 meters below grade and was completed with three thermocouple wires to measure temperature changes at 1.5 meters, 2.5 meters and 3.5 meters below grade. Subsurface soil temperatures were measured daily during the full duration (116 days) of the pilot test.

DISCUSSION OF RESULTS

Electrical Heating and Temperature Results.   Substantial temperature increases were measured within the electrode array during electrical heating. The average initial in-situ temperatures was 6 °C and increased to an average of 30 °C after 58 days, to 45 °C after 92 days and to greater than 55 °C after 116 days. The volume of soil within a two meter radius of the electrodes heated to an average temperature of approximately 70 °C within 90 days of heating.

The temperatures of the soil within the electrode array after the system was deactivated decreased at a rate of approximately 0.25 °C per day. At this rate, the in ground temperatures would not return to their original temperatures for a period of approximately 160 days (> 5 months). It is noted that the temperature patterns measured during the pilot test tracked very well with those predicted by the numerical model.

Power Consumption Results. The electrical equipment operated for a total of 108 days of the 116 days the equipment was on site. The operating uptime percentage is thus 93%. Down time was an associated cold weather factor, for example plugging of the cooling system due to freezing during cold ambient conditions. Over the period of the pilot test, the electrical equipment consumed a total of approximately 160,000 kWh of electrical energy. The monthly electrical bill was approximately $US 1,500. This includes the electrical costs for operating the soil vapour extraction equipment as well as the water pumps. The electrical heating power consumption was approximately 80% of the total.

Petroleum Hydrocarbon Removal. The subsurface soil temperatures increased to where the benzene vapour pressure was elevated by a factor of six. The multiphase extraction equipment removed a total of 200 liters of petroleum hydrocarbon from the subsurface during 132 days of extraction, for an average extraction rate of 1.5 liter/day. It is estimated that as much as 270 liters of petroleum hydrocarbons were remaining in the subsurface prior to activating the **ET-DSP** system. Therefore, over a period of 132 days, the system successfully extracted approximately **75%** of the total mass of hydrocarbons. In regions where the temperature of the soil was greater than 70 °C, the extracted mass of hydrocarbon removal was greater than 90%.

It should be noted that in the final 36 days, a larger extraction system was installed and removed hydrocarbons at a rate of 3.3 liters/day for a 43% (58% of the total removed) reduction in the total mass of hydrocarbons. The larger system could effectively lower the groundwater table within the electrode array to a level beneath the zone of affected soil. With the groundwater table lowered, air was able to pass through the previously saturated zone removing the hydrocarbons more rapidly from the soil. The soil remained saturated during the operation of the smaller system, thereby relying on the removal of dissolved-phase hydrocarbons in the groundwater and diffusion for removal which is a slower process.

Based on this data, if the extraction system had remained operating for a longer period of time (estimated to be approximately 30 to 50 days), a reduction of hydrocarbon mass of greater than 95 % may have been achieved. It should be noted that the extraction system was removed from the site due to time constraints established by the property owner and not by the client.

After electrical heating the soil remained hot for a long period of time. The rate of temperature decrease at the CFB Pilot was 0.25 °C/day. During this time bio-remediation activity increases (Newmark, et. al.) and natural attenuation of the soil is accelerated.

Economics. The monthly electrical bill was approximately $US 1,500, and was a minor factor in the overall costs. If the data acquisition costs are removed from the capital costs of the project, the capital and operating costs combine to give an overall cost of remediation of approximately $US 50 per m^3.

REFERENCES

Buettner, H. M., Daily, W. D., and Ramirez, A. L.: "Enhancing Cyclic Steam Injection and Vapor Extraction of Volatile Organic Compounds in Soils With Electrical Heating", *Proc., Nuclear and Hazardous Waste Mgmt., Spectrum 1992*, pp 1321-1324.

Buettner, H. M., and Daily, W. D.: "Cleaning Contaminated Soil Using Electrical Heating and Air Stripping", *Journal of Environmental Engineering*, August 1995, pp 580-588.

Dev, H., Stresty, G. C., Bridges, J. E., and Downey D.: "Field Test of the Radio Frequency In-situ soil decontamination process", *Proc. Superfund '88 Ninth National Conference*.

Hirata, M., Ohe, S., Nagama, K.: "Computer Aided Data Book of Vapor-Liquid Equilibria" *Elsevier*, Amsterdam, 1975.

McGee, B.C.W., Vermeulen, F. E., Vinsome, P. K. W., Buettner, M. R., and Chute, F. S.: "In-situ Decontamination of Soil", *The Journal of Canadian Petroleum Technology*, Paper No. 94-10-07, pages 15-22, October, 1994.

EPA/540/R-94/527 "IITR Radio Frequency Heating Technology, Innovative Technology Evaluation Report", *United States Environmental Protection Agency, Office of Research and Development*, June 1995

Internal Report of the U.S. Department of Energy Office of Environmental Management Office of Technology Development, "Six Phase Soil Heating - Demonstration of six phase soil heating, at M Area Savannah River Site, Aiken, SC and 300-Area Hanford Site, Richland WA", April 1995.

Editorial: "Technology and Business – Not Cleaning Up", *Scientific American*, February 1999, pp 39-40.

ORGANIC HALIDES REMOVAL FROM DYE WASTEWATER

Natalija Koprivanac (University of Zagreb, Zagreb, Croatia)
Ana Loncaric Bozic and Sanja Papic (University of Zagreb, Zagreb, Croatia)
Jarolim Meixner (DIOKI d.d. Zagreb, Zagreb, Croatia)

ABSTRACT: Wastewater originated from reactive dye processes represent a specific ecological problem due to its high content of organic halides and color besides. This type of wastewater is characterized by poor biodegradability, therefore some other treatment method should be applied prior to discharge to the sewage system. In our previous investigations coagulation/flocculation process was successfully applied as a treatment method for reactive dye wastewater decolorization. In this work, the possibility of application of AOX method in monitoring efficiency of flocculation process, detecting and measuring presence of organic chlorine in reactive dye wastewater, was investigated. Experiments, in laboratory scale, were carried out in wastewater from synthesis of anthraquinone reactive blue monochlorotriazine dye, similar to C.I. Reactive Blue 5, and model wastewaters of other representative of monochlorotriazine reactive dyes (C.I. Reactive Red 45, C.I. Reactive Green 8, C.I. Reactive Blue 49 and C.I. Reactive Blue 137). The results have shown that AOX values, as well as COD, BOD_5, TOC and IC_{50} parameters were significantly reduced. Quality of water obtained after treatment process is suitable for discharge to the sewage system.

INTRODUCTION

Organic synthetic dyes represent relatively large group of organic synthetic chemicals, which are met in practically all spheres of our daily life. Reactive dyes, mainly used in dyeing of natural fibers, are the largest product group in value terms with quarter of market and it grows as the use of natural fibbers increases (Kelshaw, 1998). Over the past two decades, manufacturers and users of these dyes have faced increasingly stringent legal regulations promulgated to safeguard human health and the environment. Wastewater originated from reactive dye processes represent a specific ecological problem due to its high content of organic halides and color besides (Willmott et al., 1998).

Investigations, in laboratory scale, of production process of monochlorotriazine anthraquinone reactive blue dye (RB), similar to C.I. Reactive Blue 5, were carried out. One of the raw material for this type of dye is cyanuric chloride (2,4,6-trichloro-1,3,5,-triazine). Wastewater from dye manufacturing units contains unreacted cyanuric chloride, chlorinated intermediates and byproducts as well as dye product which can not be completely isolated because of its high solubility. Wastewater originated from reactive dye processes is characterized by poor biodegrability, passing unaffected through conventional treatment systems. Therefore, some other treatment method should be applied prior to discharge to the sewage system (Davis, 1991; McKay, 1979; Tebbut, 1998). In our previous investigations coagulation/flocculation process was successfully applied as a treatment method for reactive dye wastewater decolorization (Koprivanac et al., 1992; Koprivanac et al., 1993; Papic et al., 2000; Koprivanac et al., 2000).

Objective.   The objective of this work was to determine the feasibility of flocculation process in organic halide removal. Experiments were carried out in wastewater from monochlorotriazine reactive dye synthesis process and model wastewaters of four other representatives of monochlorotriazine reactive dyes Process efficiency was estimate on the basis of the AOX (Absorbable Organic Halide) as well as COD (Chemical Oxygen Demands), BOD_5 (Biological Oxygen Demands), TOC (Total Organic Carbon) and IC_{50} (50 % Inhibition Concentration) parameters, determined in wastewater samples prior and after treatment with flocculation method. The goal was to achieve quality of treated water that could be discharged to the sewage system.

MATERIALS AND METHODS
        The synthesis of the monochlorotriazine reactive dye was carried out in laboratory scale and was reported elsewhere (Koprivanac et al., 2000). The wastewater after product isolation with dye concentration 1.8 g/l was used for further investigations. C.I. Reactive Red 45 (RR 45), C.I. Reactive Green 8 (RG 8), C.I. Reactive Blue 49 (RB 49) and C.I. Reactive Blue 137 (RB 137) were chosen to prepare the model wastewater of the same dye concentrations.

Flocculation. The batch device used for the wastewater treatment consisted of rapid and slow mix units The geometry of mixing device was chosen according to our previous work, where the optimum process parameters were determined Panic et al., 2000). The flocculation process was carried out applying rapid and slow mixing. A turbine impeller was used in rapid mixing at 250 rpm during 0.5 min, while slow mixing over 15 min at 20 rpm using paddle impeller. Cationic polymer flocculant (Levafloc R, By) was added at the beginning of rapid mix period and pH was adjusted using Na_2CO_3. After the slow mix operation followed by sedimentation, residual dye concentration, AOX, COD, BOD_5, TOC and IC_{50} values were determined according to the standard procedures.

AOX. AOX values were determined using DX-2000 Halide Analyzer (Dohrman) according to the ISO 9562: 1989 (E) Standard. Method involves adsorption of organic halides from the sample on to activated carbon, followed by combustion and microculometric detection (Figure 1).

COD. Determination of the COD values is based on the oxidation reaction of the oxidable substances in the sample influenced by kalium bicromate at increased temperature in sulfuric acid solution with argentum-sulphate as catalyst. The reaction is made in the closed ampoule, the oxygen quantity is measured colorimetricaly with spectrophotometer at 600nm with standard.

BOD$_5$. Determination of the BOD$_5$ values was carried out using OxiTOP system (WTW) according to the DIN 38409-52 Standard. Method is based on measuring the initial and remained quantity of dissolved oxygen in the water sample after 5 days at certain conditions.

TOC. TOC values were determined using TOC-5000 A (Shimadzu) according to the ASTM Methods D2479 and D4779. Method is based on oxidizing organic material dissolved in the water to carbon dioxide, which is then detected and quantified by nondispersive infrared detection.

IC$_{50}$. The aerobic toxicity expressed in a term of IC$_{50}$ is determined by the bacterial growth inhibition test based on turbidity measurement of the microorganism suspension in the aerobic active sludge of the tested wastewater. The test concentration corresponding to 50 % of the control is termed IC$_{50}$ and that is agreed concentration in excess of which the wastewater is not considered as harmful to the microorganism in the biological wastewater treatment facility.

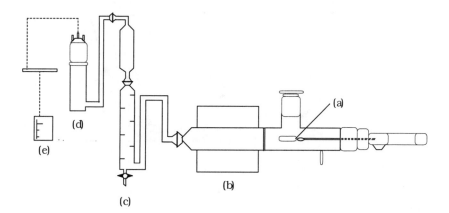

FIGURE 1. Principle of AOX measurement. (a) Sample on activated carbon (b) Furnace, 900 °C (c) Scrubber (d) Microcoulometer (e) Printer

RESULTS AND DISCUSSION
The flocculation process is influenced by the characteristics of wastewater; as temperature, pH and type and dose of flocculent used, as well as intensity and duration of mixing and geometry of mixing device. In order to obtain maximal process efficiency, optimal physical and chemical parameters should be established. Optimization of flocculation process in reactive dye wastewater treatment was carried out earlier and reported elsewhere (Papic et al., 2000).
The effect of pH values on reactive dye removal is given in Figure 2. It can be seen that polimeric flocculant operate over the wide range of pH and that the best

results regarding color removal are obtained by adjusting pH at 8.5, for all
investigated dyes contained in wastewater.

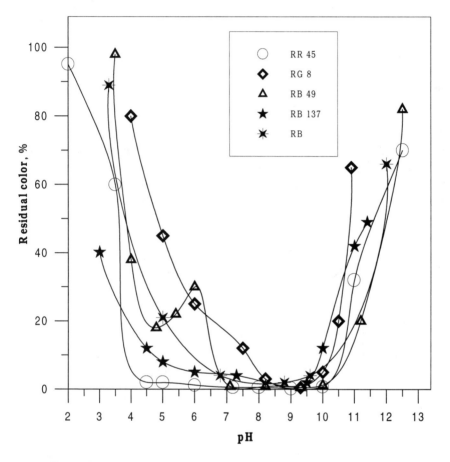

**FIGURE 2. The influence of pH on reactive dye removal from wastewater of
initial dye concentration $c_0 = 1.8$ g/L using Levafloc R ($\bullet$ = 0.25 vol %)**

Due to the nature of reactive dyes, wastewater originated from reactive dye
production and application processes are characterized by high content of organic
halides and color besides.

Organic halides are generally not found naturally in the environment. As a
group of compounds, organic halides are often very toxic and are considered
carcinogenic and mutagenic. Therefore, reactive dye processes and their waste
streams should be monitored, analyzed and treated to reduce amount of organic
halides to 0.1-1.0 mg/L, according to the regulations. AOX method is mainly used
in the control of bleaching processes in wood, paper and textile industries. In this
work AOX method was applied in monitoring efficiency of flocculation process
as a treatment method for reactive dyes wastewater. Wastewater was analyzed

prior and after flocculation treatment. Dye concentration and values of AOX, COD, BOD_5 TOC and IC_{50} parameters were determined Results are given in Tables 1 and 2.

TABLE 1. Characteristics of wastewater from synthesis of RB and model wastewater of other representatives of monochlorotriazine dyes before flocculation treatment

Dye	c (mg/L)	AOX (mg/L)	COD (mg/L)	BOD (mg/L)	TOC (mg/L)	IC_{50} (vol%)
RB	$1.8 \ 10^3$	465.8021	2760	12	2164.00	15.2
RR 45	$1.8 \ 10^3$	16.5901	2680	15	2265.00	16.0
RG 8	$1.8 \ 10^3$	12.7853	1410	3	1148.00	51.0
RB 49	$1.8 \ 10^3$	20.3141	2240	12	2274.00	14.7
RB 137	$1.8 \ 10^3$	18.4995	2850	9	2498.00	23.0

TABLE 2. Characteristics of wastewater from synthesis of RB and model wastewater of other representatives of monochlorotriazine dyes after flocculation treatment

Dye	c (mg/L)	AOX (mg/L)	COD (mg/L)	BOD (mg/L)	TOC (mg/L)	IC_{50} (vol%)
RB	4.7	0.3008	112	5	32.08	72.3
RR 45	4.1	0.2820	114	5	35.04	70.6
RG 8	2.8	0.2101	122	4	6.58	75.0
RB 49	3.3	0.2402	102	5	21.10	76.0
RB 137	3.5	0.2930	118	4	24.90	73.2

It can be seen that the values of all determined parameters have been significantly reduced after flocculation treatment. Quality of wastewater has been improved enough to be discharged to the sewage system.

Flocculation method can be successfully applied in organic halide removal from the reactive dye wastewater.

REFERENCES

Davis, J. 1991. "Improving Dye Waste Water Treatment." *American Dyestuff Reporter*. 80 (3): 19-68.

Kelshaw, P. 1998. "The future for West European dyestuff manufacturers." *JSDC*. 114: 35-37.

Koprivanac, N.,G. Bosanac, Z. Grabaric and S. Papic. 1993. "Treatment of Wastewaters from Dye Industry." *Environ. Technol.* 14: 385-390.

Koprivanac, N., J. Jovanovic-Kolar, G. Bosanac and J. Meixner. 1992. "Studies on Wastewater Decolorization by the Precipitation/Flocculation Process." *Microche. Jour.* 46: 379-384.

Koprivanac, N., A. Loncaric Bozic and S. Papic. 2000 "Cleaner production processes in the synthesis of blue anthraquinone reactive dyes." *Dyes and Pigments.*44: 33-40.

McKay, G. 1979. "Waste Color Removal From Textile Effluents." *American Dyestuff Reporter.* 68: 29-34.

Papic, S., N. Koprivanac and A. Metes. 2000. "Optimizing Polymer-Induced Flocculation Process to Remove Reactive Dyes from Wastewater." *Environ. Technol.* 21: 97-105.

Tebbut, T.H.Y. 1998. *Principles of Water Quality Control.* 5th ed., Butterworth-Heinemann, Oxford.

Willmott, N., J. Guthrie and G. Nelson. 1998. "The biotechnology approach to colour removal from textile effluent." *JSDC.* 114: 38-41.

DEGRADATIVE SOLIDIFICATION/STABILIZATION TECHNOLOGY FOR CHLORINATED HYDROCARBONS

Inseong Hwang and Bill Batchelor, Texas A&M University, College Station, Texas

ABSTRACT An experimental study was conducted to test the feasibility of degradative solidification/stabilization (ds/s) process in treating tetrachloro-ethylene (PCE) in solid phase systems. The Fe(II)-based ds/s process successfully treated PCE in a soil at the reaction rates that would not allow significant release of the contaminant in the environment. The production of hazardous intermediates was minimal in the ds/s system. A leach model was also developed that could describe the relative importance of leaching and degradation in ds/s. The first and second Damköhler numbers and dimensionless time were important parameters that determined leaching processes in wastes treated by ds/s.

INTRODUCTION

Remediation of soils contaminated with chlorinated hydrocarbons and mixtures of chlorinated hydrocarbons and metals is a major environmental challenge. Degradative solidification/stabilization (ds/s) is a potential new technology for such applications. Ds/s is a modification of conventional s/s processes that promotes degradation of organic contaminants while containing them as well as containing inorganic contaminants. As such, it combines the advantages of contaminant destruction with the low cost of conventional s/s. The economic attractiveness of conventional s/s is seen in the fact that it is the most commonly used remediation technology at Superfund sites (MacDonald, 1997). Conventional s/s can immobilize inorganic contaminants at high pH using binders such as Portland cement, fly ash, and lime. Immobilization of organics can be accomplished by physical methods such as permeability reduction.

Reductive dechlorination was studied as a degradative reaction for chlorinated hydrocarbons in ds/s. A preliminary experiment was conducted in our lab to identify an effective ds/s system for chlorinated hydrocarbons by using batch slurry reactors. The experiment found Fe(II) was a promising reductant for chlorinated hydrocarbons. Fe(II) could completely dechlorinate tetrachloro-ethylene (PCE) to non-chlorinated hydrocarbons in 10% cement slurries.

A series of experiments in batch, slurry reactors was conducted to investigate this dechlorination reaction. Results showed that PCE degradation occurred over the investigated pH range (10.6 to 13.8), which is typically observed pH range in the conventional s/s processes. PCE degradation was optimal at pH ~12.1 with a half-life of 4.8 days when Fe(II)=39.2 mM. Under this condition, 98% of PCE transformed into non-chlorinated hydrocarbons with acetylene being predominant. Production of chlorinated intermediates was minimal in Fe(II)/cement slurries. Cement was found to participate in PCE degradation reactions in an essential way, because PCE removal was minimal in the absence of cement or in the presence of solid phases such as tricalcium silicate and hydrated lime. Results of some experiments showed that a mixed-valent iron oxide such as green rust might be the species that is the reductant for PCE in Fe(II)/cement systems. These results suggest a strong potential for the development of the Fe(II)-based ds/s technology for treating contaminated soils.

Leach models are useful in evaluating the performances of the solidified wastes (Batchelor, 1998). They can simulate the overall release of the contaminants from treated materials. Leach models for ds/s systems would include description of leaching and degradation. They will be useful for evaluating conditions under which the ds/s treatment system can be applied without significant release of contaminant during the time of degradation. They will also be useful in optimizing treatment system design and conducting risk assessments for treated materials.

The objective of this research was to further evaluate the Fe(II)-based ds/s technology by investigating PCE dechlorination in solid phase systems such as would be produced by s/s treatment of soils and sediments. An additional objective was to develop a leach model that could describe the relative importance of degradation and leaching during ds/s.

MATERIALS AND METHODS

Degradation Experiments. The soil used for degradation experiments was collected from the first ~20 cm to ~40 cm of the surface at a ranch at Texas A&M University, College Station, Texas. The soil was dried and screened to particle sizes below 0.425 mm (No. 40 mesh). The soil sample consisted of 81.1% sand, 12.2% silt and 6.7 % clay. Some important physical and chemical characteristics of the sample are presented in Table 1.

TABLE 1. Surface area, organic carbon and iron contents, and pH of the sample.

Specific Surface Area (m^2/g)	Organic C (%)	Fe(II) (mg/g)	Fe(III) (mg/g)	pH
13.0	0.69	0.498	5.757	6.1

First, the sorption potential of the soil for the target compound (PCE) was tested. An isotherm experiment was conducted at pH 12.1 at a soil to water ratio of 1.29 by using 25 mL glass vials. Seven different initial concentrations of PCE (6.25, 12.5, 25, 52.5, 105, 168, 210 mg/L) were used. The equilibration time was 7 days.

PCE degradation experiments were also carried out by using the same 25 mL glass vials. Closures of the vials were designed to minimize the intrusion of oxygen and losses of PCE through volatilization. They consisted of three-layer sealing components: Teflon lined silicon septum, lead foil tape (3M, adhesive backed), and Teflon tape. All samples were prepared in an anaerobic chamber (Coy Laboratory Products). First, appropriate amounts of the soil, cement, water, and Fe(II) stock solutions were transferred to vials and were mixed. 5.3 N KOH or 5 N HCl solution was also added to maintain pH of the systems around 12.1. Two stainless steel balls of 79-mm diameter were also transferred to the vials for initial homogenization of the samples. The mass ratio of soil:cement:water was 62.5:7.5:30. Then reactions were initiated by spiking methanolic stock solutions of PCE to yield a PCE concentration of 100 mg/kg soil. After spiking, samples were mounted on a tumbler for two days to homogenize the samples. Then the samples were cured at the room temperature. At each sampling time, samples were extracted for PCE and its degradation products in 250 mL fluorine-coated plastic bottles. Samples were rapidly crushed under the water and pentane layers in the bottles. Then the bottles were capped and mounted on an orbital shaker for 6 hours for extraction. Pentane layers were analyzed for PCE, TCE, 1,1-DCE, c-DCE, t-DCE, vinyl chloride (VC), dichloroacetylene, and chloroacetylene with a GC (HP G1800 A GCD).

RESULTS AND DISCUSSION

Degradation Experiments. From the sorption experiment, a value for K_p of 0.622 L/kg was obtained. Using the f_{oc} of value for the soil (0.0069), a value for K_{oc} of 90.1 was obtained, which is about a factor of 2 to 4 lower than experimentally determined K_{oc} values of PCE for natural solids in the study of Mouvet et al. (1993).

Figure 1 shows the results of the PCE degradation experiments that tested four different Fe(II) doses. Lines in Figure 1 represent first-order fits. TCE was the only chlorinated intermediate product observed in these experiments. The concentrations of TCE did not exceed 7% of the initial concentration of PCE on a molar basis. TCE disappeared within the first 11 days in three higher Fe(II) dose experiments, but persisted throughout the experimental period in the 5.39 g Fe(II)/kg experiment. The absence of the degradation products such as DCEs and VC implies that Fe(II)-based ds/s technology may completely dechlorinate PCE in soils without production of hazardous intermediates.

Kinetics of PCE degradation was modeled by a pseudo first-order rate law. Figure 1 shows that PCE recoveries from controls steadily decreased during the experimental period. This may be due to poor extraction efficiencies in the later stages or to an intrinsic capacity of the soil to reduce PCE. To obtain pseudo first-order rate constants that are exclusively due to Fe(II) treatments, the pseudo first-order rate constant for the control was subtracted from the pseudo first-order rate constants for the reactive experiments. The rate constant for the control was 0.000952 (±29%) day^{-1}. Table 2 presents the pseudo first-order rate constants. The relationship between the rate constants and Fe(II) dose was linear with an r^2 value of 0.987. Table 2 shows that the apparent half-lives of PCE are generally within time frames allowable for ds/s technology, i.e., several months to years. Half-lives of PCE even shorter than those of the present experiments may be obtained by adding Fe(III) and/or an appropriate anion such as sulfate. Such additions were found to increase reaction rates in slurry reactors, possibly by promoting the formation of larger amounts of the reactive agent.

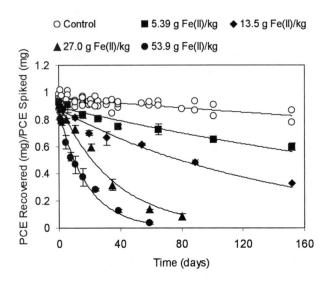

FIGURE 1. Kinetics of PCE reduction by Fe(II) in solidified soils.

TABLE 2. Pseudo first-order rate constants for PCE in the solid phase experiments

Fe(II) Dose (g/kg)	Fe(II) dose (mM)	k'_{app} [a,b] (day^{-1})	k [b,c] (day^{-1})	$(t_{1/2})_{app}$ (day)	n [d]
5.39	201	0.0021 (±37%)	0.0069 (±37%)	335	24
13.5	503	0.0062 (±24%)	0.016 (±24%)	113	18
27.0	1006	0.027 (±12%)	0.064 (±12%)	26	22
53.9	2012	0.053 (±13%)	0.13 (±13%)	13	28

[a]apparent pseudo first-order rate constants corrected for the control decay.
[b]uncertainties represent 95% confidence limits expressed in % relative to estimates for the rate constants.
[c]pseudo first-order rate constants for aqueous phase
= $(1 + K_pD)$ k'_{app} = 2.29 k'_{app}, D = soil dose = 2.08 (kg/L).
[d]number of data points.

Kinetics of PCE degradation in the soil/cement systems was much slower than that in the slurry systems. In the case of Fe(II) dose of approximately 200 mM, the reaction rate in the slurry system was 19 times higher than that in the solid phase system. The value of K_p for PCE predicts that 56% of the PCE in the solid phase experiments would be present in the soil rather than in the solution where it could be degraded. Whereas approximately 11% of the PCE was partitioned to the solid phase in the slurry experiments. As shown in Table 2, the kinetic retardation factor due to the sorption was 2.29. Other factors or processes that would potentially slow down degradation kinetics in soil systems compared to slurry systems include the following: 1) oxidation of Fe(II) by the soil, 2) interference of the soil in the formation of the reactive agent, 3) sorption of the reactive agent on the soil, and 4) less availability of cement components for the formation of the reactive agent. Identification of the reactive agent will facilitate identifying which of these factors is important.

Development of Leach Model. A model for leaching of a target compound from a waste treated by ds/s is based on a dynamic material balance within the waste form and the following assumptions:

- one dimensional transport by diffusion in a slab
- linear partitioning at equilibrium
- first-order degradation rate
- infinite bath (concentrations outside waste form remain very low)

The assumption of one-dimensional transport will be valid for any shaped waste form at early leaching times. A first-order rate form is suitable for many degradation reactions, but the rate coefficient may be a function of pH. However, use of a rate coefficient for optimum pH will provide conservative estimates of leaching. Furthermore, if the leaching front for the target compound moves more rapidly than that of the acid neutralizing component in the solid, there will be little effect of pH observed.

Using these assumptions, the material balance equation, initial condition and boundary conditions expressed in terms of dimensionless variables are as follows:

Material balance equation: $\dfrac{\partial \overline{C}}{\partial \overline{t}} = \dfrac{\partial^2 \overline{C}}{\partial \overline{x}^2} - D_{a,II}\overline{C}$ 　　　　(1)

Initial condition: $\overline{C} = 1.0$ at $\overline{t} = 0$, all $\overline{x}$

Boundary conditions: $\left(\dfrac{\partial \overline{C}}{\partial \overline{x}}\right) = 0$ at $\overline{x} = 1.0$, all $\overline{t}$

　　　　　　　　　　$\overline{C} = 0$ at $\overline{x} = 0$, all $\overline{t}$

where: $\overline{C} = \dfrac{C}{C_0}$; $\overline{x} = \dfrac{x}{L}$; $\overline{t} = \dfrac{D_e t}{RL^2}$; $D_{a,II} = \dfrac{kL^2}{D_e}$; $R = 1 + \rho_b K / \varepsilon$

The dimensionless material balance identifies two dimensionless groups that are important to understanding the way the system behaves – $D_{a,II}$ and $\overline{t}$. The product of these gives another important dimensionless group, the first Damköhler Number ($D_{a,I} = kt/R$). The material balance equation has been solved for the case of heat transfer (Carslaw and Yaeger, 1959) and the result can be applied to this problem.

$$\frac{M_t}{M_0} = 2\sum_{i=0}^{\infty} \frac{1-\exp\left(D_{all} + \left(\frac{(2i+1)\pi}{2}\right)\bar{t}^2\right)}{D_{all} + \left(\frac{(2i+1)\pi}{2}\right)} \tag{2}$$

where: M_t = amount leached at time t; M_0 = amount initially in solid.

Figure 2 shows the results of the model plotted for different value of $D_{a,II}$. This figure shows that for $D_{a,I} <0.1$, all curves are equivalent and linear. In this region, the reaction has not had sufficient time to degrade the target organic so that a model based on no reaction is reasonably valid.

$$\frac{M_t}{M_0} = \left(\frac{4D_e t}{\pi L^2 R}\right)^{1/2} \tag{3}$$

This figure also shows that the maximum fraction leached is approached as $D_{a,I}$ approaches a value of 1.0. If this value is taken to be $\pi/4$, then the maximum fraction leached can be calculated from Equation 3 using $t=\pi R/4k$.

$$\left(\frac{M_t}{M_0}\right)_{max} = \left(\frac{D_e}{L^2 R}\right)^{1/2} = D_{a,II}^{-0.5} \tag{4}$$

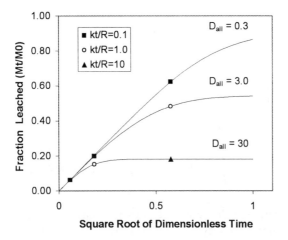

FIGURE 2. Model predictions for fraction leached as a function of dimensionless time for different values of the second Damköhler number.

The reaction coefficient needed to maintain the maximum fraction leached below a specified value depends on L (the ratio of waste volume to area exposed to leaching) and the effective diffusivity. Ranges of L, example dimensions that would give that value of L, the minimum rate coefficient required and its half-life are given in Table 3.

TABLE 3. Effect of L on minimum k for maximum fraction leached of 0.05.

L	Dimensions (m x m x m)	k (day^{-1})	($t_{1/2}$) (day)
2	40 x 40 x 5	1.7 E-3	401
0.5	4 x 4 x 2	2.8 E-2	25
0.1	0.6 x 0.6 x 0.6	6.9 E-1	1.0

The results in Table 3 show the importance of maintaining high L in order to contain contaminants for a sufficient time so that they will be effectively degraded. If the waste form remains large (e.g. L>2), then even very slow degradation reactions with half-lives in excess of one year would be fast enough to result in no more than 5% release to the environment. This release could occur over a long period of time, resulting in negligible concentrations.

CONCLUSIONS

Degradation experiments showed that Fe(II)-based ds/s technology could effectively treat PCE in soils without production of substantial amounts of chlorinated intermediates. The half-lives of PCE in the ds/s systems ranged from 13 to 335 days. This technology has a potential to be extended to treating soils or sediments contaminated with other chlorinated hydrocarbons such as poly-chlorinated biphenyls (PCBs) and dioxins. This potential would be increased further if the reactive agent of the ds/s system were identified. The leach model shows that contaminant leaching can be determined by first and second Damköhler numbers and dimensionless time. The leach model also shows that L (the ratio of waste volume to area exposed to leaching) is an important parameter that determines the time allowable for degradation reactions to remove target organics before they can leach into the environment.

ACKNOWLEGMENTS

This project has been funded entirely with funds from the United States Environmental Protection Agency (USEPA) as part of the program of the Gulf Coast Hazardous Substance Research Center. The contents do not necessarily reflect the views and policies of the US EPA nor does the mention of trade names or commercial products constitute endorsement or recommendation for use.

REFERENCES

Batchelor, B. 1998. "Leach Models for Contaminants Immobilization by pH-Dependent Mechanisms," *Environ. Sci. Technol.* 32(11): 1721-1726.

Carslaw, H.S., Yaeger, J.C. 1959. *Conduction of Heat in Solids*, 2nd Edition, Oxford University Press, London.

MacDonald, J. A. 1997. "Hard Times for Innovative Cleanup Technology." *Environ. Sci. Technol.* 31(12): A560-A563.

Mouvet, C., Barberis, D., Bourg, A. C. M. 1993. *J. Hydrol.* 149: 163-182.

Pd-CATALYZED HYDRODEHALOGENATION OF CHLORINATED COMPOUNDS: SPECTROSCOPIC ANALYSIS

N. Munakata, G. Lowry, W. Sriwatanapongse, and M. Reinhard
(Stanford University, Stanford, California)

ABSTRACT

Catalytic hydrodehalogenation is an emerging and promising technology for remediating groundwater contaminated with halogenated hydrocarbons and pesticides. The use of palladium catalysts is appealing because of the rapid reaction rates (on the order of minutes), conversion of a wide variety of compounds (e.g. ethenes, PCBs), and complete dechlorination to innocuous compounds in many cases. This work investigates changes in the catalyst surface via spectroscopic analysis, to better understand how the process might be optimized at a fundamental level. The analyses indicate that the palladium metal does undergo changes at the atomic level: initially the palladium is oxidized and present as a mixture of chloride and oxides. It forms metallic zero-valent clusters after exposure to hydrogen gas, and the palladium structure is rearranged in catalysts with reduced activity. At the larger scale, dendritic structures form on catalysts exposed to a continuous flow of water for extended periods (weeks to months). Elevated carbon and nitrogen levels on these samples indicate that the structures may be organic or biological in nature.

INTRODUCTION

Catalytic hydrodehalogenation is an emerging and promising technology for remediating groundwater contaminated with halogenated hydrocarbons and pesticides. Palladium (Pd) catalysts, for example, have been shown to catalyze hydrodehalogenation, hydrogenation, and reduction of a wide variety of compounds of environmental concern, including

- halogenated methanes: carbon tetrachloride, chloroform
- halogenated ethenes: tetrachloroethene, trichloroethene, dichloroethene, vinyl chloride
- halogenated ethanes: 1,1,2-trichloroethane, Freon-113
- halogenated benzenes: chlorobenzene, 1,2-dichlorobenzene
- PCBs: 4-chlorobiphenyl, aroclor
- PAHs: napthalene
- halogenated pesticides: 1,2-dibromo-3-chloropropane, lindane
- nitrate

(Kovenklioglu et al., 1992, Siantar and Reinhard, 1996, Schüth and Reinhard, 1998, Lowry and Reinhard, 1999, Schreier and Reinhard, 1995, Forni et al., 1997). These batch studies found extremely rapid reaction rates, implying that small reactors may be employed (e.g. in-well treatment). In addition, complete dechlorination is observed in clean systems for many of the halogenated compounds, resulting in innocuous hydrocarbon products such as ethane and methane.

In a field test, a pilot scale fixed bed Pd/H_2 reactor has been operating at Lawrence Livermore National Laboratories (LLNL) for twelve months. The reactor fits within a well bore hole and treats groundwater contaminated with 20-3000 μg/L of the following compounds: tetrachloroethene, 1,2-dichloroethane, trichloroethene, carbon tetrachloride, 1,1-dichloroethene, and chloroform. Initial conversion efficiencies were greater than 98% for each compound except chloroform which had 92% conversion and 1,2-dichloroethane which showed no measurable conversion. After 12 hours of continuous operation, conversion of some contaminants through the first bed begins to drop off; purging the reactor with 3 pore volumes of air and allowing it to it to sit for 12 hours results in almost complete activity recovery. The reactor has been operating in 12 hour cycles for over a year and activity has remained high (McNab, submitted). These results illustrate that it is possible to maintain high levels of catalyst activity for extended periods.

However, the fundamental processes which necessitate operating in 12 hour cycles are not well characterized. An improved understanding of these phenomena will facilitate optimization of the reactor design and operation. This work is a preliminary investigation into characterizing these processes through analysis of changes in the catalyst surface with exposure to waters of various compositions.

MATERIALS AND METHODS

The catalyst used in this work is a UOP catalyst consisting of dispersed Pd metal on an alumina support. The support has a surface area of 210-230 m^2/g, a pore volume of 0.82-0.96 cm^3/g, and a pore size distribution of 12-80 nm. The metal loading is 1% Pd by weight, with a dispersion of approximately 68% and a metal cluster size of approximately 20 nm. The catalyst was supplied as 1/16" (1.6 mm) spheres; these were crushed to particles of approximately 0.5 mm diameter for the column experiments, which are described below.

The catalyst samples analyzed in this work underwent one of three treatments: air exposure (fresh catalyst), batch exposure or column exposure. In the batch experiments, catalyst was exposed to one of the following aqueous salt solutions: sodium nitrate ($NaNO_3$), sodium bicarbonate ($NaHCO_3$), sodium chloride (NaCl), sodium sulfate (Na_2SO_4), sodium phosphate dibasic (Na_2HPO_4), sodium chromate (Na_2CrO_7), or sodium humics. A 40 mLbottle was filled with 20 mL of a given salt solution, and 0.4 g ofcatalyst was added. Two bottle were made of each salt solution: one was purged with hydrogen gas for 15 minutes and then capped, while the other was simply capped with no exposure to hydrogen. All bottles were then lightly agitated for approximately 6 months. At the end of this time period, all catalyst samples were tested and found to be as active for the hydrodechlorination reaction as fresh catalyst.

The operation of the column experiments is described in detail elsewhere. (Munakata, 1998). In these experiments, crushed catalyst particles were exposed continuously to a flow of TCE-amended and hydrogen-saturated water stream under one of the following conditions: deionized (DI) water and potassium nitrate (KNO_3); DI water and carbon dioxide (CO_2); DI water and sodium carbonate (Na_2CO_3); and CO_2 and groundwater (GW) from Livermore, CA. These solutions and the associated operating parameters are summarized in Table 1. In the DI/KNO_3 column, TCE removal was complete throughout the duration of the run; however, this column had significantly larger catalyst mass and it is therefore possible that the catalyst near the inlet of the column had deactivated but that the deactivation could not be detected (i.e. the remaining catalyst was still able to completely remove the TCE). The DI/CO_2 column was completely deactivated; no TCE removal was observed at the end of the experiment. Finally, the DI/Na_2CO_3 ended with approximately 75% decline in TCE removal, and the groundwater column showed a 90% decline in catalyst activity over the course of the experiment.

TABLE 1. Operating parameters of column experiments.

Solution	Catalyst Mass (g)	Running Time (d)	Loss of Activity (%)
DI/KNO_3	2.0	140	0%
DI/CO_2	0.10	14	100%
DI/Na_2CO_3	0.10	43	75
GW/CO_2	0.10	6	90

The catalyst used in the batch and column experiments was then analyzed with scanning electron microscopy (SEM), x-ray photoelectron spectroscopy (XPS), and extended x-ray absorption fine structure (EXAFS). SEM analyses were performed using a Hitachi S-2500 SEM. Samples were first coated with a Pd/Au alloy to make them conductive. The XPS analyses were performed using a Surface Science S-Probe XPS; an electron gun was used to neutralize samples and minimize charging during analysis. All binding energies were normalized to the oxygen peak at 531.4 eV, which corresponds to oxygen binding energy in alumina. EXAFS experiments were conducted using a Si 220 crystal on beam line 4-3 at the Stanford Synchrotron Radiation Laboratory at the Stanford Linear Accelerator facility. Samples were held in a Teflon cell with Mylar windows and were 2 mm thick. The Pd k-edge was measured using a fluorescence detector with Xe flow.

RESULTS AND DISCUSSION

SEM analyses were performed on three catalyst samples: fresh catalyst, catalyst exposed to DI/Na_2CO_3 for 43 days, and catalyst exposed to DI/KNO_3 for 140 days. The images, given in Figures 1 and 2, show a clean surface on fresh and batch catalysts, but signs of dendritic structures forming on both of the exposed catalysts, with more extensive formations on the DI/KNO_3 catalyst. This is consistent with biogrowth or accumulation of organics. The XPS results shown in Figure 2(a) also support this theory. Fresh catalyst or catalyst exposed in batch experiments show relatively low quantities (< ~5 atomic percent) of carbon (C) and nondetectable quantities of nitrogen (N). In contrast, all of the column catalysts showed detectable levels of N and elevated levels of C (> 10 atomic percent). Analysis of the binding energies of N (data not shown) offer further support; the binding energies indicate a reduced species of N and fall within the range of organic N.

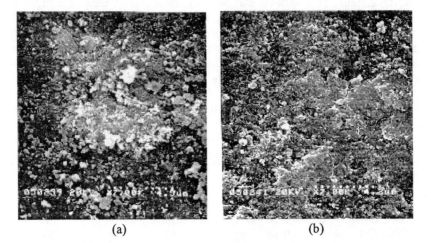

(a)   (b)

FIGURE 1. SEM images of Pd/Alumina catalyst surface at 7000X magnification, total width of each image approx. 15 micron: (a) fresh catalyst, (b) exposed to DI purged with hydrogen in batch.

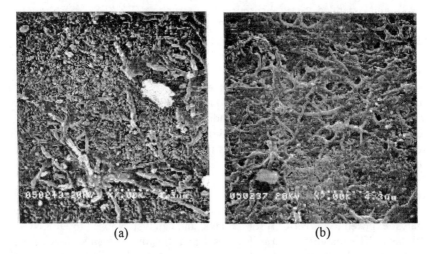

(a)   (b)

FIGURE 2. SEM images of Pd/Alumina column catalyst surface at 7000X magnification, total width of each image approx. 15 micron: (a) exposed to DI/Na$_2$CO$_3$ for 43 days, (b) exposed to DI/KNO$_3$ for 140 days.

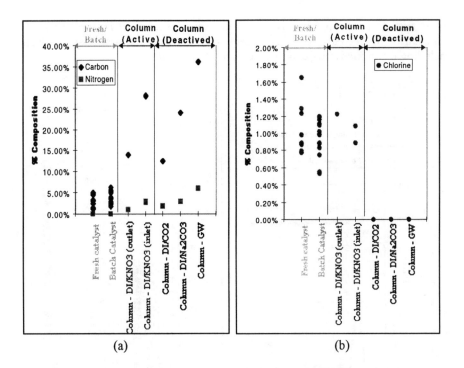

(a)   (b)

FIGURE 2. XPS Analyses of catalyst. (a) Accumulation of C and N on the surface of column catalysts. (b) Disappearance of Cl in deactivated catalysts.

EXAFS measurements were made on three catalyst samples: fresh catalyst, catalyst exposed to hydrogen gas, and catalyst used in the column experiment with the groundwater. The signals from these materials were then compared to those measured from standard materials: pure Pd metal, palladium oxide (PdO), and palladium chloride (PdCl$_2$). A comparison of the Fourier transforms of the EXAFS signals are shown in Figure 3(a). The Fourier transform signal is a function of radial distance R, in Å (not corrected for phase-shift); peaks indicate the presence of atoms at that atomic distance from the Pd atom. Of the three standard materials, O atoms bind most closely to the Pd atom, then Cl and finally Pd.

By comparing the Fourier transform of the catalyst samples to that of the standard materials, the nearest neighbors of Pd atoms in the catalyst can be determined. From Figure 3(a), it can be seen that the fresh catalyst has both O and Cl nearest neighbors, and very few Pd neighbors. Upon exposure to hydrogen gas, Pd nearest neighbors appear. After exposure to groundwater, both the Cl and O peaks diminish, indicating a relative drop in those neighbors. This can be confirmed more quantitatively using the data-fitting program Datfit. By weighting the standard material EXAFS signals and fitting them against the EXAFS signal of the catalyst samples, the relative contributions can be determined (Figure 3(b)). The Datfit results confirm the conclusions made by visual analysis of the Fourier transform signals. In addition, the loss of Cl-bound

Pd is further confirmed by XPS measurements (shown in Figure 2(b)). Cl concentrations decrease to below detection limits for all deactivated catalysts, including the groundwater exposed catalyst.

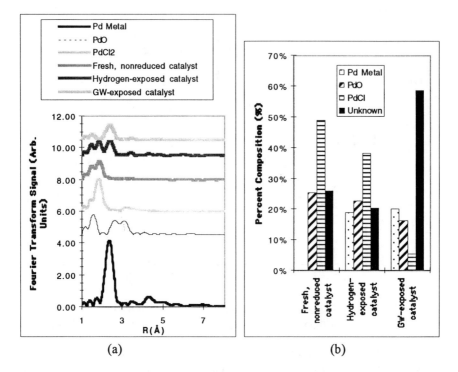

(a)   (b)

FIGURE 3. EXAFS analyses of catalyst: (a) Fourier transform of signal . (b) Composition of Pd nearest neighbors.

CONCLUSIONS

Catalysts which are exposed to flows of water for time periods between weeks and months developed dendritic structures on the surface. Samples with these structures also exhibited elevated levels of carbon and nitrogen on the surface. The nitrogen was in a reduced (i.e. probably organic) state. These results indicate that biological or organic structures formed on the catalyst surface after prolonged exposure to water flows. If the surface were to become completely coated, the contaminants of interest may be unable to reach the Pd surface, thereby reducing catalyst activity. Further research will investigate the effectiveness of maintaining catalyst activity through the prevention biogrowth or organics accumulation (e.g. by adding an oxidizing agent to the inlet stream).

EXAFS and XPS analyses indicate that the Pd in fresh catalyst is oxidized and bonded to Cl and O atoms. Upon exposure to hydrogen gas, some of the Cl-bound is converted to metallic (zero-valent) Pd clusters. Finally, catalyst which has been exposed to groundwater and has decreased catalytic activity has a very low percentage of Cl-bound Pd. This is corroborated by XPS data which indicate low Cl content on deactivated catalysts. However, the new structure of the Pd in this low-activity catalyst has not yet been established; more work is needed to determine the structure and develop methods to minimize or prevent its formation.

ACKNOWLEDGMENTS

This work was funded by the Western Region Hazardous Substance Research Center as sponsored by the United States Environmental Protection Agency under Assistance ID number R825689-01, and by a National Science Foundation Student Fellowship.

REFERENCES

Kovenkiloglu, S., C. Zhihua, D. Shah, R. Farrauto, and E. Balko. 1992. "Direct Catalytic Hydrodechlorination of Toxic Organics in Wastewater." *AICHE Journal.* 38(7): 1003-1012.

Lowry, G. V. and M. Reinhard. 1999. "Hydrodehalogenation of 1- to 3-Carbon Halogenated Organic Compounds in Water Using a Palladium Catalyst and Hydrogen Gas." *Environ Sci. and Technol.* 33(12): 1905-1910.

McNab, W. W. Jr., R. Ruiz, and M. Reinhard. 2000. "In Situ Destruction of Chlorinated Hydrocarbons in Groundwater Using Catalytic Reductive Dehalogenation in a Reactive Well: Testing and Operational Experiences." *Environ Sci. and Technol.* 34(1): 149-153.

Munakata, N., P.V. Roberts, M. Reinhard, and W. W. McNab, Jr. 1998. "Catalytic Dechlorination of Halogenated Hydrocarbon Compounds Using Supported Palladium: A Preliminary Assessment of Matrix Effects." *IAHS Publication.* 250:491-496.

Schreier, C. G. and M. Reinhard. 1995. "Catalytic Hydrodehalogenation of Chlorinated Ethylenes Using Palladium and Hydrogen for the Treatment of Contaminated Water." *Chemosphere.* 31(6): 3475-3487.

Schüth, C. and M. Reinhard. 1998. "Hydrodechlorination and Hydrogenation of Aromatic Compounds over Palladium on Alumina in Hydrogen-Saturated Water." *Appl. Cat. B: Environ.* 18: 215-221.

Siantar, D.P., C.G. Schreier, and M. Reinhard. 1996. "Treatment of 1,2-Dibromo-3-Chloropropane and Nitrate-Contaminated Water with Zero-Valent Iron or Hydrogen/Palladium Catalysts." *Water Research.* 30(10): 2315-2322.

SOLAR-POWERED PUMPING/AERATION TREATMENT OF VOLATILE ORGANICS IN GROUNDWATER

Michael Heffron
Foster Wheeler Environmental Corporation
Langhorne, Pennsylvania

Abstract: Shallow groundwater, contaminated with low concentrations of chlorinated and non-chlorinated volatile organic compounds (VOCs), was successfully recovered and treated using solar-powered pumping and aeration systems in a field-scale project conducted in July and August 1998. The systems were designed to capture and treat the outer edges of two separate groundwater plumes at the Lakehurst Naval Air Engineering Center (NAEC) site located in Ocean County, New Jersey. One groundwater plume, emanating from a former sanitary landfill, exhibited low concentrations of chlorobenzene, vinyl chloride, 1,1-dichloroethane, cis 1-2-dichloroethane, 1,2-dichlorobenzene, 1,3-dichlorobenzene, 1,4-dichlorobenzene, dichlorodifluoromethane, 1,2,3-trichlorobenzene, 1,2,4-trichlorobenzene, and benzene. The second groundwater plume, associated with a testing track used to propel jet engines for testing cable arrestor assemblies, exhibited low concentrations of trichloroethene, tetrachloroethene, and p-isopropyltoluene. Navy-surplus solar panels were used as a power source for the pump and treat systems. Solar-powered piston pumps were connected to the solar panel arrays to pump groundwater from monitoring wells through elevated sprinkler heads in order to volatilize the VOCs. The solar-powered pumping systems were designed and constructed for automatic operation. Groundwater samples were collected prior to and after sprinkler aeration to quantify the VOC removal rate. The laboratory analyses confirmed an effective VOC removal rate of 94% to non-detectable levels.

INTRODUCTION

Foster Wheeler Environmental Corporation (Foster Wheeler Environmental) was contracted by the Northern Division, Naval Facilities Engineering Command to design, and construct a solar-powered pump and sprinkler system to pump and aerate, via sprinkler, groundwater containing low concentrations of volatile organics. The outer edge of two separate groundwater contamination plumes were the focus of the groundwater pumping systems. Due to the remote location of the groundwater wells, and the requirement for low operation and maintenance costs, solar energy was chosen as the power source of the pump and treat system.

Naval Air Engineering Center (NAEC) site is located in Jackson and Manchester Townships, Ocean County, New Jersey, approximately 14 miles inland from the Atlantic Ocean (U.S.G.S., 1971). Surrounding land use is primarily undeveloped woodlands and open areas, with the closest residential area, the Borough of Lakehurst, located southeast of the facility. The U.S. Navy assumed control of the property in 1919, and it was formally commissioned Naval Air Station (NAS) Lakehurst in 1921 (U.S. EPA, 1993). The NAEC's mission is to conduct research, development, engineering, testing and system integration, limited production, and procurement for aircraft and airborne weapons systems (U.S. EPA, 1993).

The U.S. Navy's Installation Restoration Program (IRP) investigations revealed groundwater contamination at a former sanitary landfill. One groundwater plume, emanating from the former sanitary landfill, exhibited low concentrations of chlorobenzene, dichlorodifluoromethane, vinyl chloride, 1,1-dichloroethane, cis 1-2-dichloroethane, 1,2-dichlorobenzene, 1,3-dichlorobenzene, 1,4-dichlorobenzene, 1,2,3-trichlorobenzene, 1,2,4-trichlorobenzene, and benzene (U.S. EPA, 1993). The second groundwater plume, associated with a testing track used to propel jet engines for testing cable arrestor assemblies, exhibited low concentrations of trichloroethene, tetrachloroethene, and p-isopropyltoluene (U.S. EPA, 1993)

The objective of this effort was to install low-cost pumping systems in the existing monitoring wells on the outer edges of the groundwater plumes in order to remediate the low concentrations of volatile organic compounds in those wells. Solar-powered pumping systems were installed in two monitoring wells at each of the outer edges of the two separate groundwater plumes. All monitoring wells were constructed of 2-inch diameter PVC, completed at a depth 30 feet below grade. The wells were completed in fine to medium grained sand with varying amounts of trace silt and fine gravel. The depth to water in the monitoring wells ranged from 3.5 to 16.7 feet below grade. The U.S. Environmental Protection Agency has previously evaluated sprinkler irrigation as a viable volatile organic compound treatment and disposal method under the Superfund Innovative Technology Evaluation (SITE) Program (U.S. EPA, 1996)

MATERIALS AND METHODS

Solar Panels. The solar panels were constructed of photovoltaic (PV) cells, which produce small quantities of electricity when exposed to sunlight. A PV cell is a thin semiconductor wafer constructed from highly purified silicon (Pratt, 1999). The PV cells are constructed so that one side of the wafer produces a surplus of electrons and the other side produces a deficit of electrons (Pratt, 1999). When the cells are exposed to photons from sunlight, electrons are transferred from one side of the PV cell to the other by the voltage difference. An external circuit attached to the contacts establishes an electrical current. The size and efficiency of the PV cell and the amount of solar exposure determine the amount of electrical current. Since the discovery of PV cells in the early 1950's, the price of the cells has steadily come down in price, making them an affordable alternative as a power source. The price of PV cells today is approximately $6.00 per watt compared to approximately $40,000 per watt in the early 1960's (Pratt, 1999).

PV cells are wired in parallel in a PV module to increase the electrical current and produce a higher voltage. The typical PV module consists of 36 cells in series (Pratt, 1999). The modules are typically set in an aluminum frame, covered in tempered glass and waterproofed. The PV modules can be connected in series to construct solar panel arrays comprised of several modules.

The Navy supplied surplus solar panel arrays that were already integrated into aluminum frames. The PV modules were 48 inches long by 12 inches wide and 1.5 inches deep. An aluminum frame, 8 feet long by 4 feet wide, was constructed around the eight (8) PV modules, wired in series. The solar panel arrays supplied by the Navy were approximately 10-years old, and had an anticipated output of 70% of new panels based on age and condition. Each PV module produced approximately 2.1 amps and 36 watts. Each solar panel array contained 8 PV modules, therefore produced 16.8 amps or 288 watts.

In order to maximize the solar exposure, the solar panel arrays should be facing the southern direction, and the angle of the panels set to the horizontal in accordance with the time of the year (NTIS, 1995). Published data indicates that the tilt angle of the solar panel near the latitude angle will provide the most energy over a full year (NTIS, 1995). The latitude at the site was approximately 40°. Published data advises that tilt angles of + 15° (55°) in the winter and - 15° (25°) in the summer will skew the energy production for those respected seasons (NTIS, 1995). The solar panel array should be adjusted to the proper angle for the season in order to maximize the efficiency.

For the Lakehurst Project, frames were constructed to position the solar panel array above ground and at the correct angle to the sun, as well as to enable the array to be mobile. Wooden frames were constructed of treated lumber in order to support the solar panel arrays at the proper angle to the horizontal. The frames were constructed with adjustable supports to modify the position of the arrays to angles of either 25°, 40°, or 55° with respect to the horizontal. Eight-inch pneumatic wheels, on swivel coasters, were installed on one side of the frame to allow mobility of the structure.

The power requirements of the installed pumps ranged from 280 watts to 314 watts, depending upon backpressure and lift conditions. In order to meet the power requirements of the pumps, two solar panel arrays were wired in tandem to a pump at each of the recovery wells.

Solar-Powered Pumps. The pumps used for this application were Dankoff Solar Force ™ Piston Pumps (Model #3020-12PV). The piston pumps were constructed with a cast iron body, water box, oil-bath crankcase, piston assembly, gear belt drive, and a magnetic DC motor to drive the piston and gear assembly. Pressure relief valves were also installed on the pumps (set to release at approximately 80 pounds per square inch [psi]). The pumps were approximately 22 inches long by 13 inches wide, by 16 inches high and weighed approximately 80 pounds. The DC motor, powered by the solar panels, turned a gear assembly which was connected to the piston pump with a belt. The belt connected the smaller gear assembly of the motor to the larger gear assembly on the piston pump. The gear assembly of the pump drives a piston, which pulls and pushes the water. These pumps were chosen due to their lift and operation capability with a significant backpressure, durability and minimal operation and maintenance requirements.

The Dankoff Solar Force ™ Piston Pump (Model #3020-12PV) has the capability of pushing water as high as 230 feet or against 100 pounds per square inch (Dankoff, 1996). A piston pump, or displacement pump was used instead of surface rotary, jet or submersible pump because of the efficiency of the piston pump operating at a higher total head. The total head of the pumping and aeration system includes the static lift from the well, and more importantly, the discharge head needed to operate the sprinkler system with a finer droplet size. A solar-powered surface rotary pump only operates at a 15 to 25 % efficiency with a head of 15 to 60 feet (NTIS, 1995). A jet pump only operates at a 25% efficiency with a head of 60 feet (NTIS, 1995). The submersible pumps can operate with a 35% efficiency with a head of 300 feet (NTIS, 1995). The displacement pump can operate with a 45% efficiency with a head of over 300 feet. The static lift and the discharge head required for the pumping and aeration system equated to a total dynamic head of approximately 220 feet. While the lift from the wells were typically less than 25 feet, the backpressure from the sprinkler system increased the total dynamic head. The sprinkler system has a higher rate of volatilization of organic compounds with a finer mist (Spalding et al, 1994). A finer droplet size spray on the sprinkler requires the adjustment of the sprinkler that results in a higher backpressure or discharge head. The Dankoff Solar Force ™ Piston Pump is capable of a suction capacity of 25 vertical feet, and is capable of pumping up to 4.7 gallons per minute (gpm) at 96 psi (Dankoff, 1996).

Under normal operating conditions, the solar-powered pumping systems function automatically, turning on when there is sufficient sunlight to power the system, and turning off when there is not sufficient sunlight (Foster Wheeler, 1998). Based on operating conditions observed, the solar-powered pumping and sprinkler systems were capable of pumping between 3 to 4.5 gpm from the monitoring wells (Foster Wheeler, 1998).

As per the manufacturers' suggestion, the pump was sheltered from rain, dirt and direct sun, but allowed to have adequate ventilation. The pumps were placed in plastic "doghouses" which allowed the top to be removed for easy access to the pump. It is normal to expect water and some oil to drip from the rod packing of the pump (Dankoff, 1996). Oil absorbent booms were placed inside the doghouses, and holes were drilled in the bottom to allow water to drain.

Auxiliary Equipment. The solar-powered pumps were connected directly to the solar panel arrays (PV-direct). PV-direct, without using battery storage, is the most efficient way to utilize solar energy (Pratt, 1999). Approximately 20% to 25% of the energy is saved by eliminating the conversion to battery power (Pratt, 1999). The PV-direct systems must be capable of operating on the variable power output of the PV modules. The two solar panel arrays were wired, in tandem, to a Linear Current Booster (LCB™). The LCB™ is a dynamic impedance matching device designed to maximize the power transfer from the PV modules to the pump, and allow operation in lower light conditions (Sun Selector, undated). An LCB can boost the PV-direct pump output up to 40% (Pratt, 1999). A 12 VDC (30-40 amp) plug-in fuse was installed between the current booster and the pump to prevent motor overload.

In order to prevent the pumps from pumping dry, differential switches were installed on the intake line of each of the wells. The Sun Selector WLS-3 differential switch was used because of its compatibility with the LCB™. To connect the differential switch, a one-inch diameter solid piece of PVC was placed in-line on the intake hose from the well. Small holes were drilled in the PVC pipe, copper wire sensors were inserted in the PVC, and the holes sealed with silicon. The differential switch, which is water-proof, was attached to the side of the PVC and wired back to the LCB™. The differential switch was wired to shut off the pump if no water is in the intake line.

As per the manufacturers' suggestions, non-kinking hose, foot valves, and check valves were installed on the pumping system. One inch diameter black PVC hose (rated for 100 psi) was used for the intake and discharge piping of the system. The hose does not readily kink and is easy to handle. A brass strainer/foot valve was attached to the bottom of the one-inch hose and placed down the well. The strainer/foot valve allowed the pump to be primed, and keep its prime after the system is shut off. A check valve was also installed on the one-inch diameter PVC discharge hose from the pump. The check valve allowed water to be pumped from the pump to the sprinkler when operating, and closed to trap existing water in the line when the system is turned off.

Pulsating sprinklers were used to aerate the water pumped from the wells. The sprinklers were equipped with diffuser pins, adjustable distance devices, and collars to adjust sprinkler coverage areas. The sprinklers had a non-tipping base, which allowed easy mounting of the sprinklers above-grade. The sprinklers were bolted to a treated plywood base, and then bolted to a 4 inch by 4 inch treated wooden post. The wooden post was placed in the ground such that the sprinklers were elevated approximately three feet above grade. The PVC hose was connected to the sprinkler, and a wire strainer was also inserted at the sprinkler in order to prevent any small particulates from clogging the sprinkler. The diffuser pin was adjusted such that a fine spray was achieved at the sprinkler. The sprinklers were set up to pump a 360° coverage radius.

RESULTS AND DISCUSSION

Groundwater samples were collected from the solar-powered pumping/sprinkler systems at four wells (Wells IF, KI, KE, and HP) prior to and after aeration in order to determine the efficiency of the pumping/aeration process. The pumping/aeration systems, which were pumping at approximately 3.5 to 4 gallons per minute, were allowed to pump and spray for a duration that would have evacuated three to five times the volume of the water in the wells prior to sampling (Foster Wheeler, 1998). Groundwater samples were collected prior to, and after aeration. The groundwater samples were analyzed for volatile organic compounds (VOCs) using EPA SW846 method 502.2. In order to collect the water prior to aeration, one of the solar panel arrays was disconnected in order to slow the flow rate through the pump, and then the sample was obtained from the pump discharge, prior to the sprinkler. In order to collect the aerated groundwater samples, chilled, stainless steel bowls were placed on the ground surface in the spray aeration path while the systems were turned off. The circular coverage of the sprinkler was reduced by adjusting the sprinkler in order to sprinkle water over a smaller area, thus decreasing the collection time of the sprayed water. Once the water was collected in the stainless steel bowls, the water was immediately transferred to VOC vials with septum caps and placed on ice.

The groundwater samples collected prior to and after aeration revealed a VOC removal rate from 94% to non-detectable levels (Foster Wheeler, 1998). Tables 1 though 3 summarize the analytical results of the groundwater samples collected prior to and after aeration.

CONCLUSIONS

The solar-powered pumping and aeration systems were successful in their application as a low-cost, low maintenance alternative to remediate low concentrations of volatile organic compounds from the shallow groundwater aquifer. The systems were designed and

TABLE 1: Well IF: Analytical results of groundwater prior to and after aeration

	IF-98-01 Prior to Aeration (8/6/98)	IF-98-02 After Aeration (8/6/98)
Dichlorodifluoromethane	1.5 ug/l	ND
Vinyl Chloride	8.0 ug/l	ND
1,1-Dichloroethane	2.9 ug/l	ND
Cis-1,2-Dichloroethene	4.0 ug/l	ND
Chlorobenzene	6.8 ug/l	ND
1,4-Dichlorobenzene	10.5 ug/l	0.68 ug/l
Benzene	1.1 ug/l	ND

TABLE 2: Wells KI and KE: Analytical results of groundwater prior to and after aeration

	KI-98-01 Prior to Aeration (8/6/98)	KI-98-02 After Aeration (8/6/98)	KE-98-01 Prior to Aeration (8/6/98)	KE-98-02 After Aeration (8/6/98)
Trichloroethene	0.64 ug/l	ND	ND	ND
Tetrachloroethane	1.1 ug/l	ND	0.88 ug/l	ND
p-Isopropyltoluene	0.78 ug/l	ND	ND	ND

TABLE 3: Well HP: Analytical results of groundwater prior to and after aeration

	HP-98-01 Prior to Aeration (8/6/98)	HP-98-01D Duplicate of IF-98-01	HP-98-02 After Aeration (8/6/98)	HP-98-02D Duplicate of IF-98-02
Dichlorodifluoromethane	2.7 ug/l	2.5 ug/l	ND	ND
Vinyl Chloride	1.7 ug/l	1.6 ug/l	ND	ND
Chloroethane	0.50 ug/l	0.52 ug/l	ND	ND
1,1-Dichloroethane	2.0 ug/l	1.9 ug/l	ND	ND
Cis-1,2-Dichloroethene	6.4 ug/l	6.3 ug/l	ND	ND
Chlorobenzene	8.2 ug/l	7.7 ug/l	0.91 ug/l	0.92 ug/l
1,3-Dichlorobenzene	2.1 ug/l	2.0 ug/l	ND	ND
1,4-Dichlorobenzene	5.2 ug/l	5.1 ug/l	0.65 ug/l	ND
1,2-Dichlorobenzene	2.0 ug/l	1.9 ug/l	ND	ND
1,2,4-Trichlorobenzene	4.5 ug/l	6.5 ug/l	0.65 ug/l	0.60 ug/l
1,2,3-Trichlorobenzene	0.80 ug/l	1.4 ug/l	ND	ND
Benzene	0.56 ug/l	0.60 ug/l	ND	ND

constructed to operate automatically with proper solar exposure. Safety mechanisms were built into the system to ensure that the pumps were not damaged if low water level conditions occurred in the monitoring wells. The solar-powered pumping and aeration systems were demonstrated to operate automatically and have an effective groundwater VOC removal rate of 94% to non-detectable levels.

The cost of the pumping and aeration systems design, construction, installation, and laboratory analyses totaled approximately $29,000, or $7,250 per pumping/aeration setup. While the capital costs of the solar-powered pumping system may be higher than the installation of an electrical or fuel-driven pump, the operational and maintenance (O&M) costs are significantly lower. The O&M costs of the systems operations are negligible. The Dankoff Solar Force [TM] Piston Pumps have a life expectancy of 20 years. The only O&M requirements on the piston pumps is monthly checking of oil levels and occasionally replacing seals on pump (extras supplied with original purchase). The solar-power units are self-sufficient and do not require any battery backup or generators.

REFERENCES

Dankoff Solar Product. 1996. *Solar Force Piston Pump Installation and Service Manual.*

Foster Wheeler Environmental Corporation. 1998. *Analytical Results of Solar-Powered Pumping Aeration Systems, Lakehurst Naval Air Engineering Center.*

Foster Wheeler Environmental Corporation. 1998. *Operation and Maintenance Manual for Solar-Powered Pumping Systems, Lakehurst Naval Air Engineering Center.*

National Technical Information Service (NTIS) U.S. Department of Commerce. 1995. *Stand-Alone Photovoltaic Systems: A Handbook of Recommended Design Practices.*

Pratt, D., J. Schaeffer. 1999. *Solar Living Source Book, The Complete Guide to Renewable Energy Technologies and Sustainable Living.*

Spalding, R. F., M. Exner, L. Parra-Vicay and M. Burbach. 1994. "Sprinkler Irrigation: A VOC Remediation Alternative" *Journal of the Franklin Institute* 331(A): 231-241.

Sun Selector. No Date. *Installation and Operation Guide for the Linear Current Boosters.*

U. S. Geological Survey. 1971. *Lakehurst, New Jersey Quadrangle Map,* 1957. Rev. 1971, scale 1:24,000. U.S. Department of the Interior, Washington D.C.

United Stated Environmental Protection Agency. 1993. Superfund Sites-New Jersey: *Naval Air Engineering Center Record of Decision.*

United Stated Environmental Protection Agency, National Risk Management Research Laboratory. 1996. *Demonstration Plan for Sprinkler Irrigation as a VOC Treatment and Disposal Method.*

REMEDIATION OF INDOOR AIR IMPACTS DUE TO 1,1 DCE GROUNDWATER CONTAMINATION

David J. Folkes and David W. Kurz (EnviroGroup Limited, Englewood, CO)

ABSTRACT: Active soil depressurization systems were installed in 60 residential homes to control the indoor air concentrations of 1,1 dichloroethene (1,1 DCE), ranging from below detection (0.040 ug/m3) to 91 ug/m3, due to groundwater contamination. Groundwater is found at depths of 5 to 10 m, with 1,1 DCE concentrations ranging up to approximately 1600 ug/L. All systems successfully reduced indoor air concentrations below the 10^{-5} risk level, although approximately 30% of the systems required modifications, such as additional suction pits and larger fans, to achieve this level. Soil depressurizations in the range of 2.5 to 30 Pa were achieved and appear to be adequate to compensate for furnace operation and fluctuations in temperature and atmospheric pressures.

INTRODUCTION

Although the groundwater-to-indoor air pathway has been recognized for some time (e.g., Johnson & Ettinger 1991, ASTM 1995), there is little information in the literature concerning the potential for chlorinated compounds in groundwater to impact indoor air (Richardson, 1997). One reason for this may be the paucity of regulations requiring testing of indoor air in buildings over contaminated groundwater. A few states have recently developed indoor air pathway criteria for groundwater; however, these may underestimate the occurrence of indoor air impacts (Fitzpatrick and Fitzgerald 1996) and are based on models that have not been validated (Altshuler and Burmaster 1997).

For example, Massachusetts groundwater-to-indoor air criteria (310 CMR40.0932 [6]) are not applied to groundwater more than 15 ft (4.6 m) below ground surface, whereas experience at the site described herein indicates groundwater at depths greater than 6 m can significantly impact indoor air quality based on current toxicity criteria for 1,1 DCE.

Although benzene has generally received more attention than other volatile compounds in groundwater (e.g., Fischer et al., 1996), 1,1 DCE is more likely to cause indoor air impacts. The relative risk to indoor air posed by volatile compounds in groundwater is a function of the product of the Henry's Law Constant and the cancer slope factor for each compound (ASTM 1995). On this basis, risks posed by 1,1 DCE are approximately 28 times greater than benzene.

This paper is intended to fill some of the information gap on indoor air impacts caused by chlorinated compounds in groundwater, and by 1,1 DCE in particular, by presenting data on the efficacy of active soil depressurization systems in 60 residential homes.

Site Description. The site is located in Denver, Colorado, where historic use of solvents at a former manufacturing facility resulted in a groundwater 1,1 DCE plume that extends a distance of approximately 300 m under a residential neighborhood (Figure 1). The compounds of concern include 1,1,1 trichlorethane (TCA), tricholorethylene (TCE), tetrachlorethylene (PCE), and 1,1 dichloroethene (1,1 DCE). Concentrations in offsite monitoring wells are highest near the site boundary (Table 1) and decrease with distance from the site.

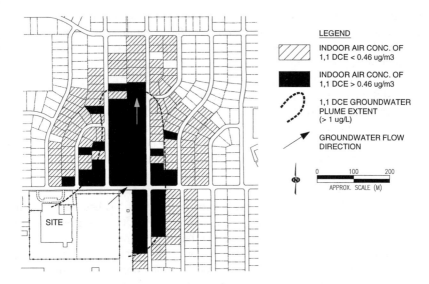

FIGURE 1. Extent of groundwater and indoor air impacts.

Table 1. Maximum Observed Concentrations in Off-Site Monitoring Wells.

Compound	Maximum Concentration (ug/L)
1,1 DCE	1600
PCE	360
1,1,1 TCA	1500
TCE	510

The site is located on fine-grained silt and clay loess deposits, underlain by weathered sandstone, siltstone and claystone deposits of the Denver Formation at depths of approximately 6 to 12 m. Fine grained soils typically extend to depths of at least 3 m. In some locations, sand and gravel alluvium is present between the loess and bedrock. Groundwater is found at depths of 5 to 6 m near the site boundary, increasing to depths of 10 m toward the northern end of the plume. Groundwater flows in the upper portion of the weathered bedrock, ultimately discharging to sand and gravel alluvium in a floodplain northeast of the site area. The downgradient edge of the solvent plume does not appear to be moving and may be controlled by dilution in the floodplain deposits.

The indoor air of 166 homes in the neighboorhood (Figure 1) was tested by the procedures described below, with 1,1 DCE concentrations ranging from below the reporting limit (0.04 ug/m3) to 91 ug/m3. Concentrations in 63 homes exceeded the Colorado Department of Public Health and Environment (CDPHE) interim action level of 0.49 ug/m3, representing an excess cancer risk of 1 x 10[-5] based on CDPHE risk equations (CDPHE 1999). All but one of these homes were in or immediately adjacent to the area overlying the groundwater plume, defined by 1,1 DCE concentrations exceeding 1 ug/L (Figure 1). Indoor air concentrations of 1,1 DCE are generally higher where groundwater concentrations are higher, but may vary by two orders of magnitude in houses over groundwater of similar concentrations. The remaining home (indoor air concentration of 0.88 ug/m3) was located approximately 300 m cross-gradient of the groundwater plume and was impacted by non-groundwater sources. Access was granted for installation and adjustment of venting systems at 60 of the homes over the groundwater plume.

DESCRIPTION OF REMEDIATION SYSTEMS

Sub-slab depressurization (SSD) systems were installed in homes with basements or slab-on-grade construction, while sub-membrane depressurization (SMD) systems were installed in homes with crawl spaces, as described below.

Sub-Slab Depressurizaton Systems.   A typical SSD installation is shown on Figure 2a. Suction pits were created below the concrete floor slabs by drilling a 9 to 10 cm diameter hole through the slab, ideally but not necessarily located near the middle of slab, then hand-excavating a void approximately 15 cm deep and 40 cm in diameter to increase the effectiveness of the depressurization sytem (e.g., see Bonnefous et al., 1992). A 76 mm diameter PVC pipe from the suction side of the fan was inserted in the hole, extending 25 mm below the bottom of the slab. The annular space between the pipe and slab was sealed with acrylic latex caulk.

(a)   **(b)**

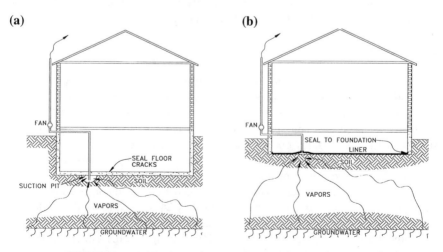

FIGURE 2. Schematic drawing of typical SSD (a) and SMD (b)

Initially, one suction point was created in the floor slab, based on information in the literature that indicates that one suction point should be sufficient to depressurize the floor area below most single family homes (e.g., EPA 1993). In finished basements, the void and pipe were placed in the utility area or in some other unfinished area, to the extent practical. The fans were installed in the attic or outside the house. Pipe exhausts were located at least 3 m above the ground and 3m from doors or windows, following EPA (1993) guidelines for radon systems. All visible cracks and joints in the floor were sealed with acrylic latex caulk.

Sub-Membrane Depressurization Systems. In homes with a crawl space, SMDs were installed by placing a 0.1 mm thick, cross-laminated polyethylene membrane or liner over the dirt floor and sealing the liner to the concrete foundation walls using acrylic latex adhesive (Figure 2b). The end of the pipe (76 mm PVC) from the suction side of the fan was inserted through a hole cut in the liner. The liner was sealed to the pipe at the penetration hole using vinyl tape to prevent loss of vacuum. When concrete footings divided the crawl space, a separate suction point was generally installed in each separate area between the footings. For homes with large crawl spaces, the fan was installed outside the house and the pipe was routed up the outside wall to exhaust above the roofline.

Fans. An inline centrifigal fan (Fantech HP190) was used to create the low pressure zone below the concrete slabs and liners. In most cases a 90 watt fan was installed, with a flow rate of 2.5 m3/min at 249 Pa. In situations where additional vacuum was desired (see Results and Discussion), a 150 watt fan with a flow rate of 5.7 m3/min at 249 Pa was installed. The fans operate 24 hours per day and each system has an oil-filled manometer in the pipe providing a visual indicator to the homeowner that the system is operating. All work was conducted by licensed contractors under mechanical and electrical permits as required by the City and County of Denver Building Inspection Department.

TEST METHODS

Summa Canister Tests. Indoor air samples were collected over a 24-hour period using an inert stainless steel container (SUMMA canister), provided by Quanterra Air Toxics Laboratory of Industrial City, CA. Each canister was laboratory cleaned, evacuated to a nominal vacuum of 0.01 torr, sealed, and shipped to the site under chain of custody documentation. The canister pressure was noted and recorded in the field at the beginning and end of the sampling event. Sample collection commenced with opening of the canister valve, which resulted in flow of air into the canister at a steady rate controlled by a regulator attached to the canister. Sample collection ceased approximately 24 hours afterwards by closing the valve. In some cases, the canisters had equilibrated with ambient atmospheric pressures when the canisters were retrieved, indicating that the sample collection period was less than 24 hours. The SUMMA canisters were shipped to the laboratory in batches under chain of custody protocols. Duplicate samples were collected at a rate of approximately 1 in 20.

SUMMA canister samples were analyzed by Quanterra in accordance with EPA Test Method TO-15 using a mass spectrometer operated in the Selective Ion Monitoring (SIM) Mode. For tests conducted after October 1998, equipment tuning procedures met the requirements of CDPHE (1999) guidelines. The SIM Mode monitors a few compounds instead of the entire mass spectra, allowing a 1,1 DCE reporting limit of 0.04 ug/m3.

Smoke Tests. Chemical smoke was used to visually verify response to installed systems at a perimeter location following EPA (1993) procedures. A small diameter test hole was drilled through the concrete slab at a location remote from the suction pit and smoke was introduced above the hole at the surface of the slab. If the response to the mitigation system was adequate, the chemical smoke would be pulled directly into the hole. If there was no response, the smoke would slowly dissipate in the air above the test hole. After testing, the holes were sealed with acrylic latex caulk.

Pressure Tests. At some homes, differential pressures across the slab or membrane were measured. Instantaneous measurements were collected using a handheld digital manometer (Dwyer Series 475-00 Mark III). The measurement capability of the unit ranges from 0 to 4 in WG (0 to 996 Pa) with an accuracy of +/-0.5% of the full-scale reading. Periodic tests over a 24-hour period were measured using three differential pressure meters (two Dwyer Series 607-1B and one Dwyer Series 607-2B) connected to a portable handheld data logger (Omega Instruments AD128). The differential pressure meters have a span of –0.25 to +0.25 in WG (-62 to +62 Pa) and –0.5 to +0.5 in WG (-124 to +124 Pa), respectively. The accuracy of the instruments is +/-0.5% of full scale.

RESULTS AND DISCUSSION

A total of 60 mitigation systems were evaluated in the study. Of these, 40 were SSDs and 20 were SMDs. The results of indoor air tests before and after system installation are shown on Figures 3 and 4, respectively. Approximately 68% (28) of the SSDs and 75% (15) of the SMDs met the CDPHE interim action level without requiring modification (i.e., with one suction point, a standard size suction pit, and a 90 watt fan). SSDs and SMDs were both able to reduce indoor air concentrations of 1,1 DCE by up to two orders of magnitude and, in some cases, almost three orders of magnitude. Therefore, the standard active soil depressurization methods used for radon mitigation (EPA 1993) appear to be adequate for the more stringent requirements of 1,1 DCE mitigation, with no significant difference in the performance of the SSDs and SMDs being observed.

The modifications or repairs that were required to bring the remaining systems into compliance with the interim action level are also consistent with radon mitigation practices, as summarized in Table 2 and described below.

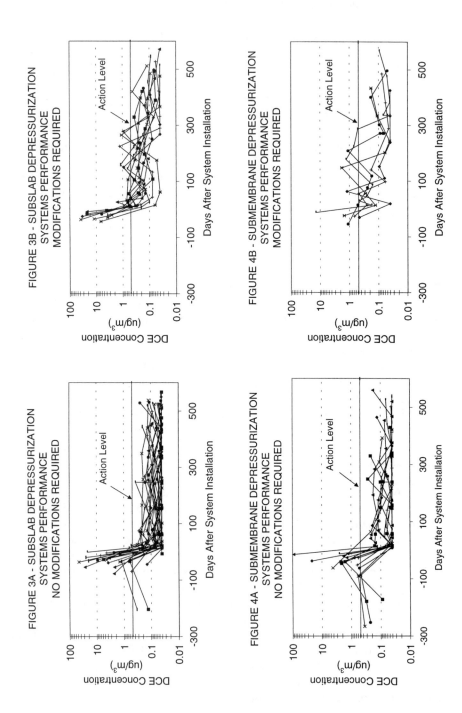

FIGURE 3A - SUBSLAB DEPRESSURIZATION SYSTEMS PERFORMANCE NO MODIFICATIONS REQUIRED

FIGURE 3B - SUBSLAB DEPRESSURIZATION SYSTEMS PERFORMANCE MODIFICATIONS REQUIRED

FIGURE 4A - SUBMEMBRANE DEPRESSURIZATION SYSTEMS PERFORMANCE NO MODIFICATIONS REQUIRED

FIGURE 4B - SUBMEMBRANE DEPRESSURIZATION SYSTEMS PERFORMANCE MODIFICATIONS REQUIRED

TABLE 2. **Modifications required to bring SSDs and SMDs into compliance with the CDPHE interim action level for 1,1 DCE.**

Modifcation	No. of SSDs	No. of SMDs
Required no modifications	28	15
Required one or more of the following actions:	12	5
• bigger suction pit below slab	6	NA
• perforated pipe below liner	NA	4
• 150 watt fan	2	3
• outside combustion air	4	0
• better membrane or crack seal	1	4
• second suction point	2	0

SSD Modifcations. In 12 homes with SSDs, the standard system (one suction point, 90 watt fan, standard size suction pit) was insufficient to achieve a negative pressure field across the entire area of the slab, likely due to low permeability soils. As a result, soil vapors in some areas could migrate through joints and cracks in the concrete into the house, particularly when the basement or house was depressurized by wind, thermal drafts, forced air furnace operation, or operation of bathroom and kitchen vents or clothes dryers. Enlargement of the suction pit, addition of a second suction pit, and/or replacement of the 90 watt fan with a 150 watt fan were sufficient, singly or in combination, as necessary, to increase the extent of the pressure field and meet the 1,1 DCE interim action level.

In four homes, a lack of outside combustion air (now required by the Denver building code in new homes) was suspected of contributing to depressurization of the basement air during furnace operation, thus overpowering the suction created by the SSD under the floor slab. The provision of outside combustion air to the furnace in these homes alleviated this condition.

In one of the SSD homes, sealing of cracks and joints in the concrete floor slab was sufficient to prevent short-circuiting of the depressurization system.

SMD Modifications. There were small gaps in seal between the liner and the foundation wall in five homes, which allowed short-circuiting of the depressurization system. This condition could be readily identified by the relaxed and slack appearance of the liner. A properly sealed liner appears to be sucked flat against the soil. In some homes, low permeability soils appear to have restricted the areal extent of the suction field below the liner. This condition was corrected in four homes by creating additional suction points below the liner, and in three cases by installing a 150 watt fan.

Impacts of Furnace Operation on System Performance. EPA (1993) recommends sub-slab and sub-membrane depressurizations of at least 0.015 in. WG (3.7 Pa) when measured in mild weather in order to maintain negative pressures during cold weather and high winds (both of which tend to reduce the air pressure in houses compared to ambient pressures). If exhaust fans are not running during the pressure test, EPA (1993) recommends a negative pressure of 0.025 to 0.035 in. WG (6.2 to 8.7 Pa) below the floor slab or membrane to ensure that operation of these appliances will not overwhelm the system. Negative pressure measurements in homes meetng the interim action level ranged from 0.02 to 0.125 in WG or 5 to 31 Pa under basement slabs and from 0.01 to 0.13 in WG. (2.5 to 32 Pa) under crawl space liners. During several 24-hour indoor air quality test periods, the vacuum at a particular location rarely fluctuated greater than 0.01 in WG (2.5 Pa) under slabs or membranes. Changes in vacuum were noticeable under slabs before and after a crack was sealed and under membranes before and after holes in the membrane/foundation interface were sealed.

REFERENCES

Altshuler, K. B. and D. E. Burmaster, 1997. "Soil Gas Modeling: The Need for New Techniques and Better Information." *Journ. of Soil Contamination.* 6(1):3-8.

American Society for Testing and Materials. 1995. *Standard Guide for Risk-Based Corrective Action Applied at Petroleum Release Sites.* E 1739 – 95.

Bonnefous, Y. C., A. J. Gadgil, W. J. Fisk, R. J. Prill, and A. R. Nematollahl. 1992. "Field Study and Numerical Simulation of Subslab Ventilation Systems." Environmental Science & Technology. 26(9): 1752-1759.

Colorado Department of Public Health and Environment. 1999. *Guidance for Analysis of Indoor Air Samples.* Hazardous Materials & Waste Management Div.

Fischer, M. L., A. J. Bentley, K. A. Dunkin, A. T. Hodgson, W. W. Nazaroff, R. G. Sextro, and J. M. Daisey. 1996. "Factors Affecting Indoor Air Concentrations of Volatile Organic Compounds at a Site of Subsurface Gasoline Contamination." *Environmental Science & Technology.* 30(10): 2948-2957.

Fitzpatrick, N. A. and J. J. Fitzgerald. 1996. "An Evaluation of Vapor Intrusion into Buildings Through a Study of Field Data." *Presented at the 11th Annual Conference on Contaminated Soils, Univerisity of Massachusetts at Amherst.*

Johnson, P. C. and R. A. Ettinger. 1991. "Heuristic Model for Predicting the Intrusion Rate of Contaminant Vapors into Buildings." *Environmental Science & Technology.* 25(8): 1445-1452.

Richardson, G. M. 1997. "What Research is Needed on Indoor Infiltration of Volatile Organic Contaminants?" *Journal of Soil Contamination.* 6(1): 1-2.

U.S. Environmental Protection Agency. 1993. *Radon Reduction Techniques for Existing Detached Houses: Technical Guidance (Third Edition) for Active Soil Depressurization Systems.* Office Research and Development. EPA/625/R-93/011

CHLORINATED HYDROCARBON TREATMENT
USING A HORIZONTAL FLOW TREATMENT WELL SYSTEM

Derek R. Ferland (Air Force Institute of Technology, Wright-Patt AFB, OH)
Kevin G. Boggs and Scott Niekamp (Wright State University, Dayton, OH)
John A. Christ (Minot AFB, ND)
Abinash Agrawal (Wright State University, Dayton, OH)
Mark N. Goltz (Air Force Institute of Technology, Wright-Patt AFB, OH)

ABSTRACT: Horizontal flow treatment well (HFTW) systems with in-well catalytic reactors, which are proposed for *in situ* destruction of chlorinated hydrocarbon contaminants dissolved in groundwater, are modeled. In an HFTW system, wells equipped with in-well catalytic reactors recirculate contaminated groundwater. No contaminated water is pumped to the surface. Due to groundwater recirculation, the contaminant makes multiple passes through the catalytic reactors, so the system achieves greater concentration reductions than would be attainable by a single-pass through the reactor. Palladium (Pd) has been found to catalyze the dehalogenation of chlorinated hydrocarbons completely and rapidly in the presence of molecular hydrogen (H_2) as the sole electron donor. Laboratory column studies were conducted to obtain kinetic parameters for dehalogenation of chlorinated ethenes in a Pd/H_2 reactor. These parameters are then incorporated into a model that accounts for both the recirculating flow through an HFTW system and the reaction kinetics of an in-well Pd/H_2 reactor. The model is used as a tool to design a system to treat trichloroethylene-contaminated groundwater at a hypothetical site.

INTRODUCTION

Organic solvents such as chlorinated hydrocarbons (trichloroethylene or TCE, perchloroethylene or PCE, etc.) are common groundwater pollutants. The widespread presence of chlorinated hydrocarbons (CHCs) as groundwater pollutants poses a significant environmental challenge due to their recalcitrance to natural biodegradation. Currently available strategies for containing migrating plumes of groundwater contaminated with CHCs are plagued by various shortcomings. The three most common strategies for containing plumes of CHC contaminated ground water are: (1) pump-and-treat, (2) funnel-and-gate, and (3) monitored natural attenuation.

Pump-and-treat systems may increase the risk of exposure because they pump contaminated groundwater to the surface and also have high operation and maintenance costs. Funnel-and-gate systems that use zero-valent metals are effective for treating groundwater contaminated with CHCs *in situ*, but can only be implemented under specific hydrogeologic conditions. Monitored natural attenuation as a strategy to deal with chlorinated contaminants is not yet generally accepted and is only applicable under specific hydrogeochemical conditions. Other disadvantages of monitored natural attenuation include complex and costly

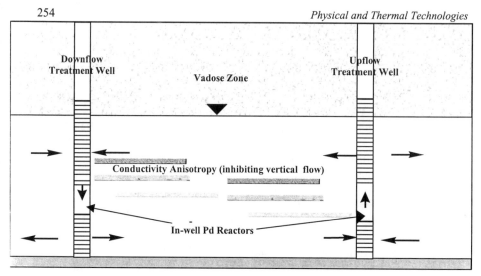

FIGURE 1. HFTW System with in-well Pd reactors.

site characterization, potential for formation of toxic by-products, long waiting period to achieve remediation goals, etc. The drawbacks of the current technologies have driven research to find new methods of containing chlorinated contaminant plumes *in situ*, with greater efficiency, and at lower costs.

It has been seen that Palladium (Pd) metal catalyzes dehalogenation of organic chemicals completely and rapidly in the presence of molecular hydrogen (H_2) as the sole electron donor (Lowry and Reinhard, 1999). The unique potential of a Pd/H_2 system to rapidly dechlorinate CHCs can be useful for rapid destruction of dissolved CHCs in groundwater. The catalytic reactions of Pd metal with CHCs include hydrodehalogenation (replacement of organochlorine by hydrogen) and hydrogenation (breaking of unsaturated carbon bonds, ultimately producing single-bonded carbons) in the presence of molecular hydrogen (Lowry and Reinhard, 1999). Under favorable conditions, these reaction pathways yield non-chlorinated, saturated hydrocarbon products that are of little or no health concern.

This work focuses on a new application of groundwater circulation wells (GCWs) called Horizontal Flow Treatment Wells (HFTWs). An HFTW system, with in-well metal Pd catalyst reactors, is being considered for active *in situ* containment of CHC plumes. An HFTW system with in-well Pd reactors combines the best features of pump-and-treat and funnel-and-gate technologies. In the catalyst-enhanced HFTW system (see Figure 1), a reactive porous medium containing a Pd metal catalyst in powder or pellet form will be placed in a reactor within the casing of the wells. The plume of contaminated groundwater captured by the HFTW system will pass through the catalytic reactor, where hydrogen gas will be added, and the toxic groundwater pollutants will be destroyed *in situ* during their brief contact with the catalyst. This system actively controls the plume of contaminated groundwater without the need to pump contaminated groundwater to the surface. In the field, it is envisioned that a line of treatment wells equipped with *in situ* catalytic reactors will be located down-gradient of a contaminant plume. Each treatment well will have two screens, one an injection screen, the other an extraction screen. Certain wells will pump in an upflow mode and others will pump in a downflow mode so that the water will circulate between the wells. Note that in the HFTW system, due to hydraulic conductivity anisotropy such as is typically seen in aquifers, groundwater flow between the injection and extraction screens of a well pair is horizontal. This is in contrast to conventional GCWs that depend on vertical flow between the injection and extraction screens of a single well. The chlorinated contaminant is treated with each pass through the in-well reactor. Due to the circulation between wells, the contaminant is treated multiple times, so that contaminant removal efficiencies (comparing contaminant concentration upgradient and downgradient of the treatment wells) can be greatly increased over the removal achieved by a single-pass of contaminated water through the reactor. McCarty *et al.* (1998) describe application of an HFTW system using *in situ* bioreactors to treat TCE-contaminated groundwater.

OBJECTIVES

Laboratory column studies were conducted to obtain kinetic parameters for dehalogenation of chlorinated ethenes in a Pd/H_2 reactor. These parameters are then incorporated into a model that accounts for both the recirculating flow through an HFTW system and the reaction kinetics of an in-well Pd/H_2 reactor. The model is used as a tool to design a system to treat trichloroethylene-contaminated groundwater at a hypothetical site.

MATERIALS AND METHODS

Bench-Scale Investigation. The reaction of Pd catalyst with several chlorinated ethenes and hydrocarbons is investigated in flow-through reactor systems. These chemicals include tetrachloroethylene (PCE), trichloroethylene (TCE), and cis-dichloroethylene (cis-DCE). A flow-through reactor investigation was carried out in a stainless steel tube (30 cm high and 5 cm internal diameter). The reactor column was packed with a mixture of clean, coarse sand and 280 g of Pd catalyst supported on 3.2 mm γ-Al_2O_3 pellets (0.5% Pd metal in Pd/Al by mass, obtained from Aldrich) resulting in 45% bulk porosity. De-ionized water was deoxygenated by continuously purging with high purity H_2 gas at atmospheric pressure, and then this H_2 saturated water was amended with an aqueous saturated solution of CHC before being introduced into the column. The influent concentration of CHC solution entering the flow-through reactor ranged between 1-30 mg L^{-1}. 2 mL aqueous samples were collected periodically in autosampler vials from the influent and the effluent sampling ports of the reactor, and they were analyzed immediately by a direct manual aqueous injection into a GC-ECD. The CHC residence time in contact with Pd catalyst was controlled by the rate of fluid flow through the reactor, and the flow varied between 40-300 mL min^{-1}. The dissolved oxygen and pH were routinely monitored using aqueous samples from the influent and effluent ports.

Modeling HFTW System with In-Well Reactor. The analytical model of Christ *et al.* (1999) was used to determine the interflow (flow of recirculating water) between a pair of upflow and downflow treatment wells. Upflow and downflow well pumping rates (Q) were assumed equal. Depths of groundwater capture by the upper and lower well screens (B), regional groundwater Darcy velocity (U) and direction (α) were all assumed known. With this information, and assuming a half-distance between the treatment wells (d), the model of Christ et al. (1999) could be used to calculate interflow (I). Using experimental values for the first-order rate constant (k) for CHC degradation in a Pd/H$_2$ reactor, with reactor porosity equal to n, and assuming a reactor volume (V), it's possible to calculate CHC degradation efficiency for a single-pass of contaminated water through the reactor (η_{sp}) using equation 1:

$$\eta_{sp} = 1 - e^{-k\frac{Vn}{Q}} \tag{1}$$

Note that the term Vn/Q in the exponent is simply the residence time of contaminated water in the reactor. Knowing η_{sp} and the interflow, the overall treatment efficiency of the HFTW system (η), which is a measure of the reduction of contaminant concentration from upgradient of the system (C_{up}) to downgradient (C_{down}), can be determined by Equation 2 (Christ et al., 1999):

$$\eta = 1 - \frac{C_{down}}{C_{up}} = \frac{\eta_{sp}}{1 - I(1 - \eta_{sp})} \tag{2}$$

Finally, the width of the contaminant plume captured by an extraction screen (CZW) may be calculated using Equation 3 (Christ et al., 1999):

$$CZW = \frac{Q}{UB}(1 - I) \tag{3}$$

RESULTS AND DISCUSSION

CHC degradation kinetics in flow-through reactors. The rate constants of CHC degradation (k_{obs}) were determined for the disappearance of a few selected chlorinated ethenes. The kinetic data for the disappearance of CHC were modeled with a simple pseudo-first-order rate law using a least-square fitting procedure, and the reactor specific half-life was calculated (Table 1). The reduction of TCE yields ethane as the only final product of reaction. Minor amounts of cis-, 1,1-, and trans-dichloroethenes (DCEs), vinyl chloride (VC) and ethene were formed as intermediates, among which cis-DCE was the dominant constituent.

The rate constant of CHC degradation, or k_{obs}, is sensitive to influent CHC concentration (Table 2), and the residence time of CHC in the reactor. The k_{obs} decreased due to increase in influent CHC concentration within the range of experimental conditions. Such decrease in k_{obs} is presumably caused by competition among CHC molecules for the reactive sites on the surface of the

TABLE 1: Degradation rate constants and half-lives for PCE, TCE and cis-DCE

Chlorinated Ethene	Influent CHC conc. (mg L^{-1})	Effluent CHC conc. (mg L^{-1})	Flow-rate (mL min^{-1})	Degradation rate constant, k_{obs} (min^{-1})	Half-life, $t_{1/2}$ (min)
PCE	1.45	0.064	65	1.05	0.66
TCE	0.97	0.009	65	1.11	0.63
cis-DCE	0.62	0.001	65	1.51	0.46

Pd/Al catalyst and the availability of dissolved hydrogen for this heterogeneous reaction. The rate constant, or k_{obs}, of TCE degradation varied also due to changes in the ambient H_2 concentration. Since molecular H_2 is the sole electron donor for the Pd-catalyzed reduction of CHC, the k_{obs} of CHC degradation is sensitive to the dissolved $[H_2]$.

TABLE 2: Effect of influent CHC concentration on degradation rate constant.

Influent PCE conc. (mg L^{-1})	Degradation rate constant, k_{obs} (min^{-1})	Influent TCE conc. (mg L^{-1})	Degradation rate constant, k_{obs} (min^{-1})	Influent cis-DCE conc. (mg L^{-1})	Degradation rate constant, k_{obs} (min^{-1})
1.45	1.05	0.97	1.10	0.62	1.51
4.33	0.88	7.52	0.92	2.59	0.85
9.61	0.74	15.40	0.80	5.04	0.69
13.74	0.63	21.20	0.73	12.80	0.33
20.30	0.51	28.70	0.63	18.70	0.13
29.30	0.37			24.75	0.094

Flow models for catalyst-enhanced HFTW. Using the flow model developed by Christ *et al.* (1999) for the specific parameter values specified in Table 3, Figures 2 and 3 were constructed.

TABLE 3: Parameters used for HFTW design

Parameter	Value
Regional groundwater Darcy velocity (U)	0.02056 m d^{-1}
Depth of groundwater capture by downflow treatment well extraction screen (B)	8 m
Direction of regional groundwater flow (α)	67.5°
First-order rate constant for TCE degradation by Pd/Al catalyst (k)	1.11 min^{-1}
Pd catalyst porosity (n)	0.3

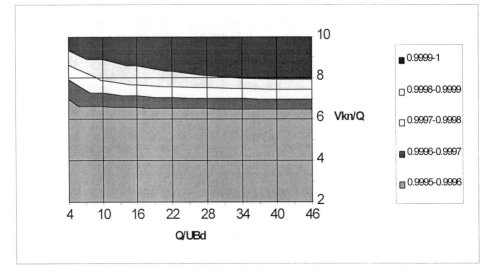

FIGURE 2. **Overall contaminant removal efficiencies (η) as a function of the dimensionless parameters Q/Ubd and Vkn/Q.**

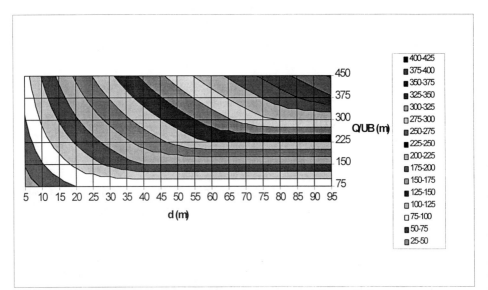

FIGURE 3. **Capture zone width (m) as a function of the half-distance between treatment wells (d) and the ratio of well flow to regional groundwater flow (Q/UB).**

The Table 3 values are typical of a TCE-contaminated aquifer at Edwards AFB (McCarty *et al.*, 1998). In this exercise, we will assume there is a contaminated groundwater plume with TCE concentrations of 15 mg L^{-1} and a width of 200 m, and that TCE concentrations downgradient of the HFTW system cannot exceed 5 μg L^{-1}. Based on the existing upgradient and required downgradient TCE concentrations, we calculate that the required overall treatment efficiency of the HFTW system (η) is 0.9997. Let us now apply Figures 2 and 3 to the upper portion of the aquifer to determine design parameters (Q, V, and d) for the downflow treatment well that meet our efficiency (η) and plume capture zone width (CZW) objectives. According to Figure 2, when η = 0.9997, the left-most Q/UBd value along the horizontal portion of the contour plot (roughly at Vkn/Q = 7) is approximately 8. This Q/UBd solution dominates other solutions on the horizontal part of the contour—that is, the other solutions result in higher values of both Q and V, but only achieve the same overall treatment efficiency. Therefore, our design will start using the value Q/Ubd = 8. The required capture zone width (CZW) is 200-m so we need to find a solution in Figure 3 above and to the right of the 200-m CZW contour. Note we want to find the solution that meets our CZW objective while minimizing Q. We also know from the previous analysis using Figure 2 that we need to be along the line Q/UB = 8d. These conditions are met at the intersection of Q/UB = 320 m and d = 40 m. Knowing U and B, we calculate Q is approximately 53 m^3/d and the CZW is about 212 m. Substituting Q back into Vkn/Q = 7 to solve for V gives us a reactor volume of 773 L. The lower part of the aquifer can be analyzed in a similar manner, to determine design parameters for the upflow treatment well. The final step is to analyze the upper and lower portions of the aquifer iteratively and redesign the system for equal flow rates (Q) and half-distances between the wells (d).

As can be seen from the above analysis, there are many values of well flow rate (Q), half-distance between treatment wells (d), and reactor volume (V) that achieve design objectives (η and CZW). On the other hand, there may be situations where no feasible combinations of Q, d, and V meet design objectives. In these cases, it becomes necessary to have more than two treatment wells in a row, or to use multiple rows of wells. Figures 2 and 3 were constructed assuming the two well system shown in Figure 1. However, using the solutions in Christ *et al.* (1999) for multiple-well rows, other figures may be constructed that may be used to design HFTW systems that attain design goals using more than two wells.

CONCLUSIONS/FUTURE WORK

The methodology presented in this study is useful for quickly screening a contaminated site to see if application of an HFTW system with in-well Pd catalysts to contain and destroy a chlorinated ethene groundwater plume is appropriate. Using the overall treatment efficiency and CZW contour plots in concert gives a designer an idea if this technology will work at a site. Based on the laboratory column studies and the modeling analysis, it appears that an HFTW system with in-well Pd catalysts has the potential for use in containing chlorinated ethene groundwater plumes.

The methodology presented in this study modeled reactor kinetics with a first-order approximation and the methodology only provides the designer with feasible solutions, not optimal solutions. Based on the laboratory results presented in Table 2, further investigation into the kinetic submodel for the Pd catalytic reactor is needed, as the reaction does not appear to be first-order. In addition, optimization techniques may be applied to the design methodology, to aid designers in selecting a solution that meets design goals while minimizing some objective function (e.g. total cost).

ACKNOWLEDGEMENT

This work was conducted with financial support from the Joint Air Force Research Laboratory/Dayton Area Graduate Studies Institute Basic Research Program. Additional financial support in the form of a Research Challenge Grant from R&SP, Wright State University, is also acknowledged

REFERENCES

Christ, J. A., M. N. Goltz, and J. Huang. 1999. "Development and application of an analytical model to aid design and implementation of in situ remediation technologies." *Journal of Contaminant Hydrology.* *37*: 295-317.

Lowry, G. V. and M. Reinhard. 1999. "Hydrodehalogenation of 1- to 3-carbon halogenated organic compounds in water using a palladium catalyst and hydrogen gas." *Environmental Science and Technology.* *33*(11): 1905-1910.

McCarty, P. L., M. N. Goltz, G. D. Hopkins, M. E. Dolan, J. P. Allan, B. T. Kawakami, and T. J. Carrothers. 1998. "Full-scale evaluation of *in situ* cometabolic degradation of trichloroethylene in groundwater through toluene injection." *Environmental Science and Technology.* *32*(1): 88-100.

FIELD DEMONSTRATION OF *IN SITU* VITAMIN B_{12}-CATALYZED REDUCTIVE DECHLORINATION

Carol S. Mowder (Dames & Moore, Bethesda, Maryland, USA)
Tim Llewellyn and Sarah Forman (Dames & Moore, Linthicum, Maryland, USA)
Suzanne Lesage, Susan Brown, and Kelly Millar (Environment Canada, Burlington, Ontario, Canada)
Don Green and Kimberly Gates (U.S. Army Garrison, APG, Maryland, USA)
George DeLong (Lockheed Martin Energy Systems, Oak Ridge, Tennessee, USA)

Abstract: Groundwater at Graces Quarters, Aberdeen Proving Ground (APG), Maryland, is contaminated with a complex mixture of chlorinated methanes, ethanes, and ethenes. The total concentration of chlorinated aliphatic hydrocarbons (CAHs) in the source area groundwater monitoring wells ranges from 4,000 to 6,000 micrograms per liter (μg/L). Based on promising laboratory-scale treatability test results, a 6-month field demonstration was conducted in 1999/2000 to evaluate using reduced vitamin B_{12} to abiotically dechlorinate the contaminants, followed by biological degradation of the daughter products. Contaminated groundwater from the site was mixed *in situ* with the vitamin B_{12} concentrate using a groundwater recirculation well as an *in situ* chemical reactor. Remedial effectiveness was evaluated through a comprehensive groundwater sampling and analysis program using a network of multiple-depth piezometers and monitoring wells in the area. The results show that the chlorinated methanes were rapidly degraded with slower degradation of the chlorinated ethanes. The chlorinated ethenes formed as daughter products are undergoing biological degradation.

INTRODUCTION

Groundwater at Graces Quarters, APG, is contaminated with volatile organic compounds (VOCs), specifically CAHs. The most frequently detected contaminants include 1,1,2,2-tetrachloroethane, carbon tetrachloride, trichloroethene, tetrachloroethene, and chloroform. The CAHs detected at the highest concentrations at the site are 1,1,2,2-tetrachloroethane and carbon tetrachloride, with peak concentrations on the order of 2,000 to 4,000 μg/L. Concentrations of trichloroethene are below 500 μg/L, and chloroform and tetrachloroethene concentrations are typically below 100 μg/L.

Based on the results of a laboratory-scale treatability study, which included the use of various microcosms and column studies, the combination of titanium citrate, vitamin B_{12}, and glucose (herein referred to as vitamin B_{12} concentrate) successfully degraded 1,1,2,2-tetrachloroethane, carbon tetrachloride, and the carbon tetrachloride daughter products within 5 minutes. Tetrachloroethene was degraded within 30 minutes. The mechanism for this degradation is abiotic reductive dechlorination utilizing the vitamin B_{12} as a catalyst for the reaction. The daughter products generated from the abiotic degradation of the 1,1,2,2-tetrachloroethane and tetrachloroethene, including trichloroethene, cis-1,2-dichloroethene, and trans-1,2-dichloroethene, did not degrade as quickly. However, the column study results showed that these compounds were subsequently degraded biologically. Vinyl chloride was also generated at low concentrations in the treatment process, but was degraded. (APG, 1999).

The laboratory-scale results were used to develop a pilot test to evaluate the use of vitamin B_{12} concentrate to treat groundwater at the site. This paper presents results from the first 14 weeks of system operation.

MATERIALS AND METHODS

The vitamin B_{12} concentrate was generated on site weekly in batch. This process involved preparing titanium oxalate (using titanium sponge and oxalic acid) followed by the addition of sodium citrate. The oxalate in solution was precipitated as calcium oxalate after the addition of calcium carbonate and calcium citrate. Sodium carbonate was added to increase the pH to above 8.0 and decrease the Eh to approximately –900 millivolts (mV). Vitamin B_{12} and glucose were then added. The vitamin B_{12} concentrate was transferred to a collapsible pillow tank (to ensure anaerobic conditions were maintained) and was continuously metered into the site groundwater. Approximately 500 gallons of vitamin B_{12} concentrate were prepared each week. Titanium citrate, the reductant, is used to reduce the cobalt atom within vitamin B_{12} and to lower the oxidation-reduction potential (ORP) of the groundwater. Glucose was added as a preferential carbon source to prevent citrate degradation by bacteria, which was found in the laboratory study to cause titanium precipitation.

A recirculation well was used to facilitate *in situ* mixing of the vitamin B_{12} concentrate and contaminated groundwater during the pilot test. A schematic of the recirculation well and vitamin B_{12} feed system is shown in Figure 1. Contaminated groundwater from the site was drawn into the lower well screen of the 10-inch diameter recirculation well and pumped above the inflatable packer which separates the upper and lower well screens. Groundwater was pumped at 2 gallons per minute (gpm) and mixed with the vitamin B_{12} concentrate, which was added at approximately 0.05 gpm. Based on the groundwater flow rate and the dimensions of the recirculation well, this system allowed an approximate residence/reaction time of 20 minutes in the well before discharging the groundwater/vitamin B_{12} concentrate through the upper well screen and into the aquifer. Based on laboratory-scale treatability study results and achievable residence time within the well, the groundwater design concentrations of the various additives were initially 4 millimolar (mM) titanium, 3 milligrams per liter (mg/L) vitamin B_{12}, and 1 gram per liter (g/L) glucose in the groundwater within the recirculation well. The glucose concentrations were subsequently lowered.

A plan view of the recirculation well and surrounding piezometers and wells are shown in Figure 2. Table 1 presents geographical information on the piezometers and monitoring wells. During system start-up, groundwater samples were collected weekly from the upper and lower well screens of the recirculation well to balance the vitamin B_{12} concentrate addition rate; these samples were

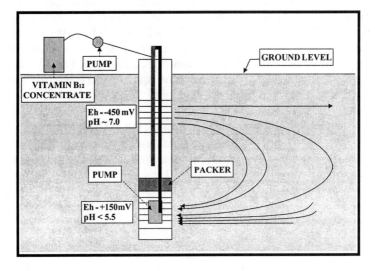

FIGURE 1. Recirculation well diagram.

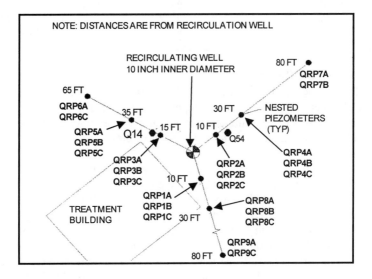

FIGURE 2. Plan view of the recirculation well, nested piezometers, and well locations.

analyzed for VOCs. Groundwater samples were collected every 2 weeks from the surrounding multiple-depth piezometers and monitoring wells and analyzed for VOCs, dissolved gases (methane, ethane, ethene, and acetylene), volatile fatty acids (VFAs - acetate, butyrate, citrate, formate, lactate, and propionate), titanium, iron, manganese, and chloride. All analyses were performed in a fixed-base laboratory. Additionally, before system start-up and after 12 weeks of operation, all of the piezometers and the adjacent wells Q14 and Q54 were sampled and analyzed in the field for redox couples (nitrate/nitrite, ferric iron/ferrous iron, sulfate/sulfide, carbon dioxide/methane) and hydrogen to evaluate the changing redox conditions of the aquifer. The piezometers and wells were monitored for Eh and pH in the field using a flow-through cell.

TABLE 1. Piezometer and Monitoring Well Information

Piezometer/Well Identification	Distance (feet) from Recirculation Well QRW1	Screened Interval Depth (feet) Below Ground Surface (bgs)
QRW1U	0	10-18
QRW1L	0	28-36
QRP1	10	13-15 (A),23-25 (B),31-33 (C)
QRP2	10	15-17 (A),23-25 (B),31-33 (C)
QRP3	15	16-18 (A),23-25 (B),31-33 (C)
QRP4	30	14-16 (A),23-25 (B),31-33 (C)
QRP5	35	14-16 (A),23-25 (B),29-31 (C)
QRP6	65	13-15 (A), 29-31 (C)
QRP7	80	13-15 (A), 31-33 (C)
QRP8	30	13-15 (A),23-25 (B),31-33 (C)
QRP9	80	13-15 (A), 31-33 (C)
Q14	18	20-30
Q54	12	24.5-34.5

RESULTS AND DISCUSSION

Due to the volume of analytical data collected and the limitations on the amount of material presented in this paper, results related to system start-up and the observed trends of contaminant concentrations over time are discussed with data graphically presented for a limited number of piezometers.

System Start-Up. During system start-up and balancing of the vitamin B_{12} concentrate addition rate, VOC samples were collected from the upper and lower well screens of the recirculation well. Although complete degradation of both carbon tetrachloride and 1,1,2,2-tetrachloroethane occurred within 20 minutes in the laboratory-scale study, the 1,1,2,2-tetrachloroethane did not degrade completely in the recirculation well. One reason for the slower contaminant degradation rate in the field is that lower titanium concentrations were achieved in the field relative to the laboratory study (2 mM titanium in the field vs. 4 mM titanium in the laboratory); the lower titanium concentrations obtained in the field were due to early difficulties encountered during mixing. As shown in Figure 3, the percent degradation of 1,1,2,2-tetrachloroethane in the recirculation well (with concurrent formation of cis- and trans-1,2-dichloroethene) varied throughout the

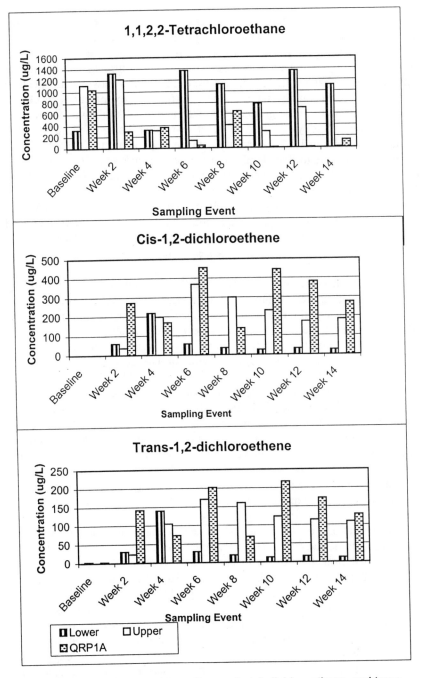

FIGURE 3. 1,1,2,2-Tetrachloroethane, cis-1,2-dichloroethene, and trans-1,2-dichloroethene concentrations in the lower and upper well screens of the recirculation well and QRP1A.

pilot test. After approximately 1 to 2 weeks of operation, the turbidity of the groundwater entering the lower well screen increased significantly and appeared to be due to fines entering the well (as opposed to biological matter). This increase in turbidity is believed to be the cause of the limited degradation of 1,1,2,2-tetrachloroethane at that time, because of interaction between chemicals in the vitamin B_{12} concentrate and the fines. After approximately 4 weeks of operation, the turbidity of the groundwater decreased. This followed an inactive period of about 10 days due to a weather-related power failure (Floyd). Subsequent variances in 1,1,2,2-tetrachloroethane degradation in the recirculation well were probably due to fluctuations in groundwater ORP and pH. As shown in Figure 3, the 1,1,2,2-tetrachloroethane continued to be degraded in the aquifer, as evidenced by the lower concentrations of 1,1,2,2-tetrachloroethane and higher concentrations of cis- and trans-1,2-dichloroethene in piezometer QRP1A. As was observed in the laboratory, carbon tetrachloride was completely degraded within the well without the accumulation of chloroform or methylene chloride, even when the groundwater turbidity increased within the well.

Contaminant Concentration Trends. Carbon tetrachloride and chloroform were completely degraded within the recirculation well. As a result, a rapid decrease in the concentrations of these constituents was observed immediately in near-by shallow Piezometers 1A and 3A; Piezometer 2A showed no significant contaminant trends due to hydrogeologic factors discussed in an accompanying paper (Forman, et. al., 2000). As the vitamin B_{12} concentrate and treated groundwater extended through the aquifer with time, a decrease in carbon tetrachloride and chloroform concentrations was also observed in the deeper "B" piezometers and those further from the well. Figure 4 presents the changes in contaminant concentrations at Piezometers 3A and 3B over time. Within the 14 weeks of operation, there were no significant changes in contaminant concentrations in the deep "C" piezometers, screened just above the confining clay layer. However, based on Eh and pH data, the reduced vitamin B_{12} concentrate had not reached the "C" depth by that time. With additional time, similar decreases in contaminant concentration should be observed in the "C" piezometers.

Concentrations of 1,1,2,2-tetrachloroethane have also decreased in the shallow piezometers, and cis- and trans-1,2-dichloroethene – the primary daughter products formed from the abiotic reaction of the vitamin B_{12} concentrate with 1,1,2,2-tetrachloroethane – initially increased. However, the concentrations of cis- and trans-1,2-dichloroethene are decreasing with time and with distance from the recirculation well. Based on the presence of VFAs, as well as the results from the laboratory-scale study, the dichloroethenes are presumed to be degraded biologically. Before injection of the vitamin B_{12} concentrate, the aquifer in the vicinity of the pilot test was essentially aerobic based on dissolved oxygen, Eh, redox, and dissolved hydrogen data. However, as expected, following injection of the vitamin B_{12} concentrate the aquifer in the vicinity of the pilot test quickly became anaerobic (Eh < -200 mV and indications of sulfate-reducing bacteria). Although degradation of the cis- and trans-1,2-dichloroethene is occurring under

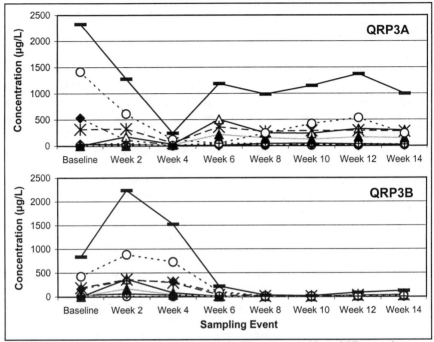

FIGURE 4. Contaminant concentrations in Piezometers 3A and 3B over time.

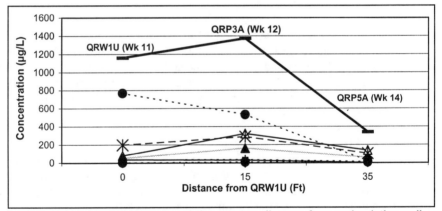

FIGURE 5. Contaminant concentrations versus distance from recirculation well.

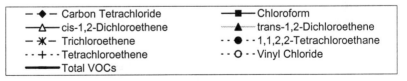

anaerobic conditions, there has been no accumulation of vinyl chloride. The lack of accumulation of vinyl chloride under the anaerobic conditions at the site may indicate that it is degraded in the presence of vitamin B_{12}.

Figure 5 presents data along the flow path from the upper well screen of the recirculation well (QRW1U) to shallow piezometer QRP3A (15 feet from QRW1U) to shallow piezometer QRP5A (35 feet from QRW1U). Based on groundwater modeling (Forman, et. al., 2000) and analytical data, it is estimated that it takes groundwater leaving the upper well screen of the recirculation well approximately 1 week to reach QRP3A and an additional 2 weeks to reach QRP5A. The data presented in Figure 5 account for the estimated time travel required between sample locations. As shown in Figure 5, 1,1,2,2-tetrachloroethane continues to degrade as it travels through the aquifer. The total VOC concentration also decreases with time and distance from the recirculation well.

One of the challenges encountered during the pilot test was the increased biological population in the vicinity of the recirculation well. After approximately 7 weeks of operation, aquifer permeability was limited as evidenced by an increased water level in the upper well screen. To limit biological growth, glucose concentrations were decreased from 1 g/L to approximately 60 mg/L; however, the aquifer permeability continued to decrease. Reformulation of the vitamin B_{12} concentrate, limited use of an anaerobic biocide, and pulsed feed of the vitamin B_{12} concentrate are being evaluated as management steps to help limit the extent of biological growth in the future.

CONCLUSIONS

The pilot test results for the first 14 weeks of operation indicate that the vitamin B_{12} concentrate effectively degrades the chlorinated methanes and ethanes in the groundwater, and that with time and distance the daughter products formed (dichloroethenes) are biologically degrading. Future efforts of the pilot test include evaluating changes to limit the extent of biological growth in the vicinity of the recirculation well.

ACKNOWLEDGMENTS

This project planned, implemented under contract, management, and oversight by AIMTech, a US Department of Energy program operated by Lockheed Martin Energy Systems, Inc., in conjunction with the APG Installation Restoration Program. Unclassified/unlimited per OPSEC No. 3368-A-4.

REFERENCES

Aberdeen Proving Ground (APG). 1999. *Final Treatability Study Report, Primary Test Area*, Appendices B and C.

Forman, S., T. Llewellyn, S. Morgan, S. Allison, R. Westmoreland, G. DeLong, F. Tenbus, D. Green, and K. Gates. 2000. "Numerical Simulation and Pilot Testing of a Recirculation Well." In proceedings for the *Second International Conference on Remediation of Chlorinated and Recalcitrant Compounds*.

CO-SOLVENT-BASED SOURCE REMEDIATION APPROACHES

Susan C. Mravik (U.S. EPA, Ada, Oklahoma)
Guy W. Sewell and A.Lynn Wood (U.S. EPA, Ada, Oklahoma)

ABSTRACT: Field pilot scale studies have demonstrated that co-solvent-enhanced *in situ* extraction can remove residual and free-phase nonaqueous phase liquid (NAPL), but may leave levels of contaminants in the ground water and subsurface formation higher than regulatory requirements for closure of a site. Various methods of improving delivery and recovery of co-solvent mixtures and of facilitating *in situ* mixing of these light remedial fluids with resident contaminants have been proposed and are being investigated. However, it is unlikely that these improvements alone will permit regulatory goals to be achieved via enhanced NAPL solubilization or mobilization. Recent laboratory and field tests have examined the feasibility and benefits of coupling co-solvent flushing with other remediation processes to achieve acceptable cleanup goals. For example, the potential for residual co-solvent to stimulate *in situ* biotreatment following partial dense nonaqueous phase liquid (DNAPL) source removal by alcohol-induced dissolution was evaluated at a former dry cleaner site in Jacksonville, Florida. Contaminant and geochemical monitoring at the site suggests that biotransformation of the tetrachloroethylene (PCE) was enhanced and significant levels of *cis*-dichloroethylene (*cis*-DCE) were produced in areas exposed to residual co-solvent.

INTRODUCTION

Chlorinated solvents were used and released to the environment in massive amounts during the 1950's, 60's, and 70's. These contaminants have migrated through the subsurface and impacted ground water at over 1000 DoD sites. Their widespread use and the physical/chemical properties of these compounds have resulted in the chloroethenes being the most commonly detected class of organic contaminants in ground water. Parent chloroethenes (PCE and TCE) can become human health hazards after being processed in the human liver or via reductive dehalogenation in the environment. This has generated a high degree of interest in efficient and cost effective technologies which can be used to remediate soils and ground waters contaminated with PCE and TCE.

The objective of this research was to demonstrate the feasibility of a treatment train approach to remediate a DNAPL contaminated aquifer. Experience has shown that conventional *pump and treat* systems are inadequate for cleaning up aquifers contaminated with DNAPL, and the highly toxic nature of these environments suppresses bioremediation. Recent advances in our understanding of the impact of organic co-solvents on NAPL behavior in porous media suggest that co-solvent-enhanced *in situ* extraction can remove residual and free-phase DNAPL. While laboratory and pilot-scale experiments have demonstrated the potential of this method for mass removal, residual amounts of solvents and contaminants are expected to remain at levels which could preclude meeting regulatory requirements. However, with the bulk of the DNAPL extracted, *in situ* biotreatment becomes a viable "polishing" procedure. *In situ* biotreatment may transform the remaining contaminants to non-hazardous compounds at a rate in excess of the rate of dissolution or displacement and at lower costs.

The efficacy of *in situ* bioremediation of solvents is usually limited by transport considerations, i.e., supplying electron donor at the appropriate levels and in conjunction with exposure to the chlorinated solvent. In this case the concurrent exposure to electron donor (co-solvent) and electron acceptor (chlorinated solvent) is facilitated by the delivery and extraction process as well as the co-solvency effect. The synergism between these abiotic and biotic processes could minimize problems associated with the individual approaches. The development of the Solvent Extraction Residual Biotreatment (SERB) technology could attenuate or eliminate the risks posed to human health and the environment by these highly contaminated sites.

MATERIALS AND METHODS

The SERB pilot demonstration was conducted at the former Sages dry cleaner site in Jacksonville, Florida where an area of PCE contamination was identified. Pre-treatment characterization of the site indicated near saturated concentrations of PCE in ground water samples collected near the source zone. Low or non-detectable levels of normal biodehalogenation daughter products were found. The zone of contamination was from 26 to 31 ft. (7.9 to 9.5 m) below ground surface and this area was targeted for remediation.

Figure 1 shows the site and location of the wells used for the SERB demonstration. The contour lines on the map represent the pre-co-solvent flush area of PCE contamination. Three injection wells (IW) and six recovery wells (RW) were placed in the source zone in July of 1998 and used for the co-solvent extraction experiment. Previously installed monitoring wells (MW) were utilized for monitoring ground water concentrations during the field experiment. Additional ground water monitoring was done with a series of 1 inch PVC wells (C), which were installed along a transect in the general direction of groundwater flow in Sept., 1998, after the co-solvent extraction test.

Ethanol was selected as the co-solvent for the *in situ* co-solvent extraction test. The alcohol flushing pilot test began on August 9, 1998, and ended on August 15, 1998. Post-test hydraulic containment began on August 15, 1998 and was discontinued on August 25, 1998, after the ethanol concentration in the treatment system influent dropped below the 10,000 mg/L termination criterion. Pre- and post-treatment partitioning tracer tests were also conducted during this time for estimation of the mass of PCE contaminant removed with the co-solvent extraction. After pumping ceased, ground water samples were collected periodically for chemical analysis. Analytes included PCE, TCE, *cis*-DCE, ethanol, methane, ethane, ethene, chloride, sulfate, and acetic acid. Approximately one year after the co-solvent extraction test, ground water samples in the source area were analyzed for dissolved hydrogen.

Sage's Dry Cleaner Site Well Location and PCE Contamination

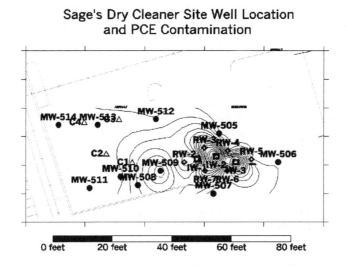

FIGURE 1. Location of ground water monitoring wells and injection/recovery wells. Pre-co-solvent extraction test contour plot of PCE contamination. PCE concentration range is 0 to 80 mg/L.

RESULTS AND DISCUSSION

Enhanced dissolution and solubilization of PCE was demonstrated as a result of the ethanol co-solvent extraction test. Analytical data from RW-7 showed that the peak PCE concentration was 80 to 90 times higher than the initial PCE concentration. In other recovery wells, the ratio of peak PCE concentration to initial PCE concentration was on the order of 30 to 40. The partitioning tracer data indicated that approximately 30.4 L of PCE was removed during the co-solvent extraction test. This is approximately 70% of the PCE mass estimated with the pre-treatment partitioning tracer test. Actual PCE concentrations monitored in the recovery well effluent indicated that approximately 41.5 L was recovered. Although there is error associated with each method of estimating PCE mass recovery, both methods showed that a significant mass of PCE was recovered with the co-solvent extraction test.

PCE concentrations in the ground water decreased immediately following the co-solvent extraction and then rebounded, as expected, to near initial concentrations. An area of PCE contamination was also detected down-gradient of the co-solvent flush when the C-wells were installed post co-solvent flush (Sept. 10-15, 1998). Monitoring wells near this location had significant concentrations of PCE in the ground water prior to the co-solvent extraction and C-well installation, but this area was not targeted with the co-solvent flood. This area of contamination may be remediated by an additional co-solvent flush in the future.

Post co-solvent extraction monitoring of the ground water indicates that degradation of the PCE contaminant is beginning to occur. Data was averaged from three of the recovery wells (RW-2, RW-3, and RW-7) where the co-solvent extraction test was conducted. This data shows the change in water chemistry over the monitoring period, where Day 1 (first data point) is pre-co-solvent extraction and Day 55 (second data point) is 1 month post-co-solvent extraction (Figures 2 and 3). Figure 2 shows the chlorinated hydrocarbon and ethanol data and Figure 3 shows the inorganic data and acetic acid and methane data.

Degradation of PCE in the area of the co-solvent extraction test is indicated by the averaged recovery well data shown in Figure 2. PCE concentrations decreased immediately following the co-solvent extraction, but then rebounded to initial concentrations. TCE concentrations remained low throughout the monitoring period. Production of *cis*-DCE began approximately 4 months post-co-solvent extraction and is an indication of reductive dechlorination of PCE. Ethanol concentrations remained relatively high during the monitoring period and served as an electron donor source.

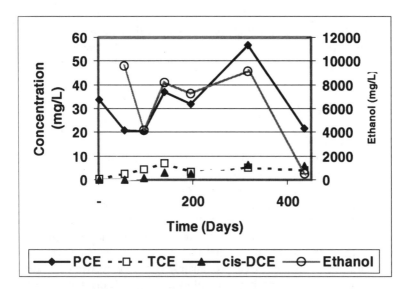

FIGURE 2. Average chlorinated hydrocarbon and ethanol concentrations in RW-2, RW-3, and RW-7 over the monitoring period.

The indication that the reductive dechlorination process is beginning is also supported by additional ground water data collected from the recovery wells (Figure 3). Chloride concentrations initially increased after the co-solvent extraction and then tailed off to near initial concentrations over time. Sulfate was utilized quite rapidly and after approximately 2.5 months sulfate concentrations decreased to approximately 5 mg/L. Acetic acid was produced immediately following the co-solvent extraction test and remained at a relatively high level over the course of the monitoring. Approximately 2.5 months after the co-solvent extraction test methane production was detected.

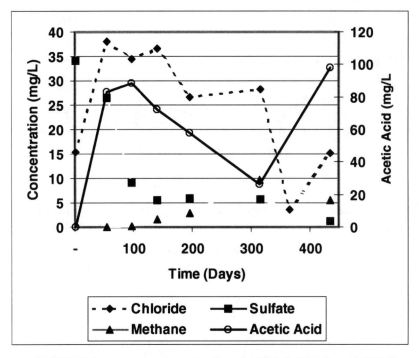

FIGURE 3. Average concentrations in RW-2, RW-3, and RW-7 over the monitoring period.

Data from MW-509, which is immediately down-gradient of the treatment zone, also indicates that reductive dechlorination is occurring (Figure 4). PCE concentrations decreased following the co-solvent extraction test and did not rebound as in the treatment zone. Ethanol concentrations increased following the co-solvent extraction test and then decreased after approximately 10 months post-extraction. Even at the lower concentrations, ethanol supplied excess electron donor to the system. TCE concentrations remained relatively constant throughout the monitoring period. After approximately 4 months post-co-solvent extraction, *cis*-DCE production began and concentrations remained low.

Additional data from MW-509 (Figure 5) followed a similar trend as the recovery wells. There was an initial increase in chloride concentrations that tailed off with time. Sulfate was utilized immediately and remained at a low level over the monitoring period. As sulfate concentrations decreased, acetic acid production began and remained relatively high. Methane production began approximately 6 months after the co-solvent extraction test.

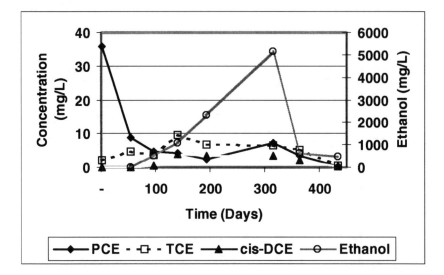

FIGURE 4. Chlorinated hydrocarbon and ethanol concentrations in MW-509 over the monitoring period.

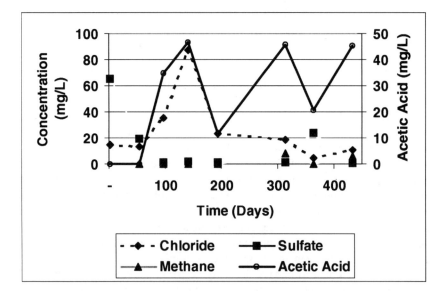

FIGURE 5. Average concentrations in MW-509 over the monitoring period.

The decrease in sulfate concentrations indicates that sulfate-reducing bacteria are active at the site. As the sulfate is reduced, more anaerobic conditions are created which can enhance the reductive dechlorination process. Methane production is another indication that the anaerobic microbial processes are enhanced and that methanogenesis is occurring. As the subsurface microorganisms adapt and become more active, reductive transformations are enhanced.

Under anaerobic conditions, ethanol can be biodegraded to acetate by incomplete oxidation or to carbon dioxide by complete oxidation. The production of acetic acid in our system indicates that the biodegradation of ethanol is beginning to occur. Hydrogen is a product of the oxidation of ethanol and is used in the dechlorination of PCE. Hydrogen analysis was conducted in the field on ground water from several wells. Production of hydrogen is higher in the area near the source area where the co-solvent extraction test was conducted (Figure 6).

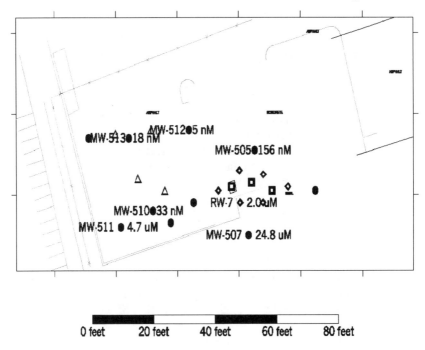

FIGURE 6. Hydrogen concentrations (nM and μM) in ground water from selected wells approximately 11 months post-co-solvent extraction test.

Depending on the extent of the oxidation of ethanol, 1 to 2 moles are required to drive complete dechlorination of PCE. The average ground water concentration of ethanol in the treatment zone after the co-solvent extraction test was 8,000 mg/L. This corresponds to 560 Kg or 12,200 moles of available ethanol. If we assume 2 moles of ethanol are required per mole of PCE for complete dechlorination, then there is 289 times the amount needed to remove the 21.1 moles of dissolved PCE in the source zone. An estimate of the total PCE in the source zone is 157.1 moles (126 moles residual PCE + 21.1 moles dissolved PCE). With the same assumption of 2 moles of ethanol per mole of PCE dechlorination, we calculate that there is 38.8 times the amount of ethanol needed for complete PCE dechlorination. While this estimate assumes no competing terminal oxidation processes, such as methanogenesis or sulfate reduction, an efficiency greater that 2% would still meet the theoretical demand.

The maximum and minimum observed rate of dechlorination based on *cis*-DCE production at the recovery wells are approximately 43.6 and 4.2 μg/liter/day, respectively. Extrapolation of these rates to a multi-step, concurrent, dechlorination process give a preliminary prediction that the dissolved phase PCE could be removed in 3 to 30 years, and that the total source zone PCE could be transformed in 24 to 240 years.

CONCLUSIONS

Co-solvent extraction was used successfully to remove a significant amount of the DNAPL contamination. The injection/extraction process aided in mixing the electron donor (ethanol) with the residual PCE so that biodegradation processes could occur. The system remains biologically active and the dechlorination products TCE, *cis*-DCE, ethene, and chloride are accumulating. High levels of acetate, methane, and hydrogen have also been detected in the treatment zone and indicate that dechlorination processes should continue. Calculated rates of dechlorination indicate that total source zone PCE transformation could take from 24 to 240 years. Monitoring of the system is continuing to better assess the reductive dechlorination process.

ACKNOWLEDGMENTS

The authors wish to acknowledge the support of Dr. Randy Sillan and Mr. Kevin Warner of Levine-Fricke Recon, Inc., Tallahassee, FL and Dr. Mike Annable of the University of Florida, Gainesville, FL for conducting the co-solvent extraction test and field sampling. We also recognize the support of SERDP, U.S. EPA-TIO and the State of Florida.

DISCLAIMER

LABORATORY INVESTIGATION OF ELECTRO-OSMOTIC REMEDIATION OF FINE-GRAINED SEDIMENTS

Nerine J. Cherepy, Dorthe Wildenschild and Allen Elsholz
Lawrence Livermore National Laboratory, Livermore, CA 94550

ABSTRACT: Electro-osmosis, a coupled-flow phenomenon in which an applied electrical potential gradient drives water flow, may be used to induce water flow through fine-grained sediments. We plan to use this technology to remediate chlorinated solvent-contaminated clayey zones at the LLNL site. The electro-osmotic conductivity (k_e) determined from bench-top studies for a core extracted from a sediment zone 36.4-36.6 m below surface was initially 7.37×10^{-10} m^2/s-V, decreasing to 3.44×10^{-10} m^2/s-V, after electro-osmotically transporting 0.70 pore volumes of water through it (195 ml). Hydraulic conductivity (k_h) of the same core was initially measured to be 5.00×10^{-10} m/s, decreasing to 4.08×10^{-10} m/s at the end of processing. This decline in permeability is likely due to formation of a chemical precipitation zone within the core. Water splitting products and ions electromigrate and precipitate within the core; H^+ and metal cations migrate toward the cathode, and OH^- from the cathode moves toward the anode. We are now exploring how to minimize this effect using pH control. The significance of this technology is that for this core, a 3 V/cm voltage gradient produced an initial effective hydraulic conductivity of 2.21×10^{-7} m/s, >400x greater than the initial hydraulic conductivity.

INTRODUCTION

Despite on-going remediation efforts utilizing a variety of technologies, fine-grained sediments contaminated with organic solvents remain recalcitrant. These contaminated fine-grained areas are sources, slowly diffusing dissolved contaminants into adjacent high-permeability zones, leading to groundwater contamination. We are exploring the use of *in-situ* electro-osmotic pumping to flush contaminants from fine-grained sediments (Cherepy, et al., 1999).

Electro-osmotic pumping of water is a known technology with applications in structural engineering for soil stabilization (Casagrande, 1949), mining (Lockhart, 1983), and remediation (Ho, et al. 1999; Acar, et al. 1995; Lageman, 1993; Probstein and Hicks, 1993). Electromigration, electro-osmosis and electrophoresis are collectively referred to as electrokinetics (Mitchell, 1993). Electrokinetic soil remediation technology employs electrodes placed in the ground with a direct current (DC) passed between them using an external power supply. Clays have a net negative surface charge, balanced in the Helmholtz double layer by exchangeable cations. Electromigration may be used for removal of metal contaminants, such as lead and cadmium (Lageman, 1993). Electro-osmosis is a secondary effect arising from electromigration of cations through the porous matrix under an applied electrical potential. The flow of current results in movement of the cations and their associated water of hydration from anode to cathode, entraining contaminants, if present in the pores, in the flow.

Electro-osmotic pumping can increase well yield in fine-grained sediments two to three orders of magnitude over flow rates achievable by hydraulic pumping alone. Contaminated water delivered to the cathode by electro-osmosis may then be mechanically pumped from the cathode well, and contaminants removed. The electro-osmotic conductivities of fine-grained clays with very low hydraulic conductivities (as low as 10^{-11} m/s) and of larger-grained sands with hydraulic conductivities of $\sim 10^{-6}$ m/s lie within the same narrow range, from 10^{-9} to 10^{-8} m^2/s-V. Thus, for soils with hydraulic conductivities so low that standard mechanical pump-and-treat technology is virtually ineffective, electro-osmotic pumping can greatly accelerate contaminant removal.

EXPERIMENTAL

A bench-top cell allowed us to make measurements of electrical (σ_e), hydraulic (k_h) and electro-osmotic (k_e) conductivities of a soil core. Electrical conductivity, σ_e, is determined using Ohm's Law, $\sigma_e = IL/EA$, where I is the current, L is the distance between electrodes, E is the voltage drop and A is the cross-sectional area of the core. Hydraulic conductivity, k_h, is obtained using Darcy's Law, $k_h = q_h L/HA$, where q_h is the hydraulic flux and H is the hydraulic pressure gradient. The electro-osmotic conductivity, k_e, is calculated using $k_e = q_e L/EA$, where q_e is the electro-osmotic flux.

Bench-top Cell. We have designed and built a test cell to measure electro-osmotic and hydraulic flow, electric current and voltage distributions (Cherepy, 1999). *In-situ* conditions are simulated by subjecting the sample to a confining pressure matching the underground stresses of the original location of the soil core. The core used in the measurements reported here was extracted from the 36.4-36.6 m depth of a well drilled for a field installation. In this area, the water table lies at 29 m, and the stresses on the core can thus be estimated to lie in the 0.21-0.42 MPa range. Therefore, all bench-top measurements were acquired with confining pressure of 30 psi (0.21 MPa). The test cell (Figure 1a) consists of a pressure vessel holding a 8.9 cm diameter by 15.2 cm long soil core (Figure 1b). The core is jacketed with a Teflon sleeve to seal against the confining pressure and to avoid short circuiting of the water flow at the circumference of the sample. Two perforated gold plated copper electrodes (anode and cathode) are placed on each end of the sample, and gold wire hoops placed around the core, 2 inches from each electrode, for use as voltage probes. The gold-plated diffusion plates are used to transfer the applied longitudinal load to the sample, as well as serving as electrodes. They are separated from the soil by a microporous membrane (Pall-RAI Electropore E40201ultra high MW polyethylene, 100 µm thick, 2 µm pores). A 0-50 V Hewlett-Packard 6633B power supply was employed in DC constant voltage mode for electro-osmotic conductivity measurements. Water is supplied to the anode side of the cell by a constant hydraulic head standpipe during electro-osmotic flow measurements and by a pressurized water vessel for hydraulic flow measurements. A narrow diameter standpipe, outfitted with a 0-1.25 psi (0-8618 Pa) pressure transducer (Validyne DP 215-50), is used to measure water outflow.

FIGURE 1A. Bench-top cell. The core is contained within the central pressure vessel, and water flow is measured as the level rises in the right hand (cathode) standpipe. For these experiments, the working electrodes were placed in direct contact with the soil.

FIGURE 1B. Core assembly. The core is jacketed in Teflon shrinkwrap, with gold-plated copper perforated electrodes at both ends, and two gold hoop voltage probes to provide information about the voltage drop along the core.

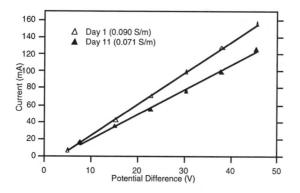

FIGURE 2. Electrical conductivity of the core is determined in the bench-top cell using Ohm's law. The current-voltage curve for the entire 15.2 cm core shows that the electrical conductivity of the core decreased during processing from 0.090 S/m at the beginning to 0.071 S/m at the end.

RESULTS AND DISCUSSION

Electrical Conductivity. In the bench-top cell, the voltage imposed between the anode and cathode (at either end of the 15.2 cm long soil core) is controlled in constant voltage mode, this voltage is referred to here as V_{14}. Further detail about the voltage drop along the core is provided by two supplemental gold hoop voltage probes at 5 and 10 cm along the core, the voltage between these probes is V_{23}. Figure 2 shows the current-voltage plots of V_{14} to be linear (V_{23}, not shown, is also linear with current). We measured a soil electrical conductivity of 0.090 S/m on day 1, decreasing to 0.071 S/m at the end of processing on day 11.

Hydraulic Conductivity. A standard technology used for remediation of organic solvent contamination at the LLNL site, "pump-and-treat", is based on pumping water through contaminated zones, extracting contaminated water, and removing the contaminants. Hydraulic flow through heterogeneous lithologies preferentially passes through sandy, permeable zones, resulting in very little penetration of clayey, fine-grained zones. It is for this reason, that we are exploring electro-osmotic pumping to specifically address the finer-grained, less permeable sediments. The core chosen for work in the bench-top cell was selected due to its high clay content, representative of the finer-grained layers in the screened zone of one of the wells. Therefore, the hydraulic conductivity measured for this core is indicative of the type of sediments we are interested in targeting for cleanup with electro-osmotic pumping, but would not be typical of a measurement between wells in a field installation.

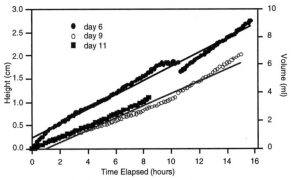

FIGURE 3. The hydraulic conductivity of the same core, measured in the bench-top cell with a 10 psi (0.069 MPa) hydraulic head gradient, shows a small decline in conductivity during electro-osmotic processing from 5.00 x 10^{-10} m/s (day 6), to 4.16 x 10^{-10} m/s (day 9), and 4.08 x 10^{-10} m/s (day 11).

Hydraulic conductivity was measured for the core in the bench-top cell using a pressure differential imposed by a pressure can, pressurized with compressed air, on the inlet side and a standpipe open to atmospheric pressure at the outlet side. Initial flow rate using a pressure gradient of 10 psi (0.069 MPa) was 0.0085 ml/min (Figure 3). This corresponds to a hydraulic conductivity for this core of $k_h = 5.00 \pm 0.36 \times 10^{-10}$ m/s. Sediments with hydraulic conductivities in this range may be considered essentially impermeable to mechanical pumping, especially when interleaving sandy layers ($k_h > 10^{-6}$ m/s) are present.

Electro-osmotic Conductivity. Electro-osmotic conductivity (k_e) measurements may be performed under controlled conditions in the bench-top cell. Five measurements taken during electro-osmotic processing are presented in Figure 4. A 3 V/cm applied voltage resulted in a $q_e = 0.082$ ml/min on day 3, decreasing to 0.038 ml/min on day 11. The electro-osmotic conductivity for the 3.5 inch diameter core calculated from these measurements declines from $k_e = 7.37 \times 10^{-10}$ m²/s-V to 3.44×10^{-10} m²/s-V after 11 days of processing. The sediment sample used is small, isolated from natural hydraulic gradients, and represents the finer-grained zones of a natural heterogeneous fabric.

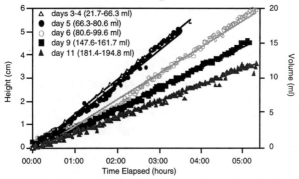

FIGURE 4. The electro-osmotic conductivity of the core, measured in the bench-top cell using a 3 V/cm voltage gradient, decreased during processing. Initially the electro-osmotic conductivity was 7.37×10^{-10} m²/s-V (day 3), then 7.26×10^{-10} m²/s-V (day 5), 6.11×10^{-10} m²/s-V (day 6), 4.33×10^{-10} m²/s-V (day 9) and 3.44×10^{-10} m²/s-V (day 11).

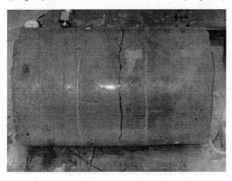

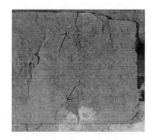

FIGURE 5. Core at the end of electro-osmotic processing. Discoloring is seen in a zone closer to the cathode (right end), with a distinct black ring at the furthest left portion of the discolored area.

FIGURE 6. The black metals deposition and white carbonate precipitate occurred at the outer perimeter of the core, in a donut-shaped area, not evenly over the cross-sectional area of the core.

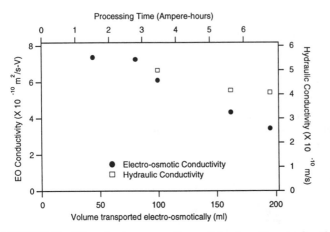

FIGURE 7. The electro-osmotic and hydraulic conductivities of the core both decreased during processing. Electro-osmotic conductivity declined by ~2x, while the decline in hydraulic conductivity was slight, ~1.2x less. The volume transported electro-osmotically, 195 ml, is 0.70 pore volumes.

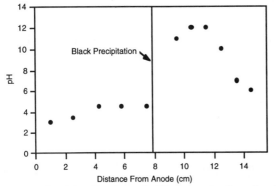

FIGURE 8. Distribution of pH along the length of the core. The anode is at the left end of the core and the cathode at the right end. The pH rises sharply at the precipitation zone.

Conclusions. We measured declines in the electrical, hydraulic and electro-osmotic conductivity of a clayey core in a bench-top test cell during electro-osmotic processing in conjunction with precipitation within the core. Metals desorbed by an acid front propagating from the anode formed a black precipitate with OH⁻ propagating from the cathode (Figures 5,6 and 8). In addition, where the base penetrated the core, carbonates precipitated. This precipitation zone developed closer to the cathode end, due to the lower mobility of OH⁻ (self-diffusion coefficient $D_0 = 52.8 \times 10^{-10}$ m²/s) relative to H⁺ ($D_0 = 93.1 \times 10^{-10}$ m²/s) (Mitchell, 1993). Since hydraulic pumping in a heterogeneous matrix will draw water primarily through the coarse-grained zones, circumventing finer-grained zones with higher contamination, mechanical pump-and-treat technology is not effective in cleanup of sediments such as that studied in the bench-top cell. We therefore hope to use electro-osmotic pumping to drive water through such sediments and clean up "impermeable" zones. From the initial electro-osmotic conductivity measured for the core in the bench-top experiments, 7.37×10^{-10} m^2/s-V, the "equivalent hydraulic conductivity" ($k_{h\text{-eq}}$) under electro-osmotic pumping (at 3 V/cm) is $k_{h\text{-eq}} = 2.21 \times 10^{-7}$ m/s. This equivalent hydraulic conductivity results in a flow through the fine-grained sediments more than 400 times greater than without the applied field ($k_h \sim 5 \times 10^{-10}$ m/s)! Electro-osmotic pumping technology offers great potential for controlled cleanup of fine-grained sediments, since flow follows the domain of the imposed electric field and passes preferentially through fine-grained sediments due to their greater electrical conductivities. It is an *in-situ* technology, requiring no excavation, nor significant chemical by-product residue. We are now testing a pH control system to minimize the formation of precipitates and we are exploring the removal efficiency of organic solvents from contaminated sediments using electro-osmotic pumping.

ACKNOWLEDGEMENTS
This work was performed under the auspices of the U.S. Department of Energy by University of California Lawrence Livermore National Laboratory under contract No. W-7405-Eng-48. Thanks to Roberto Ruiz for the cell design, and to Walt McNab for helpful discussions. We gratefully acknowledge LLNL's Laboratory Directed Research and Development Office, DOE EM-40 and EM-50 for support.

REFERENCES
Acar, Y.B., Gale, R.J., Alshawabkeh, A.N., Marks, R.E., Puppala, S., Bricka, M. Parker, R. 1995. "Electrokinetic Remediation: Basics and Technology Status", *J. Haz. Mat.,* 40, 117.

Casagrande, L. 1949. "Electro-osmosis in Soils", *Geotechnique*, I, 159.

Cherepy, N., McNab, W. Wildenschild, D., Ruiz, R. and Elsholz, A. 1999. "Electro-Osmotic Remediation of Fine-Grained Sediments", *Proc. Electrochem. Soc.: Environmental Aspects of Electrochemical Technology,* 99-39, in press.

Ho, S.V., Athmer, C., Sheridan, P.W., Hughes, B.M., Orth, R., McKenzie, D., Brodsky, P.H., Shapiro, A.M., Thornton, R., Salvo, J., Schultz, D., Landis, R., Griffith, R., Shoemaker, S., 1999 "The Lasagna Technology for In Situ Soil Remediation. 1. Small Field Test" *Environ. Sci. Technol.,* 33, 1086.

Lageman, R. 1993. "Electroreclamation: Application in the Netherlands", *Environ. Sci. Technol.,* 27, 2648.

Lockhart, N.C. 1983. "Electro-osmotic Dewatering of Fine Tailings from Mineral Processing", *Int. J. of Miner. Process.,* 10, 131.

Mitchell, J.K. 1993. *Fundamentals of Soil Behavior*, John Wiley and Sons, Inc., New York.

Probstein, R.F., Hicks, R.E. 1993. "Removal of Contaminants from Soils by Electric Fields", *Science*, 260, 498.

HYDRATION STRUCTURES AND DYNAMICS: AN ALTERNATIVE PERSPECITVE ON REMEDIATION PROCESSES

D.L. Marrin (San Diego State University, San Diego, CA and Kaua'i, HI)

ABSTRACT: As environmental scientists, our understanding of solute behavior (e.g., fate, transport, remediation) is based on macro-scale properties such as solubility, diffusivity and phase partitioning. These gross physical properties of solutes are actually a result of molecular- and atomic-scale dynamics, which are described by thermodynamic and quantum mechanical principles. The perception that water is simply a passive solvent within which contaminants react and are transported overlooks the effects of contamination on water itself, except as it may influence gross physical properties of the water. The effect of solutes on the molecular structure of water and the manner in which water is able to accommodate solutes into its rather complex network is of particular importance in assessing its behavior and degradation in aqueous environments.

INTRODUCTION

Bulk Liquid Water. Water is not a random collection of H_2O molecules, but rather an extended network that is locally interconnected by hydrogen bonds (H-bonds) linking the oxygen and hydrogen atoms of neighboring water molecules. These electrostatic H-bonds are generally considered to be distinct from the covalent bonds that connect oxygen and hydrogen atoms within the same molecule; however, a recently published study indicates that the intermolecular H-bonds have significant covalent properties (Isaacs et al., 1999). This finding suggests that electrons are shared not only between atoms of the same water molecule (intramolecular), but also among atoms of neighboring water molecules (intermolecular).

The underlying geometry of bulk liquid water is tetrahedral, owing to the ability of each water molecule to hydrogen bond with four of its neighbors. This water network is dynamic, rather than static, and re-shuffles H-bonds among neighboring water molecules at a rate of millions to trillions of times per second. The ceaseless rearrangement of H_2O molecules with respect to each other creates 3-dimensional geometries that are unique to the chemicals solvated and to the surfaces interfaced. In addition to the tetrahedral geometry of bulk liquid, water also has the ability to form larger associations that are known as clusters. Water comprising these clusters (ranging from a few to several hundred water molecules) is more structured and less dynamic than that in bulk liquid. In other words, water molecules comprising clusters re-shuffle intermolecular connections at a slower rate and are structurally more rigid (e.g., possess less orientational flexibility) than do those in the bulk liquid. These two properties of water clusters translate into slightly different physical properties than those of the bulk liquid.

Water Clusters. Water clusters exist as both pure water and as "cages" or envelopes that surround chemicals that are dissolved in water. The latter may include millions of water molecules and are of particular interest to our discussion because they constitute the structures that are responsible for the dissolution and aqueous behavior of contaminants. Essentially, water insulates solutes from its bulk tetrahedral network by forming hydration envelopes around them. Water molecules comprising the envelopes are connected to the bulk liquid via H-bonds and to the solute via H-bonds, dipole interactions or van der Waal forces, depending on the nature of the solute. Many recalcitrant organic chemicals are hydrophobic in nature and, thus, interact weakly with water comprising the envelope. Conversely, ionic and polar solutes interact more strongly with water molecules in the envelope. Within these hydration structures, solutes are contained in water with minimal (albeit unavoidable) breakage of H-bonds and the associated disruption to water's bulk network.

Water molecules comprising the hydration envelopes act as mediators for the interaction of contaminants with their "guest" compounds and with larger-scale structures such as mineral surfaces. For example, carcinogenic environmental chemicals are active, not through their direct effects on nucleic acids, but through the production of free radicals in DNA's hydration shell that, in turn, result in strand breakage. In a similar manner, sorption of contaminants on mineral surfaces is facilitated through hydration envelopes that actually mediate proton transfer and other critical processes responsible for electrostatic attraction. Physical chemists have described hydration envelopes for many common hydrophobic solutes, including volatile chlorinated/aromatic hydrocarbons and biogenic gases. Moreover, the orientation and rotational dynamics of solutes and of water molecules comprising such hydration envelopes have also been documented. These dynamics are the underlying basis of reactivity, solubility and a number of other well-known contaminant properties.

Obviously, variables such as temperature and pressure affect contaminants via changes in their molecular structure and dynamics of water comprising both hydration envelopes and the bulk liquid. However, there are less obvious phenomena such as electromagnetic radiation, certain types of fields (magnetic, vorticity) and the presence of other solutes that profoundly effect hydration envelopes and their interaction with the bulk liquid. Many of these phenomena are inherent in conventional remediation and monitoring technologies.

DISCUSSION

Hydrophobic Gases. The behavior of biogenic gases (e.g., methane, carbon dioxide, hydrogen) in aqueous systems is a major area of research as a result of its applicability to everything from assessing redox conditions to quantifying the rate at which organic contaminants are degraded in the subsurface. Those dissolved gases possessing a molecular diameter <3 angstroms (e.g., hydrogen) have a substantially different effect on the structure of water than do other solutes. This is because small gases are able to "slip through the bars" of conventional water envelopes and clathrates, affecting the thermodynamic state of water's H-bond network in a different manner than do larger gases (e.g., methane, ethane). Figure 1 shows the change in entropy and in Gibbs free energy associated with the dissolution of various biogenic and noble gases at 25° Celsius (e.g., Tenaka and

Nakanishi, 1991). Notice that hydrogen and neon (both of which have a molecular diameter <3 angstroms) display the least entropy change and a free energy change that is comparable to that of n-octane. While these small gases have a relatively small enthalpic cost associated with rearranging water molecules to accommodate them (i.e., cavity formation), they offer water almost nothing in terms of network order (i.e., decreased entropy).

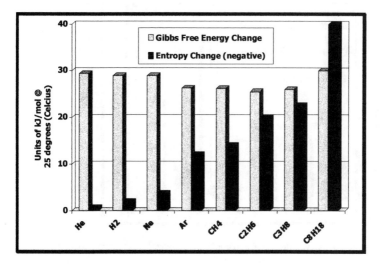

Figure 1. Gibbs free energy and entropy changes associated with solute dissolution in water.

Generally, the dissolution of a hydrophobic solute is believed to occur because the entropic gain to the water network exceeds the enthalpic cost of rearranging molecules to accommodate it. Large organic molecules are less soluble in water than are smaller ones (within a homologous series of organic compounds) because the enthalpic costs increase faster than the entropic gains. Hence, large hydrophobic compounds and very small gases are the most disruptive to water's network and, therefore, have the lowest aqueous solubilities. Figure 2 presents various physical properties of gases in water (Atkins, 1986). Notice that H_2 has both the highest diffusivity and lowest van der Waals constant of any of the gases, suggesting that it interacts with and is confined by hydration envelopes to a minimal extent. It also suggests that helium and neon are probably the best surrogates for hydrogen gas (as conservative tracers) and that these same noble gases are probably poor surrogates for light hydrocarbon and other biogenic gases, which interact with water via stronger van der Waals forces and possess lower diffusivities.

It is interesting to note that certain recalcitrant compounds seem to degrade only under highly reducing conditions, where elevated H_2 concentrations may have an effect on water comprising both the bulk liquid and the hydration shells of contaminants. In addition to the fact that H_2 is the most thermodynamically favored electron donor (i.e., the most highly reduced molecule), its affect on the structure of water itself may be another reason for its apparent rapid turnover in aqueous systems. Schuler and Conrad (1990) hypothesized that the rapid rate of H_2 oxidation (particularly at low aqueous concentrations) may be due to the presence of extracelluar enzymes that are adsorbed to natural organic matter or mineral surfaces, primarily via interactions mediated by hydration envelopes.

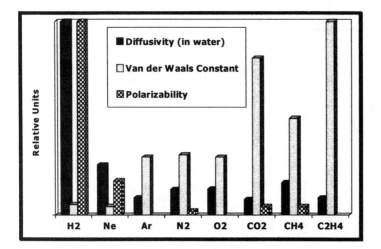

Figure 2. Relative diffusivities, van der Waal constants and polarizabilities among solutes.

Volatile Organic Compounds (VOCs). Similar to biogenic gases, the effect of hydrophobic solutes (such as VOCs) on the structure of water is related to the ordering of molecules in the hydration envelope. For this reason, hydrophobic solutes are often referred to as structure-makers, denoting that they increase the order (or decrease the entropy) of the water network immediately surrounding the VOC. Water molecules in the hydration envelope are not only more ordered (i.e., possess less orientational freedom) than those in the bulk liquid, they also rearrange or vibrate at a slower rate (Nakahara et al., 1996). This slower vibration rate is a result, or a consequence, of the fact that H-bonds in the hydration envelope are stronger than those in the bulk liquid. Although hydrophobic hydration results in an increased percentage of broken H-bonds compared to that in the bulk tetrahedral network, the H-bonds that persist are stronger (Mancera, 1996).

The effect of the hydration envelope on VOCs such as benzene is to increase the rotational movement of the solute, affecting its physical properties such as reactivity and diffusivity. Therefore, both the structure of the hydration envelope and the temperature of the water affect solute properties. As the temperature of the water rises, water molecules in the hydration envelope, as well as those in the bulk liquid, vibrate at an increasing rate until H-bonds begin to break, thus releasing solutes from their cages. This is the basis of in-situ thermal technologies, which are generally designed to either decrease sorption or increase volatilization by destroying the hydration structures that permit VOCs to remain in solution and to interact (via H-bond mediated proton transfer) with natural organic acids or mineral surfaces.

Bioremediation technologies are also dependent on hydration structures and dynamics. Both intracellular and extracellular enzymes responsible for the biological catalysis of recalcitrant compounds are fully hydrated and interact with the substrate (e.g., VOC) via their respective hydration envelopes. Hence, changes in both the contaminant hydration structure and dynamics, affecting the molecular-scale properties of the contaminant itself, influence the susceptibility to and kinetics of biodegradation. Due to the difference in the size and polarity among chlorinated ethanes, their respective hydration envelopes may be characterized by different free energies (thermodynamics). Even among this homologous group of compounds, changes in the Gibbs free energies for hydration are an order-of-magnitude less for the moderately recalcitrant 1,1,1-trichloroethane than for the readily degradable 1,1,2,2-tetrachloroethane (Paulsen and Straatsma, 1996).

In addition to organic substrates (e.g., benzene, chlorinated ethanes), the hydration of electron acceptors is also critical. Figure 3 indicates that water solvates both O_2 and CO_2, which serve as electron acceptors for aerobes and H_2-utilizing methanogens, with what are known as *Structure I* cages or clathrates (Franks, 1973). Clathrates are a specific class of water cluster that surrounds solutes with molecular diameters less than 7 angstroms; *Structure I* consists of 46 water molecules in the form of a dodecahedron (Franks, 1973). By contrast, ionic electron acceptors (e.g., nitrate, ferric iron, sulfate) interact with hydration structures predominantly via dipole or electrostatic (rather than van der Waals) forces, which reorient the surrounding polar water molecules of the bulk liquid. For example, the first hydration shell or layer surrounding the Fe^{3+} ion consists of 6 water molecules in the form of an octahedron (Degreve and Quintale, 1996). This first shell is surrounded by a second shell, consisting of 15 water molecules that H-bond with those in both the first shell and the bulk liquid.

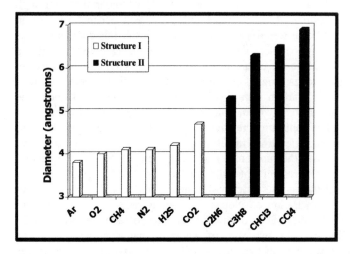

Figure 3. Clathrate structures of various aqueous solutes as a function of molecular diameter.

For both the nonpolar and ionic electron acceptors, water molecules in the hydration envelopes act as mediators and as buffers between the solute and the bulk water or other solutes. Because there is extensive H-bonding between water molecules in the hydration envelopes and the bulk liquid, changes in the bulk properties of water are translated into differences in the availability and reactivity of biodegradation reactants. Simply stated, the hydration envelopes of all solutes are affected by any re-structuring of the bulk phase.

CONCLUSIONS

There are number of ways in which the hydration structures and dynamics of water could be useful in understanding or designing remediation systems. Perhaps the simplest is the selection of surrogates that are used as conservative tracers for both organic contaminants and for biogenic gases. The increasing use of such tracers over the past five years is related to the monitoring requirements of the MNA remediation option and to the understanding of biodegradation dynamics within a contaminated plume. As previously mentioned, helium has a similar effect on the structure of water as does hydrogen gas. Similarly, argon and other moderately-sized nonpolar gases are good tracers for the common biogenic gases (e.g., O_2, CO_2, CH_4, H_2S) because they are contained within *Structure I* hydrate clathrates. The largest of molecules contained within clathrates possess a molecular diameter in excess of 5 angstroms and include C_2 and C_3 hydrocarbon gases and the chlorinated methanes. These larger compounds form *Structure II* clathrates, which consist of 136 water molecules and represent the concatenation of 4-, 5- and 6-sided polygons (Franks, 1973). Compounds possessing a molecular diameter greater than 7 angstroms are hydrated by customized envelopes rather than by specific clathrate structures.

Clathrates are currently being utilized in the mitigation of the greenhouse gas, carbon dioxide, and seem to be very stable under low temperatures and high pressures. The formation of CO_2 hydrates either in deep ocean basins or in engineered processes is being tested as a method to sequester this greenhouse gas from the atmosphere. In essence, CO_2 clathrates are being entombed at the bottom of the oceans. While the environmental impacts of such a technology are poorly understood, the process does pose the prospect that other clathrate formers (e.g., methylene chloride, carbon tetrachloride) could be sequestered under the appropriate conditions, which may or may not be applicable to the environment.

Another remedial application related to hydration dynamics is related to the effects of electromagnetic (EM) radiation on bulk water, hydrate structures and solutes. Table 1 lists the molecular and atomic effects of EM frequencies within various ranges. The reason that radio and microwave frequencies are used in heating water is that they cause water molecules to vibrate and reorient in both the bulk liquid and the hydration envelopes. While IR radiation is sufficient to distort the H-bonds of water, UV radiation is capable of actually breaking its covalent bonds. Similarly, the exposure of water to magnetic fields has been shown to re-structure water such that number of clusters (water only) increases relative to the tetrahedral bulk phase. H-bond network effects (e.g., electron rearrangements) induced by magnetic or electric fields are due to the slight polarization of water molecules and depend on the exact location of a molecule in the chain or cluster (Hermansson and Alfredsson, 1999). Magnetic fields actually induce current flow that affects the solute's molecular rotation and orientation, both of which are related to its chemical properties. As presented on Figure 2, the biogenic and light hydrocarbon gases (particularly H_2) possess a polarizability much greater than that of water itself.

Table 1. Effects of EM radiation within various frequency ranges.

FREQUENCY (Hz)	EFFECT
10^7 to 10^{11} **RADIO/MICROWAVES**	Rotations & vibrations (molecular orientation)
10^{12} to 10^{15} **IR/UV/VISIBLE LIGHT**	Electronic shell excitation (molecular distortion)
10^{16} to 10^{19} **X-RAYS**	Electron excitation (quantum effects)
$>10^{20}$ **COSMIC RAYS**	Nuclear excitation

The purpose of this paper is not to suggest specific technologies, but to consider contaminants (and associated remediation processes) from the perspective of their effects on the molecular structure and dynamics of water. Perhaps a greater understanding of the structure/dynamics of water will provide insights into methods of investigating, tracing and remediating recalcitrant contaminants that otherwise would not be obvious.

REFERENCES

Atkins, P. 1986. *Physical Chemistry.* W.H. Freeman & Co., New York.

Conrad, R. 1999. "Contribution of hydrogen to methane production and control of hydrogen concentrations in methanogenic soils and sediments." *FEMS Microbiol. Ecol. 28*: 193-202.

Degreve, L. and C. Quintale. 1996. "The interfacial structure around ferric and ferrous ions in aqueous solution: the nature of the second hydration shell." *J. Electroanalytical Chem. 409*: 25-31.

Franks, F. (Ed.). 1973. *Water: A Comprehensive Treatise*, Vol. 2. Plenum Press, New York.

Hermansson, K. and M. Alfredsson. 1999. "Molecular polarization in water chains." *J. Chem. Phys. 111*: 1993-2000.

Isaacs, E., A. Shukla, P. Platzman, D. Hamann, B. Barbiellini and C. Tulk. 1999. "Covalency of the hydrogen bond in ice: a direct X-ray measurement." *Phys. Rev. Lett. 82*: 600-603.

Mancera, R. 1996. "Hydrogen-bonding behavior in the hydrophobic hydration of simple hydrocarbons in water." *J. Chem. Soc., Faraday Trans. 92*: 2547-2554.

Nakahara, M., C. Wakai and Y. Yoshimoto. 1996. "Dynamics of hydrophobic hydration of benzene." *J. Phys. Chem. 100*: 1345-1349.

Paulsen, M. and T. Straatsma. 1996. "Chlorinated ethanes in aqueous solution: parameterization based on thermodynamics of hydration." *Chem. Phys. Lett. 259*: 142-145.

Schuler, S. and R. Conrad. 1990. "Soils contain two different activities for oxidation of hydrogen." *FEMS Microbiol. Ecol. 73*: 77-84.

Tenaka, H. and K. Nakanishi. 1991. "Hydrophobic hydration of inert gases: thermodynamic properties, inherent structures, and normal-mode analysis." *J. Chem. Phys. 95*: 3719-3727.

TREATMENT OF CHLORO-HYDROCARBON CONTAMINATED GROUNDWATER BY AIR STRIPPING

Kuyen Li, Sung-Woon Chang, and Chung-Ching Yu
Lamar University, Beaumont, Texas

ABSTRACT: Air stripping of chloro-hydrocarbon contaminated groundwater was studied by using a 5-foot by 4-inch I.D. glass column packed with 12.7 mm Raschig ceramic rings. The contaminated groundwater was simulated by using a local residential well water spiked with six chloro-hydrocarbons: trichloro-thylene (TCE), dichloro-methane (DCM), 1,2-dichloroethylene (DCE), 1,2-dichloroethane (DCA), chlorobenzene (Cl-Bz), and 2-dichloroethyl ether (DCEE). Experimental results indicated that all the six chemicals could be removed easily from the groundwater by air stripping except DCEE. The measured over-all mass transfer coefficients were found in good agreement with those model predictions except for DCEE, at which the Sherwood's predictions were significantly higher. In general, water flowrate affects the value of over-all mass transfer coefficient more than the other factors. However, for the low-volatile chemical such as DCEE, the air flowrate effect was found to be significant. In order to remove DCEE efficiently, high air-water loading ratio has to be utilized. At this high air flowrate the cooling effect could be significant. An air stripping system of multiple columns set up in series could be more effective than a single column for the removal of DCEE.

INTRODUCTION

Groundwater supplies are extremely important. They account for nearly half of our drinking water and a large portion of the water used for irrigation. They are also vulnerable to contamination and difficult to clean up once contaminated. During the past decade, control of the pollution of ground water has been one of the highest research priorities of the USEPA. Choloro-hydrocarbons are the most concerned contaminants for groundwater because they are heavier than water, stable to the biodegradation, and carcinogenic.

Air stripping has been accepted as one of the best available technologies for removing volatile organic contaminants from groundwater (Li 1997). However, technical and economical considerations do not favor this application to contaminants of low volatility. For effective removal of a solute in air stripping process, the stripping factor, S, must be kept above 3 (Kavanaugh and Trussel 1981). Since the stripping factor is defined as $S = (H/P)(G/L)$, where P is the system pressure (atm), the molar air-to-water flow ratio, G/L, has to be maintained very high for contaminants of low H values. This high air-to-water flow ratio may cause a significant cooling effect in the stripping column (Yu 1990). As a result of temperature drop the Henry's constant drops and therefore the removal efficiency of the air stripping drops too. Thus, for effective removal of low volatile chemicals in an air stripping tower, the air-to-water ratio must be very large and so the power required to run the air will be tremendous. Due to this reason, the Water Pollution Control Federation's Groundwater Committee has suggested that the air-stripping process is most economical for compounds which have Henry's constants higher than 50 atm.

Ball et al. (1984) showed that correlation equation was a valuable tool to predict an overall mass transfer coefficient. Robert et al. (1985) and Lamarche et al. (1989) assessed the performance of three correlation models such as Onda, Shulman, and Sherwood by comparing the model predictions with experimental results. They concluded that the Onda's model, which is based on two-film theory, provided the best overall fitting.

Objective. the objectives of this study are (1) to obtain experimental data of removal percents of six chemicals: trichloro-thylene (TCE), dichloro-methane (DCM), 1,2-dichloroethylene (DCE), 1,2-dichloroethane (DCA), chlorobenzene (Cl-Bz), and 2-dichloroethyl ether (DCEE); (2) to determine the overall mass transfer coefficient which is one of the important design parameters; and (3) to evaluate the performance of the mass transfer prediction models.

MATERIALS AND METHODS

A schematic diagram of the experimental set up is shown in Figure 1. Air was pumped by an air compressor (AC) through a saturator (C2) to adjust the humidity before entering the stripping column (C1). The stripping column was constructed with 5-ft by 4-inch I.D (152.4 cm by 10.16 cm I.D.) Pyrex conical column. The column is packed with 12.7 mm ceramic Rashig rings and the packing height is 1.13 m. Two Rosette-type PVC water redistributors (RD) were placed inside the column to reduce wall effects. A 3-mm thick brace plate with 60% hole area was used to

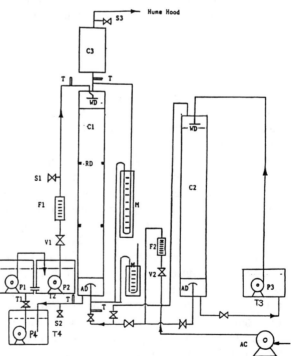

FIGURE 1. Experimental setup for packed column air stripping

support the packing. Water was distributed a half-inch above the packing by a 24-spoked acrylic distributor which has a diameter of 2.5 inches. A section of 1-inch copper tubing was placed 2 inches below the supporting plate to distribute air.

The adsorption column (C3) was half-filled with 14-mesh granular activated carbon and placed above the stripping column to reduce air pollution from the effluent air. Simulated ground water was contained in two 30-gallon fiber glass tanks (T1 and T2) which were connected together to increase the capacity of liquid supply. The effluent liquid was stored in a 50-gallon PVC tank (T4). The effluent liquid water was reused in the experiment. Air and water flowrates were measured by two separate rotameters (F2 and F1) which were calibrated before the experimental measurements.

Water Sampling and Analysis. All the chemicals (purchased from Aldrich) was chemical pure and was used without further purification. The ground water was obtained from a local well. Contaminated aqueous solution was prepared using methanol as the spiking agent. The concentrations of the contaminants were measured by GC-ECD (Varian 3500) together with a purge-and-trap concentrator (Tekmar LSC 2000), using a Supelco Vocol glass capillary column (60 m x 0.75 mm I.D. with 1.5 μm thick film) as the separation column.

A 40-ml glass bottle with TFE septum was used to collect water sample. To prevent possible contamination of the sample bottle due to adsorption of chemicals, sampling bottles were washed with hot water and then rinsed with methanol. Samples were collected at different time ranges of 30, 40, 50, and 60 minutes. The result of sample analysis showed that the steady state had been reached after 40-minute run. Before collecting sample, sampling ports were flushed and fresh samples were obtained.

Data Analysis. The equation used in this data analysis has been derived (Treybal 1980) and is given below.

$$Z = HTU \times NTU \tag{1}$$

Here, Z is the packing height, HTU is called the height of transfer unit and is defined as

$$HTU = L/(aK_L C_0) \tag{2}$$

and NTU is so called the number of transfer unit and is defined as

$$NTU = (S/(S-1)) \, ln\{[(x_{A1}/x_{A2})(S-1)+1]/S\} \tag{3}$$

where S is the stripping factor $(S = (H/P)(G/L))$, G and L are the molar flowrates of air and water solution, respectively, aK_L is the overall mass transfer coefficient, and x_{A1}/x_{A2} is the liquid molar fraction ratio of the contaminant A at the bottom to the top of the stripping tower. Once the values of Z, S, L, C$_0$, and x$_{A1}$/x$_{A2}$ are obtained, the overall mass transfer coefficient, aK$_L$, can be calculated from Equation (1).

RESULTS AND DISCUSSION

The removal percentages of the six chloro-hydrocarbon contaminants at different air and water flowrates and different temperatures are listed in Table 1.

TABLE 1. Removal percentages at different operating conditions

Temp C	L mol/(m²s)	G mol/(m²s)	TCE Removal %	Cl-Bz Removal %	DCEE Removal %	1,2-DCE Removal %	DCM Removal %	DCA Removal %
31.5	118	4.2	93	83	-	86	87	85
31.5	118	8.67	94	86	-	87	88	90
29.1	118	14.5	94	87	-	88	89	90
29.1	118	21	95	89	-	89	90	92
29.6	205	4.23	82	77	7.5	82	81	72
29.4	205	8.73	80	81	12	85	86	81
30.7	205	14.5	87	83	13	89	88	85
30.7	205	20.9	89	87	17	90	89	87
29.8	291	4.23	84	71	11	83	79	68
29.6	291	8.23	85	77	12	86	85	75
28.7	291	14.85	84	79	14	87	86	80
28.5	291	20.9	85	80	15	87	86	85
24.7	119	4.3	92	83	8	88	85	54
25.2	119	8.85	93	86	9	88	89	69
26.0	119	14.7	93	88	11	89	91	79
25.5	119	21.2	94	87	15	90	92	82
25.3	205	4.29	91	65	7	82	79	74
24.8	205	8.87	92	76	8	82	84	84
24.8	205	14.7	92	78	9.5	84	86	86
24.9	205	21.3	92	87	14	87	87	88
23.8	291	4.31	84	72	7.5	80	72	66
24.3	291	8.88	86	78	8	81	77	75
25.0	291	14.7	86	78	10.5	83	83	79
25.0	291	21.2	89	81	11	85	86	84
19.3	121	4.38	87	77	8	87	81	66
20.4	120	9	89	82	9	88	87	75
19.7	121	15	89	86	15	89	88	79
20.5	120	21.6	90	88	18	91	90	84
20.1	205	4.4	85	73	7.5	74	74	54
19.6	205	9	87	81	8	80	81	69
20.3	205	15	88	82	9	82	82	70
20.1	205	21.6	88	83	12	84	84	79
20.5	291	4.4	84	53	7	78	68	87
21.5	291	9	86	66	7.5	80	77	84
21.1	291	14.9	86	65	9.4	82	80	87
20.4	291	21.6	86	74	12	84	83	89

The above removal percentages can be plotted versus air-water molar loading ratio, G/L, and the results are shown in Figure 2 for three average temperatures (20, 25, and 30 °C with a fluctuation of ± 2 °C).

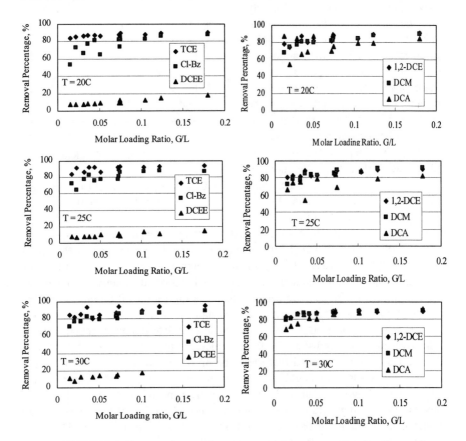

FIGURE 2. Removal percentage versus air-water molar loading ratio

It can be seen from Figure 2 that all the six chloro-hydrocarbon contaminants, except DCEE, can be removed from the contaminated groundwater easily. This may be realized from the Henry's law constants which are shown in Table 2. DCEE has the smallest Henry's law constant among these

TABLE 2. The Henry's law constant in the unit of atm

Temp C	TCE	DCE	Cl-Bz	DCM	DCA	DCEE
20	575	267	200	145	55	1.0
25	630	295	225	178	61	1.2
30	685	324	250	215	70	1.4

six chloro-hydrocarbon contaminants. In order to remove DCEE from the groundwater more effectively, a high air-water loading ratio has to be used. A special experimental set of high air-water loading ratio was performed for the stripping of DCEE and the results are shown in Figure 3. At the beginning, the removal percentage increases linearly with the air-water loading ratio. However, when the air-water molar loading ratio is higher than about 0.6 the removal percentage starts to deviate away from the linear and approaches to a constant, as can be seen from Figure 3. At this high air flow rate (the air-water molar loading ratio of 1.43 is equivalent to volume loading ratio of 1,800), the cooling effect of the water becomes significant if the relative humidity of the air is not at its saturation (Yu,

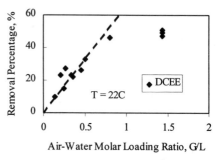

FIGURE 3. Air stripping of DCEE at high air-water loading ratio.

1990). It was estimated in his thesis that the water interface temperature could be 10 °C lower than the inlet bulk temperature of the water. When the water interface temperature drops the removal percentage drops too. Therefore, for achieving higher removal percent of DCEE, it is more advantageous to use two stripping columns with each air-water molar loading ratio of 0.6 than to use one stripping columns with air-water molar loading ratio of 1.2.

Equations (1) through (3) were used to calculate the experimental values of the over-all mass transfer coefficient, aK_L, from the experimental data in Table 1. The results indicated that the value of aK_L increases as the temperature increases. This increase could be due to an increament of the Henry's law constant (as shown in Table 2) and an increment of molecular diffusivity (Treybal, 1980). In general, water flowrate affects the value of the over-all mass transfer coefficient more than the air flowrate. For an example of TCE, the value of aK_L increases from 0.004 to 0.0045 sec^{-1} when the air flowrate increases from 1.8 to 8.8 ft^3/min (4.67 to 22.8 $mole/m^2s$), giving only 10% increase. However, as the water flowrate increases from 0.275 to 0.675 gpm (118.9 to 291.8 $mole/m^2s$), the value of aK_L increases from 0.004 to 0.0094 sec^{-1}. This increment is more than double.

For the low-volatile chemical such as DCEE, the air flowrate effect was found to be significant. For an example of the water flowrate of 0.12 gpm (51.88 $mole/m^2s$), the value of aK_L increases from 3 x 10^{-4} to 7 x 10^{-4} sec^{-1} as the air flow rate increases from 7.0 to 18 ft^3/min (18.17 to 46.72 $mole/m^2s$).

The overall mass transfer coefficient, aK_L, can be estimated by the models proposed by Sherwood and Holloway (1940), Shulman et al (Treybal, 1980), and Onda et al (1968). The predicted aK_L values were plotted versus the observed values of aK_L in Figure 4 for the Sherwood's model, in Figure 5 for the Shulman's model, and in Figure 6 for the Onda's model. It can be seen from these figures that, in general, the Onda's model probably gives the best prediction comparing with the experimental data obtained in this study. The Shulman's model tends to over-predict the values of aK_L except for DCEE. The Sherwood's model predicted higher values of aK_L, especially for the case of DCEE. This could be due to the fact that the Sherwood's model does not take the gas phase resistance into consideration.

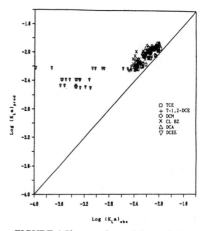

FIGURE 4 Sherwood's model correlation

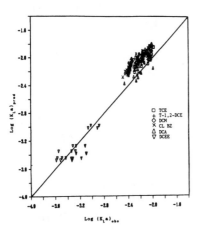

FIGURE 5. Shulman's model correlation

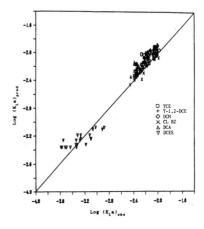

FIGURE 6. Onda's model correlation

CONCLUSIONS

All the six chloro-hydrocarbon contaminants tested here can be removed easily from the groundwater by air stripping except DCEE. The measured over-all mass transfer coefficients are in good agreement with those model predictions except for DCEE, at which the Sherwood's predictions were significantly higher. In general, water flowrate affects the value of over-all mass transfer coefficient more than the other factors. However, for the low-volatile chemical such as DCEE, the air flowrate effect was found to be significant. In order to remove DCEE efficiently, high air-water loading ratio has to be utilized.

ACKNOWLEDGMENT

This project has been funded partially with funds from the United States Environmental Protection Agency (USEPA) as part of the program of the Gulf Coast Hazardous Substance Research Center. The contents do not necessarily reflect the views and policies of the USEPA nor does the mention of trade names or commercial products constitute endorsement or recommendation for use.

REFERENCES

Ball, W. P., M. D. Jones, and M.C. Kavanaugh. 1984. "Mass Transfer of Volatile Organic Compounds in Packed Tower Aeration," *J. WPCF* 56(2):127.

Kavanaugh, M.C., and R.R. Trussell. 1980. "Design of Aeration Towers to Strip Volatile Contaminaants from Drinking Water," *J. AWWA* 72(12): 684.

Lamarche, P., and R. L. Droste. 1989. "Air-Stripping Mass Transfer Correlations for Volatile Organics," *J. AWWA* 81(1): 78

Li, K. 1997. *"Air Stripping,"* Section 1.8 in the Handbook of Separation Techniques for Chemical Engineers*, edited by P. A. Schweitzer, McGraw-Hill Company 1-393

Onda, K., H. Takeuchi, and Y. Okumo. 1968. "Mass Transfer Coefficients Between Liquid and Gas Phases in Packed Columns," J. Chem. Engrg. Japan. 1: 56.

Roberts, P. V., G. D. Hopkins, C. Munz, and A. H. Riojas. 1985. "Evaluating Two-Resistance Models for Air Stripping of Volatile Organic Contaminants in a Counter-Current Packed Column," *J. Environ. Sci. & Technol.* 19(2): 164.

Sherwood, T. K. and F. A. L. Holloway. 1940. "Performance of Packed Towers - Liquid Film Data for Several Packings," *Trans. Amer. Inst. Chem. Engrs.* 36: 39.

Treybal, R. E. 1980. *"Mass Transfer Operations,"* McGraw-Hill Book Company, New York, New York.

Yu, C. C. 1990. *"Nonisothermal Air Stripping for Low-volatile Organic Contaminants in Ground Water,"* Master thesis, Lamar University, Beaumont, Texas

REMEDIATION OF CONTAMINATED SOILS BY COMBINED CHEMICAL AND BIOLOGICAL TREATMENTS

Shelley A. Allen (Colorado State University, Fort Collins, Colorado)
Kenneth F. Reardon (Colorado State University, Fort Collins, Colorado)

ABSTRACT: In a laboratory-scale project, soils obtained from two different sites were used to study the combination of chemical oxidation and biodegradation. The primary contaminants in Soil #1 and Soil #2 were polycyclic aromatic hydrocarbons (PAHs) and pentachlorophenol (PCP). The specific chemical oxidation process utilized in this study involved the addition of a Fenton's reagent (ferrous iron and hydrogen peroxide) to generate hydroxyl free radicals (•OH). Once chemically oxidized, biodegradation by indigenous microorganisms was used to more completely degrade the contaminants. The effects of different extents of chemical oxidation were tested by applying 0, 1, and 2% (w/w) hydrogen peroxide, with and without added nutrients. The different treatments were analyzed for pollutant levels and microbial population size. The addition of even a small amount of hydrogen peroxide produced better degradation than bioremediation alone. The 1 and 2% hydrogen peroxide treatments resulted in similar final extents of degradation of phenanthrene and anthracene, both 3-ring compounds. Fluoranthene, however, was degraded to a lower final concentration by the 1% hydrogen peroxide treatment. Although the concentration of microorganisms decreased significantly after the hydrogen peroxide was added, the microbial population rebounded within one week. The addition of nutrients increased the microbial growth rate. These results have been used to design a pilot-scale treatment study.

INTRODUCTION

Chemical oxidation and bioremediation are two remediation techniques commonly used to treat contaminated soil. Bioremediation can be an economically sound method of treating easily degradable compounds (Marco, 1997). However, many pollutants are not effectively degraded by bioremediation. In contrast, chemical oxidation technologies are considerably more expensive but successfully degrade contaminants that are biologically recalcitrant (Heinzle, 1992; Tang, 1997). The aim of this work was to identify the optimal degree of chemical oxidation as defined by the overall extent of degradation. Thus, the impact of the chemical oxidant on the microbial population was an important concern.

The chemical oxidant used in this experiment was Fenton's reagent, a combination of hydrogen peroxide and ferrous iron. The Fenton's reagent process is an advanced oxidative process that generates hydroxyl free radicals (•OH) and has been shown to be very effective in treating biologically recalcitrant compounds (Mokrini, 1997).

The contaminants chosen for this experiment were polycyclic aromatic hydrocarbons (PAHs) and pentachlorophenol (PCP). These contaminants are widely used in the wood treatment industry to protect wood products from rotting.

PCP and many PAHs are prevalent priority pollutants regulated by the U.S. Environmental Protection Agency. They are often slowly biodegraded and the larger PAHs are biologically recalcitrant.

Laboratory-scale experiments were performed to help predict the outcome of pilot-scale experiments combining chemical oxidation and bioremediation on the same soils. The goals of the laboratory experiments were to identify an acceptable range of hydrogen peroxide addition, to determine the time scales for each treatment phase, and to verify that significant changes in total microbial mass could be detected.

MATERIALS AND METHODS

Two soils were used for this experiment. Soil #1 was a sandy loam from a PCP disposal site that was contaminated with PCP only. Soil #2 was a silty loam from a wood treatment facility contaminated with PCP and several PAHs ranging from two to five rings. The indigenous soil microorganisms were used for the biological degradation in both experiments.

The soil experiments were performed in Erlenmeyer flasks closed with a rubber stopper. Each batch of soil was treated with different levels of hydrogen peroxide solution and 5% (w/w) iron aggregate, the ferrous iron source used in this experiment. Water was added to bring the water level to 4% (w/w) to ensure adequate soil moisture for the microorganisms. The soil was then incubated at 20 °C to allow biodegradation to take place.

125 g of Soil #1 were placed in 250-mL Erlenmeyer flasks and treated with 0 and 1% (w/w) hydrogen peroxide. 1000 g of Soil #2 were placed in 1000-mL Erlenmeyer flasks and treated with 0, 1, and 2% (w/w) hydrogen peroxide. One treatment of 1% hydrogen peroxide also was enhanced with 31 mL of a nutrient solution containing 845 mg/L KH_2PO_4, 2186 mg/L K_2HPO_4, 1063 mg/L NH_4Cl, 97.5 mg/L $MgSO_4$, 25 mg/L of $MnSO_4$-H_2O, 5 mg/L of $FeSO_4$-$7H_2O$, 1.5 mg/L $CaCl_2$-$2H_2O$, and trace amounts of H_2SO_4.

PAH concentrations in the soil were determined using a modification of EPA Method 8270. Both acid and base/neutral extractions were performed on 20 g of soil in methylene chloride and water. This extract was then rinsed through a solid phase extraction apparatus with florisil cartridges to remove humics. The remaining extract was concentrated down to 1 mL under a nitrogen atmosphere and analyzed by gas chromatography (GC). The GC used was a HP 5890 Series 2 GC with an HP 5971A mass spectrometer. The column used was a HP-5 Trace Analysis Column.

The microbial concentration was measured by the phospholipid fatty acid (PLFA) assay (Findlay, 1997; Cavigelli, 1995). This assay quantifies the concentration of fatty acid methyl esters (FAMEs) found in the sample. From the amount of FAMEs present, a qualitative comparison of microbial concentrations can be found. Extraction and interpretation were performed as described by White et al. (1979).

RESULTS AND DISCUSSION

The experiment using Soil #1 demonstrated more extensive PCP degradation from the 1% hydrogen peroxide treatment followed by biodegradation than biodegradation alone (Figure 1a). A significant increase in the PCP concentration can be seen three hours after addition of the Fenton's reagent. This peak is thought to be produced by breakage of soil particles by both mixing and the generation of oxygen bubbles, which would increase the availability and dissolution of PCP.

The addition of hydrogen peroxide initially decreased the concentration of FAMEs and therefore the microbial population (Figure 1b). However, by the end of the six-day experiment, the microorganisms recovered from the initial shock and the cell concentrations increased significantly.

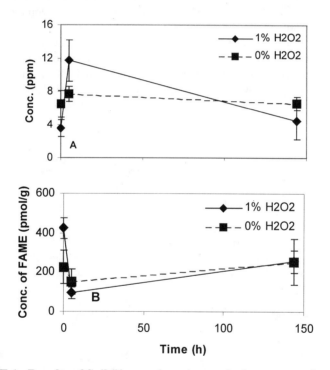

FIGURE 1. Results of Soil #1 experiment over six days comparing 0 and 1% hydrogen peroxide additions. (A) Degradation of PCP. (B) Microbial concentration profiles.

The experiment using Soil #2 produced similar results. The treatments with hydrogen peroxide oxidation followed by biodegradation degraded the PAHs considerably more than biodegradation alone (Figures 2a-c). The 1 and 2% hydrogen peroxide treatments resulted in similar final extents of degradation of phenanthrene and anthracene, both 3-ring compounds. Fluoranthene, however, was degraded to a lower final concentration by the 1% hydrogen peroxide treatment. Though this result is unexpected, it is hypothesized that the fluoranthene-degrading microorganisms are more sensitive to the hydrogen peroxide, and therefore the biodegradation of fluoranthene is limited by the 2% hydrogen peroxide addition.

The PLFA analysis produced the expected results. The 1 and 2% hydrogen peroxide additions produced an initial drop in the microbial population. However, the microbial population increased to initial values within one week, then continued to increase, surpassing the original microbial mass at the end of two weeks. Also, the results clearly confirmed that the addition of nutrients to the soil doubled the microbial growth rate over the two-week experiment. Since no data were gathered on the PAH levels of the nutrient-amended soil, it cannot be determined whether the increased microbial growth correlated with increased PAH degradation. However, this was most likely the case as the contaminants are the major carbon source in the soil.

The PLFA analysis also revealed some information about the nature of the microorganisms in Soil #2. Although the majority of the microorganisms were found to be aerobic, some desulfobacters were detected (Kieft et al., 1997). Furthermore, stress indicators were found in the soils. The levels of these stress indicators increased with higher concentration of hydrogen peroxide added, which was expected.

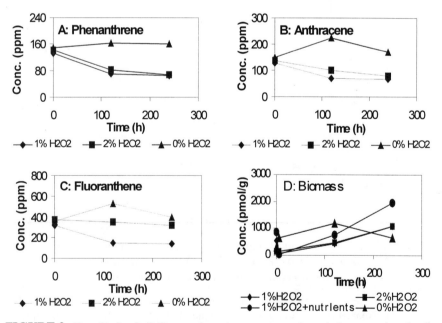

FIGURE 2. Results for Soil #2 experiment over two-week period comparing the 0, 1, and 2% hydrogen peroxide additions. (A) Phenanthrene degradation. (B) Anthracene degradation. (C) Fluoranthene degradation. (D) Microbial concentration profiles for 0, 1, and 2% hydrogen peroxide and nutrient-enhanced 1% hydrogen peroxide treatments.

CONCLUSIONS

The combination of chemical oxidation and biodegradation proved to degrade PCP and PAHs in the two soils tested better than biodegradation alone. The two hydrogen peroxide treatments resulted in similar final extents of degradation. Also, as expected, the microbial population decreased upon addition of hydrogen peroxide, but increased significantly within a week. Nutrient addition doubled the growth rate of the microorganisms.

The results from these experiments are currently being applied to pilot-scale experiments using the same soils. The goals of the pilot-scale experiments are to study the degradation achieved through the addition of 0, 1, 2, and 4% hydrogen peroxide with 5% iron aggregate. Nutrients are being added to all the soils, since this experiment demonstrated such a marked increase in microbial growth upon nutrient enhancement. For these large-scale experiments, a Microenfractionator™ – a highly efficient, large-scale mixer – is being incorporated to break up as well as homogenize the soil. Along with the PAH and PLFA levels, other factors are being studied in this large-scale experiment: the mutagenicity level of the soil is being measured by the Ames Assay (Ames, 1983), the microbial community changes are being examined by Denaturing Gradient Gel Electrophoresis (Ovreås, 1997), and the water content of the soil is being measured by a combination of pressure plates and soil moisture assays.

ACKNOWLEDGEMENTS

The authors are grateful for funding from the Colorado Institute for Research in Biotechnology and H&H Eco Systems, Inc. We also thank Terry and Ron Horn from H&H Eco Systems, Inc. for supplying Soil #2, the use of the Mini-Microenfractionator™, and their continued technical assistance. Finally, we thank David Stewart and Paul Stone from Stewart Environmental Consultants Inc. for supplying Soil #1.

REFERENCES

Ames, B.N. and D.M. Maron. 1983. "Revised Methods for the Salmonella Mutagenicity Test." *Mutation Research.* 113: 173-215.

Cavigelli, M.A., G.P. Robertson, and M.J. Klug. 1995. "Fatty Acid Methyl Ester (FAME) Profiles as Measures of Soil Microbial Community Structure." *Plant and Soil.* 170: 99-113.

Findlay, R.H. and F.C. Dobbs. 1993. "Quantitative Description of Microbial Communities Using Lipid Analysis." In P.F. Kemp, et al. (Eds.), *Handbook of Methods in Aquatic Microbial Ecology*, pp. 271-280. Lewis Publishers, Boca Raton.

Heinzle, E., et al. 1992. "Integrated Ozonation-Biotreatment of Pulp-Bleaching Effluents Containing Chlorinated Phenolics." *Biotech. Prog.* 8: 67-77.

Kieft, T.L., E. Wilch, K. O'Conner, D.B. Ringelberg, D.C. White. 1997. "Survival and Phospholipid Fatty Acid Profiles for Surface and Subsurface Bacteria in Natural Sediment Microcosms." *Applied and Env. Microbiology.* 63(4): 1531-1542.

Marco, A., S. Esplugas, and G. Saum. 1997. "How and Why Combine Chemical and Biological Processes for Wastewater Treatment." *Wat. Sci. Tech.* 35: 321-327.

Mokrini, A., D. Ousse, and S. Esplugas. 1997. "Oxidation of Aromatic Compounds with UV Radiation/Ozone/Hydrogen Peroxide." *Wat. Sci. Tech.* 35(4): 95-102.

Ovreås, L., et al. 1997. "Distribution of Bacterioplankton in Meromictic Lake Saelenvannet, as Determined by Denaturing Gradient Gel Electrophoresis of PCR-Amplified Gene Fragments Coding for 16S rRNA." *Appl. Environ. Microbiol.* 63(9): 3367-3373.

Tang, W., et al. 1997. "TiO2/UV Photodegradation of Azo Dyes in Aqueous." *Solutions. Environ. Technol.* 18: 325-332.

CHEMICAL OXIDATION SOURCE REDUCTION AND NATURAL ATTENUATION FOR REMEDIATION OF CHLORINATED HYDROCARBONS IN GROUNDWATER

Michael J. Maughon (NAVFAC South Division, N. Charleston, South Carolina)
Clifton C. Casey (NAVFAC South Division, N. Charleston, South Carolina)
J. Daniel Bryant (Geo-Cleanse International, Kenilworth, New Jersey)
James T. Wilson (Geo-Cleanse International, Kenilworth, New Jersey)

ABSTRACT: In-situ chemical oxidation may be an effective element of source reduction and cost-effective plume management strategies for groundwater impacted by chlorinated volatile organic compounds (CVOCs). Monitored natural attenuation (MNA) and other complimentary technologies for downgradient plume areas with lower CVOC concentrations are more feasible if coupled with source reduction. In-situ chemical oxidation has been effectively implemented for source reduction at NSB Kings Bay and NAS Pensacola. At NSB Kings Bay, 23 injectors were installed and 12,063 gallons of hydrogen peroxide were injected over 24 days during two treatment events. Source area CVOC concentrations were reduced by an average of 98%. The most contaminated well experienced a reduction from 9,074 µg/L total CVOCs (including 8,500 µg/L PCE) to 65 µg/L (>99%reduction), with little or no rebound 13 months after treatment when total CVOCs measured 85 µg/L. As a result, the State of Georgia approved MNA as a final remedy for the downgradient plume and terminated a long-term pump-and-treat program. At NAS Pensacola, 15 injectors were installed and 10,127 gallons of hydrogen peroxide were injected over 11 days during two treatment events. Source area CVOC concentrations were reduced by an average of 95%. The most contaminated well experienced a reduction from 2,608 µg/L total CVOCs (including 2040 µg/L TCE) to <1 µg/L with relatively little rebound eight months after treatment when total CVOCs measured 198 µg/L. As a result, the Florida Department of Environmental Protection (FDEP) approved termination of an existing pump-and-treat system, and the RCRA permit is being revised to establish MNA as the final remedy.

INTRODUCTION

CVOC remediation of soil and groundwater is one of the primary Department of Defense environmental needs. Slow dissolution of CVOCs from dense non-aqueous phase liquids (DNAPLs) in concentrated source areas can cause a widespread and persistent groundwater problem. One of the most rapid and cost-effective methods to address CVOC remediation is to aggressively attack the DNAPL source area. After the source area DNAPL and persistent groundwater CVOC source is eliminated, residual CVOCs in downgradient groundwater may rapidly attenuate due to natural biological and chemical degradation processes. Regulatory acceptance of MNA as a final remedy is more feasible if a concentrated source area has been remediated.

As part of an overall plume management strategy implemented by Southern Division NAVFAC, in-situ chemical oxidation has been evaluated as a key source reduction technology, which can be complimented by MNA, enhanced bioremediation, or other technologies for remediation of downgradient plume areas with lower CVOC concentrations. Southern Division NAVFAC has recently utilized the Geo-Cleanse® Process, an in-situ chemical oxidation method utilizing a pressurized injection of Fenton's reagent, for CVOC source area reduction at two sites: Site 11 at Naval Submarine Base Kings Bay, Georgia, and the Wastewater Treatment Plant (WWTP) at Pensacola Naval Air Station, Florida. At each of these sites, an assessment of the natural attenuation capacity of the aquifer indicated CVOC plumes should collapse within a few years if concentrations in the source areas were substantially reduced.

SITE 11, NSB KINGS BAY, GEORGIA

Site Background and Characterization. The site is a former municipal landfill located on the edge of the base. Site investigations identified a groundwater CVOC plume migrating off base with low concentrations of cis-1,2-dichloroethene detected under an adjacent residential neighborhood. As an interim measure a pump-and-treat system was constructed with extraction wells along the downgradient installation boundary. The lithology of the shallow soils is typified as fine quartz sand with interbedded silty and clayey fine sands. Groundwater is located approximately 6 ft below land surface. The efficiency of natural attenuation processes was assessed over a two-year time period as part of the remedial investigation and feasibility study. Results indicated that ongoing microbiological processes, reductive dechlorination, were very effective at remediating the plume. However, given the relatively high concentrations of CVOCs in the source area there was not enough distance for the processes to prevent the remaining low concentrations of daughter products from reaching the neighborhood. Based upon an analysis of the natural attenuation capacity (Chapelle and Bradley, 1998), the State of Georgia agreed to terminate the pump and treat system and accept MNA as the final remedy if total CVOC concentrations of <100 µg/L were achieved within the source area.

Tetrachloroethylene (PCE) and its degradation products trichloroethene (TCE), cis-1,2-dichloroethene (CIS), and vinyl chloride (VC) were identified at the site. Maximum CVOC concentrations observed within the treatment area prior to injection were at monitoring well KBA-11-34 where 8,500 µg/L PCE, 550 µg/L TCE, and 24 µg/L CIS was measured (Table 1; only monitoring wells shown, but most of the injectors were also sampled). The PCE data exceed 1% solubility of PCE, indicating the potential presence of an unknown DNAPL mass. Vertical contaminant profiles prior to injection detected no contamination outside of the 30-40 ft depth interval.

Project Design, Implementation and Results. Based upon the interpreted plume area and depth, an array of 23 injectors was installed within and surrounding an area of concern (Figure 1). The injectors were all screened between 27 and 42 ft to encompass the targeted depth interval. A total of 12,063 gallons of 50% hydrogen peroxide and a similar volume of ferrous iron catalyst solution were injected over the course of 17 days during two treatment periods in November 1998 and February 1999. Groundwater samples were collected for up to 6 months after treatment. All sampled locations within the target treatment area yielded total CVOCs less than the 100 μg/L target, and held those levels for >30 days following treatment. The average reduction in total CVOCs in all monitoring points within the treatment area was to 32 μg/L from 3,028 μg/L (99%). KBA-11-34, the most contaminated monitoring well, has maintained CVOCs less than 100 μg/L for 13 months since treatment. Wells outside of the treatment area yielded no change or a decrease in CVOC concentrations.

Current Status. The State of Georgia allowed the pump-and-treat system to be terminated based upon the injection results and the ongoing natural attenuation of the downgradient plume. Since the February 1999 treatment, the scope of work has been expanded to treat other identified source areas. One injector location outside and upgradient of the initial treatment area, I-14 (Figure 1), yielded 3,500 μg/L PCE (3,800 μg/L total CVOCs) following the February 1999 treatment. Additional delineation work was conducted and identified a very concentrated source area with up to 130,000 μg/L PCE, and contamination extending from approximately 10 to 40 ft below grade. A hot spot excavation was performed and several 15-gallon containers of solution having high concentrations of PCE were removed. This source area was then treated with Fenton's reagent in January 2000. Preliminary results indicate a maximum residual of 1,700 μg/L PCE, and a follow up treatment is planned.

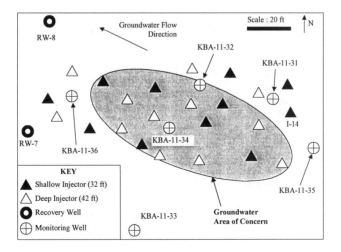

FIGURE 1. NSB Kings Bay Site 11 site layout map.

TABLE 1. CVOC results (µg/L) for NSB Kings Bay Site 11

Location	Date	PCE	TCE	CIS	VC
KBA-11-31	10/30/98	<1	1	7	<1
	11/6/98	1	<1	<1	<1
	11/24/98	8	7	2	<1
	12/22/98	12	12	6	<1
	1/27/99	14	15	4	<1
	2/18/99	23	17	5	<1
	3/15/99	15	12	7	<1
	5/20/99	69	57	73	5
KBA-11-32	10/30/98	<1	110	54	4
	11/13/98	<1	<1	<1	<1
	11/24/98	4	9	3	<1
	12/22/98	7	13	5	<1
	1/27/99	9	17	5	<1
	2/18/99	6	8	3	<1
	3/15/99	11	11	4	<1
	5/20/99	18	13	<1	<5
KBA-11-33	10/30/98	<1	<1	<1	<1
	11/24/98	<1	<1	<1	<1
	12/22/98	<1	<1	<1	<1
	1/27/99	<3	<1	<1	<1
	3/15/99	<1	<1	<1	<1
	5/20/99	<3	<1	<1	<5
KBA-11-34	10/30/98	8,500	550	24	<1
	11/13/98	180	30	<5	<5
	11/24/98	200	19	<1	<1
	12/22/98	87	6	<1	<1
	1/27/99	83	7	<1	<1
	2/18/99	9	<1	<1	<1
	3/15/99	62	3	<1	<1
	5/20/99	83	3	<1	<5
	7/26/99	24	<1	<1	<1
	2/11/00	83	2	<1	<1
KBA-11-35	10/30/98	<1	<1	<1	<1
	11/6/98	<1	<1	<1	<1
	11/24/98	<1	<1	<1	<1
	12/22/98	<1	<1	<1	<1
	1/27/99	<3	<1	<1	<1
	3/15/99	<1	<1	<1	<1
	5/20/99	<3	<1	<1	<5
KBA-11-36	10/30/98	5	360	60	<1
	11/24/98	17	55	3	<1
	12/22/98	370	44	2	<1
	1/27/99	380	44	2	<1
	2/18/99	3	<1	3	<1
	3/15/99	51	3	<1	<1
	5/20/99	90	10	<1	<5

NAS PENSACOLA WWTP

Site Background and Characterization. The site is at the location of a former sludge drying bed within the wastewater treatment plant (WWTP), which was previously operated as an industrial wastewater treatment plant. Initial site investigations under RCRA in 1986 indicated chlorinated compounds and other contaminants had been released from sludge drying beds and a surge pond. In 1987 soil source areas were removed and a groundwater extraction system was installed to treat contaminated groundwater and prevent migration to Pensacola Bay. In the mid-1990's the extraction system was optimized to focus recovery efforts on the area near monitoring well GM-66 (Figure 2), which had persistently high levels of TCE averaging about 3,000 to 4,000 µg/L. In 1997 groundwater monitoring at the site was revised to include an ongoing assessment of natural attenuation. Results indicated that natural attenuation processes were effective with TCE and its daughter products being completely destroyed within 250 feet downgradient primarily as a result of reductive dechlorination (Chapelle and Bradley, 1999).

To accelerate remediation of the site and eliminate operation and maintenance costs associated with the pump and treat system, source reduction was planned in 1998. The groundwater source area near monitoring well GM-66 was further delineated and characterized with direct push sampling at 5-foot depth intervals at 8 locations. Two of the sampling locations were converted to temporary monitoring wells, USGS-5 and USGS-6. Groundwater was encountered at 5 ft below land surface. The lithology of the shallow soils is typified as white to light brown, fine to medium quartz sand underlain by relatively impermeable clay at approximately 40 ft below grade.

TCE and its biological degradation products, CIS and VC, were identified as the primary contaminants. Maximum CVOC concentrations in the source area were observed at GM-66/66R where 3,600 µg/L TCE, 520 µg/L CIS, and 63 µg/L VC was measured in July 1998 (Table 2). Chlorinated aromatic petroleum hydrocarbons (primarily chlorobenzenes) and other chlorinated aliphatics were also detected at much lower concentrations. Vertical contaminant profiling within the source area detected only naphthalene at depths shallower than 30 ft and indicated that CVOC contaminants were concentrated at a depth of 35 to 40 feet, just above the confining clay layer. The CVOC data distribution (Figure 2) indicated a former sludge drying bed was the most likely source.

Based on the site-specific contaminants of concern, the hydrogeologic parameters, and the area to be treated, in-situ chemical oxidation using the Fenton's Reagent method was selected as the technology for source reduction. The FDEP, which had allowed the pump and treat system to be temporarily turned off during the second year of the natural attenuation assessment, allowed the pump and treat system to remain off during the technology demonstration.

TABLE 2. CVOC results (µg/L) for the NAS Pensacola WWTP Site.

Location	GM-66 / 66R			USGS-5			USGS-6		
Date/CVOC	TCE	CIS	VC	TCE	CIS	VC	TCE	CIS	VC
Jul-98	3,600	520	63	-	-	-	-	-	-
8/20/98	2,440	403	115	1,420	169	49	743	369	976
12/1/98	2,040	430	138	2,460	218	31	773	344	595
12/9/98	3,000	510	53	1,700	200	14	-	-	-
Samples collected during and after Phase I injection (Dec 7-12, 1998)									
12/10/98	180	16	<1	<1	<1	<1	-	-	-
12/11/98	200	31	2	830	103	6	-	-	-
12/12/98	130	10	<1	<1	<1	<1	-	-	-
12/29/98	485	42	<5	1,820	140	11	825	391	575
Jan-99	190	ND	ND	-	-	-	-	-	-
1/27/99	415	27	4	2,100	<5	6	926	<5	218
4/7/99	460	38	<1	2,700	210	17	890	380	250
Samples collected during and after Phase II injection (May 11-17, 1999)									
5/11/99	310	27	<1	1,600	180	7	500	280	180
5/12/99	<1	<1	<1	960	130	7	45	27	18
5/13/99	510	37	<1	1,900	170	11	81	46	38
5/14/99	330	33	2	<1	180	19	590	340	250
5/15/99	370	77	27	120	16	5	550	350	250
5/17/99	1	<1	<1	110	<1	<1	<1	<1	<1
5/20/99	<1	<1	<1	79	10	1	32	14	6
6/21/99	<5	<1	<1	100	20	3	110	49	21
7/15/99	86	8	<5	-	-	-	-	-	-
1/11/00	180	18	<5	-	-	-	-	-	-

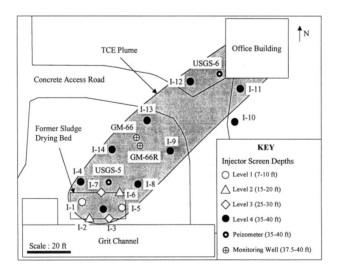

FIGURE 2. NAS Pensacola WWTP site layout map.

Project Design, Implementation and Results. Based upon the interpreted plume area and the presumed source being the former sludge drying bed, 6 injectors were installed within the former sludge drying bed in a vertical array from 11 to 31 ft and 8 injectors were installed at 35-40 ft to target the downgradient plume area. This provided additional sampling points for source area delineation.

A total of 4,089 gallons of 50% hydrogen peroxide and a similar volume of ferrous iron catalyst solution were injected over the course of 5 days in December 1998. Groundwater samples were collected daily during the injection and 4 rounds were collected over a 4-month period following injection. An 81% reduction in total CVOCs was achieved and maintained in GM-66 over that period, but USGS-5 and USGS-6 experienced rebound with post-treatment TCE concentrations slightly higher than pre-treatment levels (Table 2). Post-treatment data evaluation indicated that naturally occurring dissolved ferrous iron concentrations were found to be exceptionally high (up to 975 mg/L) within the treatment area prior to the Geo-Cleanse® injection. Because dissolved iron serves as a catalyst in the Fenton's reagent method, the elevated iron concentrations limited the radius of influence and decreased treatment effectiveness. Accordingly, the injection program was modified to include phosphoric acid to buffer and slow the reaction and increase the effective radius of influence. Post-treatment sampling also indicated that the screen in monitoring well GM-66 was apparently damaged, so GM-66R was installed directly adjacent with identical construction. Modifications also included installing an additional source area injector, I-15, at a depth of 35-40 feet. The second injection phase was then conducted in May 1999.

An additional 6,038 gallons of 50% hydrogen peroxide and a similar volume of catalyst were injected during the second injection phase over 6 days. Groundwater was sampled daily during the treatment and again 30 days after treatment. Post-treatment samples indicated that CVOCs at GM-66R, which measured 2,608 µg/L before initial treatment, were reduced to below detection levels. USGS-5 exhibited a 97% reduction in CVOCs from 2,709 µg/L to 90 µg/L, and USGS-6 had a 97% reduction from 1,712 µg/L to 52 µg/L. Thirty days after treatment CVOCs at GM-66R were still below detection levels. USGS-5 and USGS-6 had exhibited only slight rebound to 123 µg/L and 180 µg/L, respectively. Subsequent data collected during ongoing semi-annual RCRA monitoring indicates CVOCs at GM-66R rebounding slightly to 94 µg/L two months after treatment and to 198 µg/L eight months after treatment.

Current Status. The goal of the in-situ chemical oxidation treatment at the NAS Pensacola site was to substantially reduce the source area concentrations to ensure the protectiveness and timeliness of MNA as the final remedy. Based on the successful results of the source reduction effort and data indicating continued natural attenuation of the downgradient plume, the FDEP has allowed the pump and treat system to be permanently discontinued. The RCRA permit is currently being revised to establish MNA as the final remedy for the site with a contingency plan for additional corrective action if warranted by any future migration of contaminants to sentinel wells.

CONCLUSIONS

A systems engineering approach to groundwater contamination problems, i.e., applying the right technology to the appropriate area of the plume given the site specific conditions, can yield substantial cost savings, accelerate cleanup, and ensure protection of downgradient receptors. For the site conditions at NSB Kings Bay, Site 11, and NAS Pensacola, WWTP Site, in-situ chemical oxidation using a Fenton's Reagent method has proven to be a successful and economical technology for aggressively reducing groundwater source area concentrations of CVOCs. Coupled with analyses of the natural attenuation capacity, the substantial source reduction achieved at each site has 1) ensured the protection of downgradient receptors, 2) ensured the effectiveness and timeliness of MNA as the final, polishing remedy, 3) eliminated the need for existing pump-and-treat systems, and 4) substantially reduced long-term O&M costs.

REFERENCES

Chapelle, F. H., and P. M. Bradley. 1998. "Selecting Remediation Goals by Assessing the Natural Attenuation Capacity of Groundwater Systems." *Bioremediation Journal.* 2: 227-238.

Chapelle, F. H., and P. M. Bradley, October 1999. "Natural Attenuation of Chlorinated Ethenes and Chlorinated Benzenes, Wastewater Treatment Plant, NAS Pensacola"

2000 AUTHOR INDEX

This index contains names, affiliations, and book/page citations for all authors who contributed to the seven books published in connection with the Second International Conference on Remediation of Chlorinated and Recalcitrant Compounds, held in Monterey, California, in May 2000. Ordering information is provided on the back cover of this book.

The citations reference the seven books as follows:

2(1): Wickramanayake, G.B., A.R. Gavaskar, M.E. Kelley, and K.W. Nehring (Eds.), *Risk, Regulatory, and Monitoring Considerations: Remediation of Chlorinated and Recalcitrant Compounds.* Battelle Press, Columbus, OH, 2000. 438 pp.

2(2): Wickramanayake, G.B., A.R. Gavaskar, and N. Gupta (Eds.), *Treating Dense Nonaqueous-Phase Liquids (DNAPLs): Remediation of Chlorinated and Recalcitrant Compounds.* Battelle Press, Columbus, OH, 2000. 256 pp.

2(3): Wickramanayake, G.B., A.R. Gavaskar, and M.E. Kelley (Eds.), *Natural Attenuation Considerations and Case Studies: Remediation of Chlorinated and Recalcitrant Compounds.* Battelle Press, Columbus, OH, 2000. 254 pp.

2(4): Wickramanayake, G.B., A.R. Gavaskar, B.C.Alleman, and V.S. Magar (Eds.) *Bioremediation and Phytoremediation of Chlorinated and Recalcitrant Compounds.* Battelle Press, Columbus, OH, 2000. 538 pp.

2(5): Wickramanayake, G.B. and A.R. Gavaskar (Eds.), *Physical and Thermal Technologies: Remediation of Chlorinated and Recalcitrant Compounds.* Battelle Press, Columbus, OH, 2000. 344 pp.

2(6): Wickramanayake, G.B., A.R. Gavaskar, and A.S.C. Chen (Eds.), *Chemical Oxidation and Reactive Barriers: Remediation of Chlorinated and Recalcitrant Compounds.* Battelle Press, Columbus, OH, 2000. 470 pp.

2(7): Wickramanayake, G.B., A.R. Gavaskar, J.T. Gibbs, and J.L. Means (Eds.), *Case Studies in the Remediation of Chlorinated and Recalcitrant Compounds.* Battelle Press, Columbus, OH, 2000. 430 pp.

2000 KEYWORD INDEX

This index contains keyword terms assigned to the articles in the seven books published in connection with the Second International Conference on Remediation of Chlorinated and Recalcitrant Compounds, held in Monterey, California, in May 2000. Ordering information is provided on the back cover of this book.

In assigning the terms that appear in this index, no attempt was made to reference all subjects addressed. Instead, terms were assigned to each article to reflect the primary topics covered by that article. Authors' suggestions were taken into consideration and expanded or revised as necessary to produce a cohesive topic listing. The citations reference the seven books as follows:

2(1): Wickramanayake, G.B., A.R. Gavaskar, M.E. Kelley, and K.W. Nehring (Eds.), *Risk, Regulatory, and Monitoring Considerations: Remediation of Chlorinated and Recalcitrant Compounds.* Battelle Press, Columbus, OH, 2000. 438 pp.

2(2): Wickramanayake, G.B., A.R. Gavaskar, and N. Gupta (Eds.), *Treating Dense Nonaqueous-Phase Liquids (DNAPLs): Remediation of Chlorinated and Recalcitrant Compounds.* Battelle Press, Columbus, OH, 2000. 256 pp.

2(3): Wickramanayake, G.B., A.R. Gavaskar, and M.E. Kelley (Eds.), *Natural Attenuation Considerations and Case Studies: Remediation of Chlorinated and Recalcitrant Compounds.* Battelle Press, Columbus, OH, 2000. 254 pp.

2(4): Wickramanayake, G.B., A.R. Gavaskar, B.C.Alleman, and V.S. Magar (Eds.) *Bioremediation and Phytoremediation of Chlorinated and Recalcitrant Compounds.* Battelle Press, Columbus, OH, 2000. 538 pp.

2(5): Wickramanayake, G.B. and A.R. Gavaskar (Eds.), *Physical and Thermal Technologies: Remediation of Chlorinated and Recalcitrant Compounds.* Battelle Press, Columbus, OH, 2000. 344 pp.

2(6): Wickramanayake, G.B., A.R. Gavaskar, and A.S.C. Chen (Eds.), *Chemical Oxidation and Reactive Barriers: Remediation of Chlorinated and Recalcitrant Compounds.* Battelle Press, Columbus, OH, 2000. 470 pp.

2(7): Wickramanayake, G.B., A.R. Gavaskar, J.T. Gibbs, and J.L. Means (Eds.), *Case Studies in the Remediation of Chlorinated and Recalcitrant Compounds.* Battelle Press, Columbus, OH, 2000. 430 pp.

A

α-ketoglutarate-dependent cleavage **2(7):**229
abiotic release date **2(7):**181
acetate **2(4):**23, 107, 389, 437
acid mine drainage, *see* mine waste
acid-enhanced degradation **2(7):**33
acridine orange (AO) **2(6):**233
actinomycetes **2(4):**455

activated carbon **2(6):**257, 315
advanced oxidation technology (AOT) **2(6):**201, 209, 217, 225, 233, 241, 249; **2(7):**25
aeration **2(5):**237
air monitoring **2(1):**207
air sparging, *see* sparging
air stripping **2(5):**293